CONFORMAL MAPPING ON RIEMANN SURFACES

CONFORMAL MAPPING ON RIEMANN SURFACES

Harvey Cohn

Distinguished Professor of Mathematics
City University of New York

Dover Publications, Inc.
New York

Published in Canada by General Publishing Company, Ltd., 30 Lesmill Road, Don Mills, Toronto, Ontario.
Published in the United Kingdom by Constable and Company, Ltd., 10 Orange Street, London WC2H 7EG.

This Dover edition, first published in 1980, is an unabridged and corrected republication of the work first published in 1967 by McGraw-Hill Book Company.

International Standard Book Number: 0-486-64025-6
Library of Congress Catalog Card Number: 80-65863

Manufactured in the United States of America
Dover Publications, Inc.
180 Varick Street
New York, N.Y. 10014

DEDICATED TO
BERNICE

PREFACE

The subject matter loosely called "Riemann surface theory" is one of the most valuable bodies of knowledge within mathematics for a student to learn. It has a permanent place both within pure mathematics, where it marks the crossroads of modern mathematics, and within applied mathematics as long as complex variables will serve as tools. Such a subject should be made available to the senior undergraduate or beginning graduate student with a minimal preparation consisting of, say, an ordinary semester of complex analysis. This is best done before he learns topology, functional analysis, modern algebra, or any one of a dozen recent branches of mathematics which were guided by the remarkable development of Riemann surface theory.

Such course material can be made available to the mathematics student at the desired level only in the spirit of compromise. The need for compromise is evident from the very size and age of Riemann surface theory, and if further proof were required of the need for compromise, one should merely recall the multitude of "nothing but" attitudes to which the material is prone in speech and literature. Riemann surface theory is variously described as *nothing but* "the theory of algebraic curves," or "elliptic operators," or "harmonic differentials," or "two-dimensional orientable manifolds," or "valuations on sheaves," or (why not) "information theory"! Clearly, we are dealing with a subject which is the mirror of changing mathematical culture and which has the special value to the student of synthesizing much of his knowledge as no other branch of mathematics can.

It is therefore necessary to compromise unwillingly with rigor and completeness, and more sadly with historical motivation and elegance, to even get the material between two covers. Still, the teacher who uses this book will probably be called upon also to compromise by further deletions, in order to cram the material into *one semester*.

The book consists of five parts: (I) Review, (II) Definitions, (III) Derivation of Existence Theorems, (IV) Existence Proofs, (V) Classical Applications, and two historically necessary appendixes for student reference. The main omissions desired in order to cover the book in *one semester* might be some of the more algebraic content and the details of Part IV (on existence proofs). Here the existence proofs are rather fully presented in the spirit of E. Picard's "Traité d'Analyse," but since they have been superseded by more advanced minimal techniques, these proofs could be truncated (except for the brief section 11-3 on the Schottky double). The rest of the material has not been superseded even where imbedded in more modern concepts. For a one-semester course, portions involving modular functions and algebra may be omitted. For a two-semester course, no omissions should be necessary, and the appendixes can be included and even elaborated with supplementary reading on direct minimal methods.

Perhaps the main guideline in selecting and omitting material is that the material is meant to serve as a desirable course for all mathematicians, regardless of their special interests. My original interest was in the physical applications, and I later returned to the material because of its necessary role for algebra and number theory; it is therefore hoped that the book will provide "equal time" as warranted by the material.

My professional debts are too numerous to recount. I have benefited as a student from the lectures of R. Courant and L. V. Ahlfors and have later benefited by professional associations with M. Schiffer and S. Bergman. The fact that their outlook is not adequately reflected in the text is only evidence that the present undertaking still lacks some desirable appendixes. Many students have helped over the last ten years in reading five successive preliminary versions of the work, including Roy Lippmann, Terry Heil, and particularly Bill Woodruff. Dr. Stoddart Smith assisted in reading the galley proofs.

Harvey Cohn

CONTENTS

Preface *vii*

PART ONE Review of Complex Analysis 1

Introductory Survey

Chapter 1. *Analytic Behavior* *3*

Differentiation and Integration

1-1. Analyticity *3*
1-2. Integration on curves and chains *5*
1-3. Cauchy integral theorem *7*

Topological Considerations

1-4. Jordan curve theorem *10*
1-5. Other manifolds *12*
1-6. Homologous chains *15*

Singular Behavior

1-7. Cauchy kernel *19*
1-8. Poles and residues *22*

Chapter 2. *Riemann Sphere* *25*

Treatment of Infinity

2-1. Ideal point *25*

2-2. Stereographic projection 26
2-3. Rational functions 28
2-4. Unique specification theorems 31

Transformation of the Sphere

2-5. Invariant properties 33
2-6. Möbius geometry 36
2-7. Fixed-point classification 39

Chapter 3. Geometric Constructions 46

Analytic Continuation

3-1. Multivalued functions 46
3-2. Implicit functions 51
3-3. Cyclic neighborhoods 54

Conformal Mapping

3-4. Local and global results 57
3-5. Special elementary mappings 60

PART TWO Riemann Manifolds 65

Definition of Riemann Manifold through Generalization

Chapter 4. Elliptic Functions 67

Abel's Double-period Structure

4-1. Trigonometric uniformization 67
4-2. Periods of elliptic integrals 71
4-3. Physical and topological models 74

Weierstrass' Direct Construction

4-4. Elliptic functions 77
4-5. Weierstrass' ℘ function 80
4-6. The elliptic modular function 83

Euler's Addition Theorem

4-7. Evolution of addition process 89
4-8. Representation theorems 92

Chapter 5. Manifolds over the z Sphere 96

Formal Definitions

5-1. Neighborhood structure 96
5-2. Functions and differentials 100

Triangulated Manifolds

5-3. Triangulation structure 103
5-4. Algebraic Riemann manifolds 108

Chapter 6. Abstract Manifolds 116

6-1. Function field on $\mathfrak{M}$ 116
6-2. Compact manifolds are algebraic 120
6-3. Modular functions 122

PART THREE Derivation of Existence Theorems 131

Return to Real Variables

Chapter 7. Topological Considerations 133

The Two Canonical Models

7-1. Orientability 133
7-2. Canonical subdivisions 136
7-3. The Euler-Poincaré theorem 141
7-4. Proof of models 144

Homology and Abelian Differentials

7-5. Boundaries and cycles 151
7-6. Complex existence theorem 156

Chapter 8. Harmonic Differentials 160

Real Differentials

8-1. Cohomology 160
8-2. Stokes' theorem 166
8-3. Conjugate forms 171

Dirichlet Problems

8-4. The two existence theorems 177
8-5. The two uniqueness proofs 180

Chapter 9. Physical Intuition 184

9-1. Electrostatics and hydrodynamics 184
9-2. Special solutions 188
9-3. Canonical mappings 196

PART FOUR Real Existence Proofs 201

Evolution of Some Intuitive Theorems

Chapter 10. Conformal Mapping 203

10-1. Poisson's integral 203
10-2. Riemann's theorem for the disk 210

Chapter 11. Boundary Behavior 219

11-1. Continuity 219
11-2. Analyticity 228
11-3. Schottky double 235

Chapter 12. Alternating Procedures 242

12-1. Ordinary Dirichlet problem 242
12-2. Nonsingular noncompact problem 248
12-3. Planting of singularities 256

PART FIVE Algebraic Applications 263

Resurgence of Finite Structures

Chapter 13. Riemann's Existence Theorem 265

13-1. Normal integrals 265
13-2. Construction of the function field 270

Chapter 14. Advanced Results 275

14-1. Riemann-Roch theorem 275
14-2. Abel's theorem 284

Appendix A. Minimal Principles 297

Appendix B. Infinite Manifolds 307

Table 1: Summary of Existence and Uniqueness Proofs 315

Bibliography and Special Source Material 316

Index 319

CONFORMAL MAPPING ON RIEMANN SURFACES

PART I
REVIEW OF COMPLEX ANALYSIS

INTRODUCTORY SURVEY

We first undertake a critical review of introductory complex function theory for the purpose of examining the "ideal" case of the Riemann sphere (or the ordinary plane with a "point at infinity" adjoined). We make every effort to think of this two-dimensional plane (or sphere) as being represented by a single complex variable, say z.

Here we find that, in a certain sense, those functions of z which "belong in a natural way" to the sphere are *rational*. This is significant from two seemingly opposite viewpoints:

(*a*) *Algebraic.* The rational functions $f(z)$ can first of all be handled formally, through high-school-level identities in z, without any apparent concern for the magnitudes of z or $f(z)$ or for analysis (limiting operations, differentiation, integration, etc.).

(*b*) *Analytic.* The rational functions $f(z)$ are now determined by their (limiting) magnitude where they become infinite. This so-called "singular" behavior determines the functions completely, as though "normal" (analytic) behavior were unimportant.

We find that to fully appreciate this condition aesthetically, we must explore the z plane as a two-dimensional (real) xy plane, as is done in Part III. Such considerations were first introduced as "conformal mapping" based on strong ties with applications; but the proper arena for such investigations is the Riemann manifold, which is introduced in Part II and is examined in terms of intuitive physical and topological techniques in Part III. In Part IV we show how to define real functions on manifolds analogously with the algebraic and analytic points of view outlined here. In Part V we return to the use of a single complex variable z.

Thus Part I presents (as a review) the course in microcosm.

CHAPTER 1
ANALYTIC BEHAVIOR

At first we restrict our interest to functions which are well behaved where defined. All well-behaved functions are "equally" well behaved under this restriction, for assuming differentiability in a neighborhood of a point, we find a valid power series to exist.†

DIFFERENTIATION AND INTEGRATION

1-1 Analyticity

Let us start with the concept of a real function $u(x)$ defined in an interval I: $(a < x < b)$. There is a certain interest in "degrees of regularity" of the function $u(x)$ as represented by the smoothness of the curve or its tangent, etc. Thus we consider restrictions of $u(x)$ to classes **C**, **C**′, **C**″, . . . , $\mathbf{C}^{(\infty)}$, **A** as follows:

C: $u(x)$ is continuous for all x in I.

C′: $u'(x)$ exists and is continuous for all x in I.

C″: $u''(x)$ exists and is continuous for all x in I.

$\mathbf{C}^{(\infty)}$: $u^{(n)}(x)$ exists and is continuous for all x in I regardless of n.

A: $u(x)$ is representable by a convergent Taylor series about each x in I.

It is clear, a priori, that each restriction implies the preceding ones, and Exercise 1 will indicate that each restriction is satisfied by fewer functions than the preceding one.

What happens when we introduce $\sqrt{-1}$ (or i)? We write the independent variable z in terms of two independent real variables x and y, as

† In this *review* chapter, the elements are presented in an order which is certainly *not suitable for learning the elementary theory* but rather for stressing the viewpoint of the later text. The definitions are given mostly when there is either ambiguity of usage or unusual significance, since the reader is expected to have an elementary reference at hand. In many cases the definitions will be justified only because of the reader's preliminary knowledge of results in the elementary theory.

(1-1) $$z = x + iy$$

and we write the dependent variable symbolically as $w(z)$; and in terms of two real functions $u(x,y)$ and $v(x,y)$ we write $w(z)$ as

(1-2) $$w(z) = u(x,y) + iv(x,y)$$

Thus in real terms a complex function is a two-dimensional "surface" $[x,y,u(x,y),v(x,y)]$ in a real four-dimensional (x,y,u,v) space. We *could* speak of smoothness of $u(x,y)$ and $v(x,y)$ in terms of tangents to this four-dimensional surface, but this type of smoothness has no interest for us. We are considering the stronger condition of differentiability of w in terms of z.

We call a function $w(z)$ *analytic* in a region† $\mathcal{R}$ if the derivative $w'(z)$ exists‡ for each point of the region. The function is then called *analytic at each point* of the region. We do not define analyticity at a point except by defining it in a surrounding region. (Synonyms for "analytic" are "regular" and "holomorphic.")

THEOREM 1-1 *An analytic function in $\mathcal{R}$ has derivatives of all orders in $\mathcal{R}$ and is representable as well by a convergent power series expansion at each point of the region within any circular disk centered at that point and lying in the region.*

This theorem is practically the contents of an entire course in complex variables! Its importance is evident since it essentially vitiates all regularity distinctions once the derivative is presupposed. The proof of this theorem is effected by the rather ingenious well-known *integration* procedure which leads to the existence of power series.

Strangely enough, despite the fact that Theorem 1-1 does not mention integration, it has been proved classically§ only by integration. Perhaps then, rather than become overly impressed by this matter of semantics, we should do better to conclude that, from a more mature point of view, differentiation and integration are not so very different but are part of, say, some generalized study of linear operators.

† A *region* is defined as an open set which is arcwise connected. Some authors call this a *domain*, but we use the term "domain" to mean a region with part (or all) of its boundary.

‡ We recall that in relation to the xy plane, such differentiability requires the existence of a double (real) limit since we define $w'(z)$ or dw/dz as the limit of $[w(z + h) - w(z)]/h$ as the single (complex) variable $h \to 0$. The Cauchy-Riemann conditions, so vital to the initial understanding of the subject, are deferred until Part III in order to stress our effort to believe that the complex variables enjoy an existence largely—if not wholly—independent of their real and imaginary parts!

§ Modern revisionism is reflected in work such as that of Whyburn (see Bibliography), but we generally ignore developments outside of our direct line of motivation.

EXERCISES

1 Show how to form real functions $u(x)$ with all degrees of regularity desired above. {*Hint:* Show $U_t(x) = [\exp(-1/x^2)]/x^t$ vanishes as the real variable $x \to 0$ for any positive integer t; show $U_0(x)$ [with $U_0(0) = 0$] serves as a function with all orders of derivative but no power series representation at the origin.}

2 Find a function $w(z) = u(x,y) + iv(x,y)$ with a complex derivative at every point on the real axis but which is not analytic in any neighborhood of a real point. [*Hint:* Use $U_0(y)$.]

3 Show that $\overline{f(\bar{z})} = g(z)$ is analytic at $\bar{z}_0$ whenever $f(z)$ is analytic at z_0.

1-2 Integration on Curves and Chains

We introduce integration by defining a *smooth curve segment* $\mathcal{K}$ by differentiable functions parametrized by arc length s

$$\mathcal{K}: \quad x = x(s) \qquad y = y(s) \qquad 0 \leq s \leq l \tag{1-3}$$

Here $dx^2 + dy^2 = ds^2$ and l is the length in usual elementary fashion. A *piecewise smooth curve* (or more simply a *curve*) is the union of any finite set of curve segments meeting end to end, i.e., in the manner described formally as follows: If $\mathcal{K}_1$ and $\mathcal{K}_2$ are two curve segments with

$$\begin{aligned} \mathcal{K}_1: \quad x &= x_1(s) \qquad y = y_1(s) \qquad 0 \leq s \leq l_1 \\ \mathcal{K}_2: \quad x &= x_2(s) \qquad y = y_2(s) \qquad 0 \leq s \leq l_2 \end{aligned} \tag{1-4}$$

then if $x_1(l_1) = x_2(0)$ and $y_1(l_1) = y_2(0)$, we join $\mathcal{K}_1$ to $\mathcal{K}_2$, writing the sum (purely formally) as

(1-5)

$$\mathcal{K}_1 + \mathcal{K}_2: \quad x = \begin{cases} x_1(s) \\ x_2(s - l_1) \end{cases} \qquad y = \begin{cases} y_1(s) & \text{when } 0 \leq s \leq l_1 \\ y_2(s - l_1) & \text{when } l_1 < s \leq l_1 + l_2 \end{cases}$$

It is clear how any number of segments $\mathcal{K}_1 + \mathcal{K}_2 + \mathcal{K}_3 + \cdots + \mathcal{K}_n$ can be joined to form a curve $\mathcal{K}$. We shall use $-\mathcal{K}$ to denote† the curve $\mathcal{K}$ parametrized in reverse fashion (with $l - s$ replacing s if l is the length of $\mathcal{K}$). We shall say qualitatively the distinction between $\mathcal{K}$ and $-\mathcal{K}$ is its *orientation*. Thus $\mathcal{L} = \mathcal{K} - \mathcal{K}$ would denote the tracing of $\mathcal{K}$ followed by the retracing of $\mathcal{K}$ with opposite orientation. Here $\mathcal{L}$ is not "zero"; it is a curve twice as long as $\mathcal{K}$ but one which returns to its origin.

The definition of the integral can now be effected in terms of a complex parametric form of our curve $\mathcal{C}$ as $x + iy = z(s)$. If $f(z)$ is an analytic function in a region $\mathcal{R}$ and if $\mathcal{C}$ is in the region $\mathcal{R}$, then we define the integral by the Riemann sum as follows:

$$\int_{\mathcal{C}} f(z)\, dz = \lim_{\delta \to 0} \sum_{j=1}^{m} f(\zeta_j)(z_j - z_{j-1}) \tag{1-6}$$

† The symbol $|\mathcal{K}|$ accordingly denotes the point set in question, often called the "arc."

Here $\mathcal{C}$ has length l and $0 = s_0 < s_1 < \cdots < s_m = l$ is a partition of l with maximum interval $(s_j - s_{j-1}) = \delta$ and $z_j = x(s_j) + iy(s_j)$; and ζ_j is an arbitrary point on the curve segment from z_{j-1} to z_j. The existence of the limit is shown in elementary texts.

Now when we consider the integral (1-6), we become aware of the fact that retracing the path of integration nullifies the integral; for example,

$$\int_{\mathcal{K}-\mathcal{K}} = \int_{\mathcal{K}} - \int_{\mathcal{K}} = 0$$

We therefore introduce a purely abstract concept to make manifest this almost trivial property.

We consider the formal additive group of finite linear combinations of curves with integral coefficients. This group contains expressions such as

$$\mathcal{L} = n_1\mathcal{K}_1 + n_2\mathcal{K}_2 + \cdots + n_r\mathcal{K}_r \qquad (n_i \text{ positive or negative integers})$$

where $\mathcal{K}_1, \mathcal{K}_2, \ldots, \mathcal{K}_r$ need not even have a point in common. Indeed any curve $\mathcal{K}_i$ can have some retraced portions, or for that matter, $\mathcal{K}_i$ might be of the type $\mathcal{K}$-$\mathcal{K}$. We say that two such formal expressions constitute the same *chain* if they can be identified by cancellation and regrouping in obvious fashion. Thus a 0- (null) chain is defined (and it is not identified with any particular point). We can, of course, speak of an obvious correspondence between the adjoining of curve segments in (1-5) and the addition of the corresponding chains. The correspondence is not biunique; two different curves, for example, 0 and $\mathcal{K}$-$\mathcal{K}$, might determine the same chain. Sometimes the chains $\mathcal{L}$ are called "1-chains" because of the dimensionality of curves.

The purpose of all this formalism is the following idea: We can write

$$\int_{\mathcal{C}} f(z)\,dz = (\mathcal{C}, f(z)\,dz) \tag{1-7}$$

and we can think of the above expression as being bilinear (or linear separately) in the differential† $f(z)\,dz$ and the chain $\mathcal{C}$. The essence of the bilinearity is our ability to write

$$(\mathcal{C}, f_1(z)\,dz) + (\mathcal{C}, f_2(z)\,dz) = (\mathcal{C}, [f_1(z) + f_2(z)]\,dz) \tag{1-8a}$$

and at the same time

$$(\mathcal{C}_1, f(z)\,dz) + (\mathcal{C}_2, f(z)\,dz) = (\mathcal{C}_1 + \mathcal{C}_2, f(z)\,dz) \tag{1-8b}$$

regardless of whether or not $\mathcal{C}_1$ adjoins $\mathcal{C}_2$.

These seemingly trivial ideas will be brought to the foreground now in preparation for Part III where they play a vital role. In the meantime,

† For now, we could write the integral as $(\mathcal{C}, f(z))$ if we wished, but we want to have a notation completely compatible with that of Part III where a more reasonable justification for speaking in terms of *differentials* will be developed.

our language will be sufficiently flexible for us to regard the symbol $\mathcal{C}$ as either a curve or a chain as the occasion requires.

The important integral estimate acquires a special symmetry when we use the bilinear form (1-7)

$$\left| \int_{\mathcal{C}} f\, dz \right| = |(\mathcal{C}, f\, dz)| \leq l(\mathcal{C}) \sup_{\mathcal{C}} |f| \tag{1-9}$$

where $l(\mathcal{C})$ is the length of $\mathcal{C}$ and $\sup_{\mathcal{C}} |f|$ is the supremum of $|f(z)|$ for z on $\mathcal{C}$. Here $\mathcal{C}$ designates a curve (not a chain).

EXERCISES

1 Show that two piecewise smooth curves can have an infinitude of isolated points of intersection. In fact this is true even for convex segments (segments where d^2y/dx^2 is positive on each curve). [*Hint:* For $-1 \leq x \leq 1$ consider the curves $y_1(x) = 20x^2 + x^5 \sin (1/x)(x \neq 0)$, $y_1(0) = 0$, and $y_2(x) = 20x^2$.] Also verify that they satisfy the smoothness condition **C**$''$.

2 Prove the integral estimate (1-9) by direct appeal to the definition (1-6).

1-3 Cauchy Integral Theorem

We now think of curves in terms of the boundary of a region. We define a *closed curve* $\mathcal{K}$ as one for which $x(0) = x(l)$, $y(0) = y(l)$ where l is the length of $\mathcal{K}$. We define a *simple curve* as a curve with no self-intersections except in the case of a closed curve which meets itself at its end points. Symbolically, if l is the length and if $0 \leq s_1 < s_2 \leq l$, then $(x(s_1),y(s_1)) = (x(s_2),y(s_2))$ only (possibly) when $s_1 = 0$, $s_2 = l$.

Let us say that a region $\mathcal{R}$ "is bounded by a curve $\mathcal{K}$" (or $-\mathcal{K}$) if the set of boundary points† of $\mathcal{R}$ consists of the point set $|\mathcal{K}|$. This is a concept we shall surely have to enlarge, but for the time being, it enables us to express theorems on integration.

THEOREM 1-2 (*Cauchy-Goursat*) *Let $f(z)$ be analytic in a region $\mathcal{R}^*$ which contains a simple closed curve $\mathcal{C}$ and a subregion $\mathcal{R}$ for which $\mathcal{C}$ is the boundary; then*

$$\int_{\mathcal{C}} f(z)\, dz = 0 \tag{1-10}$$

This theorem was proved by Cauchy (1825) under a (really innocuous) variant of the concept of analyticity, requiring that $f'(z)$ not merely exist in $\mathcal{R}$ but also be continuous there. Goursat (1900) proved the stronger result by a proof which also had the advantage of keeping the argument in terms of *complex* quantities (avoiding the explicit use of $f = u + iv$, $dz = dx + i\,dy$, etc.).

† *A boundary point of a set* is defined as a point any neighborhood of which contains a point belonging to the set and a point not belonging to the set.

Although the sequence of proofs is somewhat interrupted, this is a good place to insert certain corollaries to Theorem 1-2.

COROLLARY 1 *If* z_1 *and* z_2 *are two points interior to* $\mathcal{R}$, *as defined in the above theorem, then* $\int_{z_1}^{z_2} f(z)\,dz$ *can be defined independently of the curve joining* z_1 *to* z_2 *as long as the curve lies in* $\mathcal{R}$.

COROLLARY 2 (*Morera's theorem*, 1886) *If* $f(z)$ *is a continuous function of* z *in* $\mathcal{R}$ *and if*

$$\int_{\mathcal{C}} f(z)\,dz = 0 \tag{1-11}$$

about any closed curve $\mathcal{C}$ *in* $\mathcal{R}$, *then* $f(z)$ *represents an analytic function in* $\mathcal{R}$.

Let us recall the proof of this last result. On the basis of Corollary 1, we can define the unique function

$$F(z) = \int_a^z f(\zeta)\,d\zeta \tag{1-12}$$

where a is an arbitrarily fixed point of $\mathcal{R}$ and z is a variable point of $\mathcal{R}$. Then $F'(z)$ exists [$=f(z)$ by Exercise 2] in $\mathcal{R}$. Hence $F(z)$ is analytic, and (by Theorem 1-1) so is its derivative $f(z)$, which completes the proof. The naïvety of this device is in the characteristic vein of classical complex variable theory.

COROLLARY 3 *If* $f_n(z)$ *is a sequence of analytic functions converging uniformly to a limit function* $f(z)$ *in a region* $\mathcal{R}^*$ *as* $n \to \infty$, *then* $f(z)$ *is analytic in* $\mathcal{R}^*$.

PROOF In any subregion $\mathcal{R}$ (of $\mathcal{R}^*$) bounded by a simple closed curve $\mathcal{C}$, we apply Morera's theorem to the following result (based on uniform convergence and Theorem 1-2):

$$\int_{\mathcal{C}} f(z)\,dz = \lim_{n\to\infty} \int_{\mathcal{C}} f_n(z)\,dz = \lim_{n\to\infty} 0 = 0 \tag{1-13}$$

Q.E.D.

Of course, if $f(z)$ is analytic merely on $\mathcal{C}$, a closed curve, it certainly does not follow that $\int_{\mathcal{C}} f(z)\,dz = 0$. The best known counterexample, of course, is

$$\int_{\mathcal{C}} \frac{dz}{z} = 2\pi i \tag{1-14}$$

where $\mathcal{C}$ is a simple closed curve "surrounding the origin" in the sense that the angle subtended at the origin by $\mathcal{C}$ increases by 2π with the parametrization. To see (1-14) as a triviality, it suffices to look upon the integrand dz/z as† $d \log z$ and write

(1-15a)
$$\int_{\mathcal{C}} \frac{dz}{z} = \int_{z \text{ on } \mathcal{C}} d \log z = \log z \Big|_{\mathcal{C}} = \log |z| + i \arg z \Big|_{\mathcal{C}} = i \arg z \Big|_{\mathcal{C}} = 2\pi i$$

meaning that if z is parametrized as $z(s) = x(s) + iy(s)$ by a continuous variation of s from 0 to l (the length),

(1-15b)
$$\log z \Big|_{\mathcal{C}} = \log z(l) - \log z(0) = 2\pi i$$

Here we note that $d \log z$ is always single-valued *as a differential* despite‡ the fact that the indeterminacy of $\log z$ can be any number of the type $2\pi im$ for m a positive or negative integer.

Equation (1-14) is important enough to be sometimes stated as a theorem. For our purposes, it is superseded by Cauchy's residue theorem (see Sec. 1-8 below). For any curve $\mathcal{C}$ (simple or not) and any point a not on $\mathcal{C}$, we note that it enables us to define the following function:

(1-16)
$$N(\mathcal{C};a) = \frac{1}{2\pi i} \int_{\mathcal{C}} \frac{dz}{z - a} = \frac{1}{2\pi i} \int_{z \text{ on } \mathcal{C}} d \log (z - a) = \frac{1}{2\pi} \arg (z - a) \Big|_{\mathcal{C}}$$

This is called the "winding number" of $\mathcal{C}$ with respect to a. All we can say for now is that *when* $\mathcal{C}$ *is closed*, the values of $N(\mathcal{C};a)$ are *integers* (because the indeterminacy of the *argument* is in multiples of 2π).

The expression $N(\mathcal{C};a)$ is simultaneously a function of $\mathcal{C}$ and a (although the term "functional" is traditionally more appropriate). Clearly, N depends continuously on a and $\mathcal{C}$ unless a reaches $\mathcal{C}$. This means that if a family of curves $\mathcal{C}_t$ (none passing through a) were parametrized continuously by parameter t, then $N(\mathcal{C}_t;a)$ would vary continuously with t. Finally, when $\mathcal{C}$ is closed, N takes on discrete values $0, \pm 1, \pm 2, \ldots$. *Thus the winding number for a closed curve* $\mathcal{C}$ *about a point* a *remains constant, as* $\mathcal{C}$ *and* a *vary continuously, as long as* a *is never on* $\mathcal{C}$.

Later on we shall see that the logarithm function essentially characterizes the geometry of the plane (rather than conversely)! Indeed, we generalize the plane to "other geometrical entities" and the logarithm function to so-called "abelian integrals," but only for the plane will we be able to speak of anything like a winding number.

† We assume $\exp z$ to be defined by power series and $\log z$ to be defined as its inverse. Trigonometric functions are defined (as in elementary texts) through series from Euler's relation $\exp iz = \cos z + i \sin z$; and from it $\log z = \log |z| + i \arg z$ (referring to the real logarithm and argument).

‡ This is incidentally the first step in thinking of single-valued integrands or differentials as a device to broaden the concept of single-valued functions to such "multivalued" integrals.

EXERCISES

1 Let $f(z) = x - iy$. Test Morera's theorem. Also show directly that $f'(0)$ does not exist.

2 (*a*) ("Fundamental" theorem of calculus.) In Morera's theorem show $F'(z) = f(z)$ by taking the limit of $[F(z+h) - F(z)]/h - f(z) = \int_z^{z+h} [f(\zeta) - f(z)]\, d\zeta/h$.
(*b*) Show that Morera's theorem is valid if in the hypothesis the arbitrary closed curve $\mathcal{C}$ is restricted to an arbitrary triangle.

3 Show that the definition of a region can be taken equivalently as an *open set which is arcwise connected* (such that any two points are connected by an *arc* lying in the set).

TOPOLOGICAL CONSIDERATIONS

1-4 Jordan Curve Theorem

We have asserted that Theorem 1-1 on differentiation depends on Theorem 1-2 on integration. Equally remarkable is the fact that Theorem 1-2 (as well as its relation to Theorem 1-1) depends on another concept not yet stated, namely, the idea that a boundary of a region (although defined as a set of boundary points) actually, in some sense, *encloses* a region.

Let a *smooth* curve segment $\mathcal{C}$ be part of the boundary (set) of a region $\mathcal{R}$. Let $\mathcal{C}$ be parametrized as $z(s) = x(s) + iy(s)$. Let us say that $\mathcal{R}$ *lies to the left* of $\mathcal{C}$ for the given parametrization if for each point $z(s)$ of $\mathcal{C}$ the normal to $\mathcal{C}$, which is given by

$$n(s) = -y'(s) + ix'(s) \tag{1-17}$$

is directed into $\mathcal{R}$. More precisely, this means that the point

$$Z(\varepsilon) = z(s) + \varepsilon n(s) \tag{1-18}$$

lies in $\mathcal{R}$ for all $\varepsilon > 0$ sufficiently small and no $\varepsilon < 0$ sufficiently small numerically. In symbols, $\mathcal{R}$ lies to the left of $\mathcal{C}$ if for every s there exists a real number $\Delta(s) > 0$, such that

$$\text{if } 0 < \varepsilon < \Delta(s) \text{ then† } Z(\varepsilon) \in \mathcal{R} \text{ and } Z(-\varepsilon) \notin \mathcal{R} \tag{1-19}$$

The left-hand normal, incidentally, is called the "inner" normal to $\mathcal{C}$ with respect to $\mathcal{R}$.

Let $\mathcal{R}$ be a region for which the set of boundary curves can be parametrized as disjoint simple closed curves $\mathcal{C}_1, \ldots, \mathcal{C}_n$ finite in number. (For now, we thereby bar regions with isolated boundary points or "slits.") Then the boundary consists of the set $|\mathcal{C}_1| + \cdots + |\mathcal{C}_n|$. Customarily

† We use the usual paraphernalia of set theory: $\cap$ (intersection), $\cup$ (union "and/or"), $\in$ (belongs to), etc. The closure of a set $\mathcal{S}$ is $\bar{\mathcal{S}}$ ($= \mathcal{S} \cup$ boundary points of $\mathcal{S}$). Also $c(\mathcal{S})$ is the complement of set $\mathcal{S}$. The inclusion $\subset$ is strict ($\subseteq$ is used for nonstrict inclusion). Negations $\notin$ and direction of inclusion, $\supset$, $\supseteq$, etc., are self-explanatory as they occur.

$\mathcal{R}$ is called "n-tuply-connected" ("simply-connected" for $n = 1$). Also $\mathcal{R}$ is called a "disk" for $n = 1$, an "*annulus*," for $n = 2$.

If, in accordance with the given parametrization of each $\mathcal{C}_k$ ($1 \leq k \leq n$), $\mathcal{R}$ lies to the left† of $\mathcal{C}_k$ at each point, we say that the boundary of $\mathcal{R}$ (written $\partial\mathcal{R}$) is given formally as a "chain"

$$\partial\mathcal{R} = \mathcal{C}_1 + \cdots + \mathcal{C}_n \tag{1-20}$$

Otherwise expressed, we say that $\partial\mathcal{R}$ as given in (1-20) "surrounds" any point of $\mathcal{R}$.

THEOREM 1-3 (*Jordan curve theorem*, 1887) *Every simple closed curve* $\mathcal{C}$ *in the* z *plane determines a finite region* $\mathcal{R}$ *for which*

$$\partial\mathcal{R} = \mathcal{C} \qquad \text{or} \qquad \partial\mathcal{R} = -\mathcal{C} \tag{1-21a}$$

If $\partial\mathcal{R} = \mathcal{C}$ *we say that* $\mathcal{C}$ *is "positively" oriented.*

COROLLARY 1 *The complement of a simple closed curve* $\mathcal{C}$ *consists of two disjoint regions, the finite region* $\mathcal{R}$ *and an infinite region* $\mathcal{S}$.

COROLLARY 2 *If* $\mathcal{R}$ *is an n-tuply-connected region with boundary set consisting of the disjoint simple closed curves* $\mathcal{C}_1, \mathcal{C}_2, \ldots, \mathcal{C}_n$, *then with proper choice of signs*

$$\partial\mathcal{R} = \pm\mathcal{C}_1 \pm \cdots \pm \mathcal{C}_n \tag{1-21b}$$

Thus in (the Cauchy-Goursat) Theorem 1-2 it would have been superfluous to state that $\mathcal{C}$ is the boundary of a subregion $\mathcal{R}$ if $\partial\mathcal{R}^*$ were a simple closed curve also. Now actually, in most problems, it would not be necessary to know the Jordan curve theorem since inequalities can give the region and the boundary. In general, however, a simple closed curve must be thought of as a "maze threading its way" throughout the plane, hence it is not obvious at first glance whether a point near the maze is interior or exterior to $\mathcal{C}$.

A proof of the Jordan curve theorem is a standard part of topology and will not be reproduced here. We consider only piecewise smooth curves so that the full strength of topology is not called into play.

The relationship between the Jordan curve theorem and complex variable theory is essentially established through the winding number $N(\mathcal{C};a)$ defined in (1-16).

THEOREM 1-4 *If* $\mathcal{C} = \partial\mathcal{R}$ *for a simple closed curve* $\mathcal{C}$, *then*

$$N(\mathcal{C};a) = \begin{cases} 1 & \text{if } a \in \mathcal{R} \\ 0 & \text{if } a \notin \mathcal{R} \text{ and } a \notin \mathcal{C} \end{cases} \tag{1-22}$$

† We ignore corner points on $\partial\mathcal{R}$, of course, in the test of "leftness" given in (1-19); the right- and left-hand sides can be determined by approximating such corners.

(*A sketch of the proof of Theorem* 1-4 *using the Jordan curve theorem is given as Exercises* 2 *to* 4.)

Thus Theorem 1-2 can be restated in terms of $N(\mathcal{C};a)$.

THEOREM 1-5 *Let $f(z)$ be analytic in a region $\mathcal{R}^*$. Consider a simple closed curve $\mathcal{C}$ in $\mathcal{R}^*$ such that $N(\mathcal{C};a) = 0$ for all $a \notin \mathcal{R}^*$. Then*

$$\int_{\mathcal{C}} f(z)\,dz = 0$$

This statement of the Cauchy integral theorem is the work of Ahlfors (1953).

EXERCISES

Assume the Jordan curve theorem (Theorem 1-3) for the following exercises.

1 Deduce the corollaries of Theorem 1-3. [*Hint:* For Corollary 2 consider the finite region $\mathcal{R}_t$ determined by each simple closed curve $\mathcal{C}_t$ (such that $\partial\mathcal{R}_t = \pm\mathcal{C}_t$).] Show how to express $\mathcal{R}$ explicitly in terms of $\mathcal{R}_t$ by means of the $\cup$, $\cap$, and c (complement) symbols.

2 Prove that $N(\mathcal{C};a) = 0$ for the simple closed curve $\mathcal{C}$ and a point a sufficiently far from the origin, such that for every $Z \in \mathcal{C}$, $|Z - a| > l(\mathcal{C})/2\pi$. (*Hint:* Use the integral estimate of Sec. 1-2 above.) Prove that $N(\mathcal{C};a) = 0$ for $a \in \mathcal{S}$, the exterior region of $|\mathcal{C}|$.

3 Prove that $N(\mathcal{C};a)$ changes by 1 as the variable point a crosses the curve $\mathcal{C}$, say, from the exterior region on $\mathcal{S}$. For convenience, let point a cross $\mathcal{C}$ along a normal at point $A \in \mathcal{C}$. [*Hint:* Let $\mathcal{C}_1$ be a "small" segment of $\mathcal{C}$ containing A, and let $\mathcal{C}_2$ be the remainder of $\mathcal{C}$ ($= \mathcal{C}_1 + \mathcal{C}_2$).] Then, writing

$$N(\mathcal{C};a) = N(\mathcal{C}_1;a) + N(\mathcal{C}_2;a)$$

show that $N(\mathcal{C}_1;a)$ must change by 1 and $N(\mathcal{C}_2;a)$ must change by 0, as a crosses $\mathcal{C}$ at A. [*Hint:* The hard part is to show $N(\mathcal{C}_1;a)$ changes by 1; but consider two positions of a symmetric with respect to a chord on $\mathcal{C}_1$.]

4 Complete the proof of Theorem 1-4.

5 Prove Theorem 1-5.

1-5 Other Manifolds

To appreciate the role of the Jordan curve theorem, it is desirable to imagine a suitable situation where it does not hold.

We introduce the concept of an n-dimensional *manifold* $\mathfrak{M}$ as a connected point set in, say, k ($\geq n$) dimensional euclidean space which "looks like n-dimensional euclidean space at each point." Technically speaking, this means that n (curvilinear) coordinates can be introduced in the neighborhood of any point of $\mathfrak{M}$ to create a one-to-one correspondence with a neighborhood of the origin in n-dimensional euclidean space such that the correspondence and its inverse are continuous.† For example, a region of

† Such a correspondence is called a *homeomorphism*.

a plane is trivially a two-dimensional manifold because it is open, and a unit circle in the plane is a one-dimensional manifold. A dihedral angle formed by half planes in three-dimensional euclidean space is capable of being opened "flat" and hence is a two-dimensional manifold.

To think of some "nonmanifolds," take the union of two intersecting lines, say the real and imaginary axes (positive and negative). Here, a neighborhood of their point of intersection is a set consisting of four intersecting rays. It is not possible to construct a homeomorphism of it with a neighborhood of the origin† in one-dimensional space, e.g., with $-\varepsilon \leq x \leq +\varepsilon$. We might also take the closed disk $\mathfrak{D}$ $|z| \leq 1$. It consists of two manifolds, the (open) disk $|z| < 1$ and the circle $|z| = 1$, but $\mathfrak{D}$ is, of course, not open; e.g., the points on $|z| = 1$ do not have the (full) neighborhoods in the closed disk as required by the definition.

We use the generic term "variety" to describe a geometric set which is not a manifold. The term "algebraic variety" is used if the set is described by an algebraic system of equations (such as the union of the x and y axes which is described by $xy = 0$).

We do not write any more formal definitions at this point since *analytic* manifolds will be considered in some detail later on. We can, however, intuitively consider some two-dimensional manifolds imbedded for convenience in three-dimensional euclidean space, vis-à-vis the validity of the Jordan curve theorem. Actually, returning to a simple closed curve $\mathcal{C}$ in the plane, we are able to write $\pm\mathcal{C} = \partial\mathcal{R}$ for some $\mathcal{R}$ in Theorem 1-3 essentially because of two specifications:

(*a*) The left-hand ε neighborhood of $\mathcal{C}$ is definable as being distinct from the right-hand ε neighborhood of $\mathcal{C}$ (*orientability*).

(*b*) The left-hand ε neighborhood of $\mathcal{C}$ cannot be joined to the right-hand ε neighborhood of $\mathcal{C}$ without crossing $\mathcal{C}$ (*separation*).

There are two well-known manifolds which show how Jordan's theorem might fail on either specification.

MÖBIUS STRIP

Consider the domain $\mathcal{S}$ given by coordinates (in the xy plane)

$$\begin{aligned} 0 &\leq x < l \\ -a &< y < a \end{aligned} \tag{1-23}$$

We "identify" the points $(0,-y)$ and $(l,+y)$. This process of identifying points is discussed in some detail only in the theory of manifolds (Part II). For the time being let us think of representing the domain $\mathcal{S}$ as a thin paper strip (see Fig. 1-1a) and the ends ($x = 0$ and $x = l$) are tied after a 180° twist [identifying $(0,-y)$ with $(l,+y)$].

† For example, a homeomorphic image of this neighborhood could lie on only two of the four rays at the intersection.

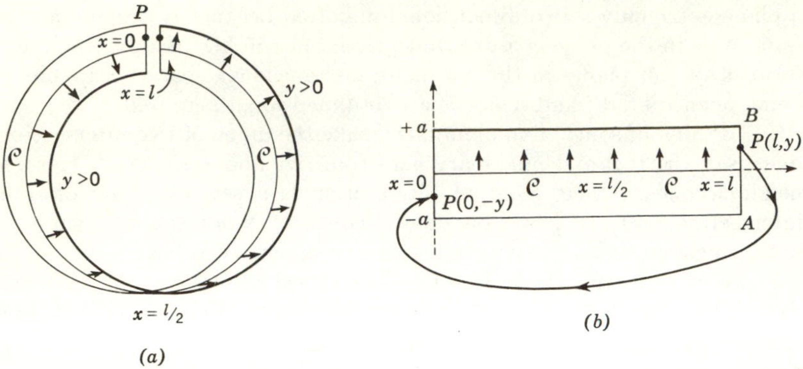

Fig. 1-1 *Möbius strip.* The left-hand ε neighborhood of $\mathcal{C}$ can be joined to the right-hand ε neighborhood of $\mathcal{C}$ by going "to the right" of $x = 1$ (see long arrow).

The curve (in the xy plane again)

$$\mathcal{C}: \quad 0 \leq x \leq l \qquad y = 0 \tag{1-24}$$

becomes a closed curve, but it does not provide the left-hand boundary of any region $\mathcal{R}$. To see this, note that the left-hand side of $\mathcal{C}$ contains only points where $y > 0$, but these points still "connect with" points where $y < 0$ after we go from $x = l$ to $x = 0$ (as shown by the long arrow in Fig. 1-1b).

Here the difficulty is that the right- and left-hand sides of $\mathcal{C}$ are connected on completing a circuit of $\mathcal{C}$. We shall ultimately see that this situation can be completely avoided in complex variable theory.

TORUS

Consider the surface of revolution of, say, the circle

$$(\xi - a)^2 + \zeta^2 = b^2 \qquad a > b > 0 \tag{1-25}$$

about the ζ axis in (real) (ξ,η,ζ) space. Then, these loci

$\mathcal{C}$: azimuthal circles (on ζ = const)

$\mathcal{C}^*$: longitudinal circles $\left(\text{on } \dfrac{\xi}{\eta} = \text{const}\right)$

constitute simple closed curves which still do not serve as boundaries of any region.

Here we have orientability, however. If we define the (inner) left-hand normals to $\mathcal{C}$, they point to only one side of $\mathcal{C}$ (unlike the case of the Möbius strip), but the curve $\mathcal{C}^*$ joins the left-hand side of $\mathcal{C}$ to the right-hand side of $\mathcal{C}$ "through the hole in the torus." Nevertheless, we can speak of a right-hand versus a left-hand ε neighborhood of $\mathcal{C}$ (see Fig. 1-2b).

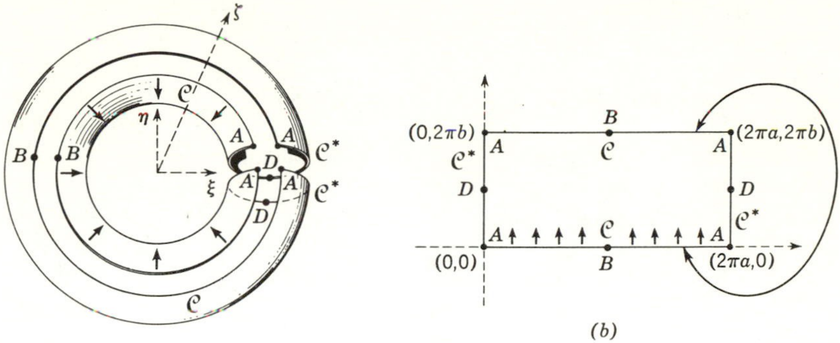

(a) (b)

Fig. 1-2 *Torus.* The normals to $\mathcal{C}$ point out the left-hand ε neighborhood, but $\mathcal{C}^*$ joins them to the right-hand ε neighborhood. The torus (*a*) is cut along $\mathcal{C}$ and $\mathcal{C}^*$ (so that there are two replicas of each) and the torus is flattened out (*b*).

To see orientability in terms of "flattening out to a plane," we can imagine the torus cut along $\mathcal{C}$ and $\mathcal{C}^*$ and then imagine it opened up and laid out flat on the rectangle measuring, say, $2\pi a$ by $2\pi b$ [allowing intuitively for some slight "stretching" (see Fig. 1-2*a*)]. The curve $\mathcal{C}$ appears as both the upper and lower side of the rectangle identified (see long arrow in Fig. 1-2*b*). The curve $\mathcal{C}^*$ appears as both the right- and left-hand side of the rectangle identified. The points B and D show the identification; the point A is "the same at all four corners." In other words, the torus can be laid out as the rectangle and opposite sides identified without twisting; the Möbius strip involves twisting (in some intuitive fashion).

1-6 Homologous Chains

The term "topological" will be used rather loosely to mean "pertaining to properties of sets, functions, etc., which are invariant under one-to-one continuous mappings with continuous inverses." (Such mappings are called "bicontinuous," "topological," "homeomorphic," etc.) This interpretation of the term† is in close keeping with analysis.‡

† More classically, topology (analysis situs) had been interpreted to mean the study of relative positions, although more recently it is being interpreted to mean the study of certain algebraic operations, particularly "homology" sequences, based on geometrical constructs. These are barely hinted at in Chaps. 7 and 8.

‡ On the whole, we shall pursue proofs in topology only to the extent that they have an obvious interest for complex analysis. This leaves the reader much occasion to consult with literature in the Bibliography. We shall use the "double standard of rigor" that proofs "on manifolds" are consummated when they are reduced to "obvious situations in the plane." Thus, in referring to Fig. 1-2*a*, the reader is not expected to see that $\mathcal{C}^*$ joins the right- and left-hand sides of $\mathcal{C}$, but he is expected to see (and believe) this remark in reference to Fig. 1-2*b*.

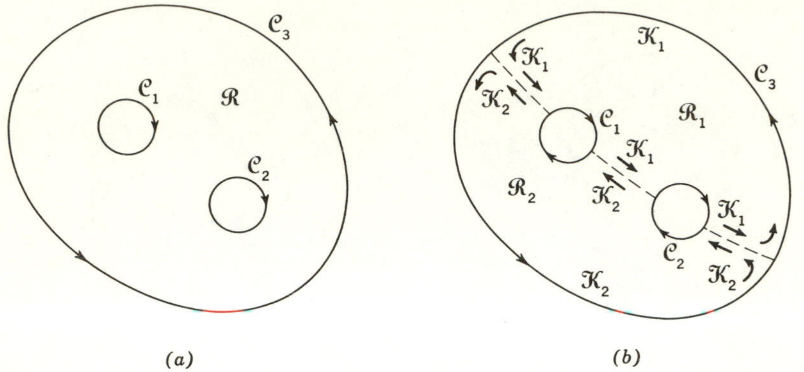

Fig. 1-3 Illustration for remarks on homology.

Let us view a very familiar situation in the light of topological concepts by referring to Fig. 1-3. Here a function $f(z)$ is analytic in some region $\mathcal{R}^*$ (not marked in the diagram), but $\mathcal{R}$ is a region whose closure lies interior to $\mathcal{R}^*$. Then we would say

(1-26)
$$\int_{\mathcal{C}_1} f(z)\,dz + \int_{\mathcal{C}_2} f(z)\,dz + \int_{\mathcal{C}_3} f(z)\,dz = 0$$

For proof, we would divide $\mathcal{R}$ into $\mathcal{R}_1$ and $\mathcal{R}_2$ (as shown in Fig. 1-3b), and in this manner we would take

$$\int_{\mathcal{K}_1} f(z)\,dz = \int_{\mathcal{K}_2} f(z)\,dz = 0$$

where $\mathcal{K}_1$ is the boundary of $\mathcal{R}_1$ and $\mathcal{K}_2$ is the boundary of $\mathcal{R}_2$. Then we would show that the chain $\mathcal{K}_1 + \mathcal{K}_2$ is equivalent to $\mathcal{C}_1 + \mathcal{C}_2 + \mathcal{C}_3$ (by the cancellation of retraced portions in usual elementary fashion).

In effect, we are asserting $\partial\mathcal{R}_1 + \partial\mathcal{R}_2$ is equivalent to $\partial\mathcal{R}$. In doing this elementary proof, we find ourselves thinking of $\mathcal{R}$ as $\mathcal{R}_1 + \mathcal{R}_2$. It is quite convenient to formalize this.

Let $\mathcal{R}^*$ be some region of the plane. Consider a (finite or infinite) family of subregions $\{\mathcal{R}\}$ such that

(a) Each $\mathcal{R} \in \{\mathcal{R}\}$ is bounded by a finite number of simple closed curves.

(b) $\{\mathcal{R}\}$ is closed under intersection [if $\mathcal{R}_1 \in \{\mathcal{R}\}$ and $\mathcal{R}_2 \in \{\mathcal{R}\}$, then $\mathcal{R}_1 \cap \mathcal{R}_2 \in \{\mathcal{R}\}$].

(c) $\{\mathcal{R}\}$ is contained with closures in $\mathcal{R}^*$; that is, if $\mathcal{R} \in \{\mathcal{R}\}$, then $\overline{\mathcal{R}} \subset \mathcal{R}^*$.

We next define† formal chains of regions or 2-chains:

(1-27a) $$\mathcal{S} = m_1\mathcal{R}_1 + m_2\mathcal{R}_2 + \cdots + m_r\mathcal{R}_r$$

by taking finite integral combinations of regions in $\{\mathcal{R}\}$. We consider chains to be *equivalent* if they can be identified by a composition and cancellation process which ignores one-dimensional continua. For example, if $\mathcal{R}_1$ and $\mathcal{R}_2$ compose $\mathcal{R}$, as in Fig. 1-3b, and $\mathcal{R}_1$ and $\mathcal{R}_2$ do not intersect, then we can write $\mathcal{R} = \mathcal{R}_1 + \mathcal{R}_2$ or $\mathcal{R}_1 = \mathcal{R} - \mathcal{R}_2$, ignoring the common boundary. (This is a situation analogous to the ignoring of points of intersections when we assert the equivalence of the 1-chains referred to in Sec. 1-2.)

If we let $\mathcal{S}$ denote a 2-chain, then we can define

(1-27b) $$\partial\mathcal{S} = m_1\,\partial\mathcal{R}_1 + m_2\,\partial\mathcal{R}_2 + \cdots + m_r\,\partial\mathcal{R}_r$$

as the *boundary* of $\mathcal{S}$. Indeed the form of $\partial\mathcal{S}$ is unaffected by equivalence operations; this amounts to realizing that (in Fig. 1-3b) we can write $\partial(\mathcal{R}_1 + \mathcal{R}_2) = \partial\mathcal{R}_1 + \partial\mathcal{R}_2$ because the (common) retraced portion of $\partial\mathcal{R}_1$ and $\partial\mathcal{R}_2$ will cancel. (This in turn is due to the fact that $\partial\mathcal{R}_1$ and $\partial\mathcal{R}_2$ are traced so as to keep $\mathcal{R}_1$ and $\mathcal{R}_2$ on the left.)

Thus, for two chains $\mathcal{S}_1$ and $\mathcal{S}_2$,

(1-27c) $$\partial(\mathcal{S}_1 + \mathcal{S}_2) = \partial\mathcal{S}_1 + \partial\mathcal{S}_2$$

Now Cauchy's integral theorem can be restated as

THEOREM 1-2 (bis) *Let $f(z)$ be analytic in $\mathcal{R}^*$ and let $\mathcal{S}$ be a 2-chain in $\mathcal{R}^*$. Then $\int_{\partial\mathcal{S}} f(z)\,dz$ vanishes or*

(1-28) $$(\partial\mathcal{S}, f(z)\,dz) = 0$$

The idea is that we have already established (1-28) for $\mathcal{S}$ equal to $\mathcal{R}_i$ and therefore the *additive property* (1-27c) gives us (1-28).

Often it is convenient to say that two 1-chains $\mathcal{L}$ and $\mathcal{M}$ are *homologous* or

$$\mathcal{L} \sim \mathcal{M} \qquad \text{if } \mathcal{L} - \mathcal{M} = \partial\mathcal{S}$$

for some 2-chain $\mathcal{S}$. Thus, any boundary $\partial\mathcal{S}$ is homologous to zero; that is, $\partial\mathcal{S} \sim 0$ for all 2-chains $\mathcal{S}$. We can then rewrite the Cauchy integral theorem as follows:

(1-29) $$(\mathcal{L}, f(z)\,dz) = (\mathcal{M}, f(z)\,dz) \qquad \text{if } \mathcal{L} \sim \mathcal{M}$$

† The chains of curves (or 1-chains) of Sec. 1-2 were easier to define because we were not concerned with the possibility that two well-behaved curves can have an infinitude of isolated points of intersection. (See Exercise 1 of Sec. 1-2.) In Chap. 7, we shall avoid such questions by using the "simplicial structure" of the manifold (in this case $\mathcal{R}$).

The idea that $\mathcal{L} \sim \mathcal{M}$ is essentially that the chain $\mathcal{L} - \mathcal{M}$ $(=\partial\mathcal{S})$ can be "filled in" by a region† or 2-chain $\mathcal{S}$. Thus, returning to Fig. 1-3*a*, we say $-\mathcal{C}_3$ is homologous to $\mathcal{C}_1 + \mathcal{C}_2$ which means the indicated chains can be filled in by the region $\mathcal{R}$. Likewise, $\mathcal{C}_1 \sim -\mathcal{C}_2 - \mathcal{C}_3$, etc.

The ideas introduced in Sec. 1-2 and here do not reach fruition until analogs are developed for "differentials on a manifold" in Chaps. 7 and 8. But proceeding in a purely formal way, let us observe that the boundary operator ∂ has a natural application to 1-chains: If $\mathcal{K}$ is a curve going from P to Q we can write formally

(1-30*a*) $$\partial\mathcal{K} = Q - P$$

Then it can be easily seen that the composition law holds

(1-30*b*) $$\partial(\mathcal{K}_1 + \mathcal{K}_2) = \partial\mathcal{K}_1 + \partial\mathcal{K}_2$$

when $\mathcal{K}_2$ continues where $\mathcal{K}_1$ leaves off. Thus, for a 1-chain, $\mathcal{L} = \Sigma m_i\mathcal{K}_i$ to within the usual equivalence (cancellation) operations.

The "fundamental" theorem of calculus states that

(1-31) $$f(Q) - f(P) = \int_{P(\mathcal{K})}^{Q} f'(z)\,dz$$

It can be restated in the neat formal manner:

(1-32) $$(\partial\mathcal{K},f) = (\mathcal{K}, df\,(z))$$

if we make the obvious definition, $(\partial\mathcal{K},f) = f(Q) - f(P)$ according to (1-30*a*). Actually (1-32) holds for 1-chains $\mathcal{L} = \Sigma m_i\mathcal{K}_i$ (instead of $\mathcal{K}$) if we define $(\partial\mathcal{L},f) = \Sigma m_i(\partial\mathcal{K}_i,f)$ in the usual manner. The relation (1-32) seems frivolous, but it indicates the "transfer of the boundary operator" from $\mathcal{K}$ to f (where it becomes the differential). This process will be reexamined (in Sec. 8-2) on manifolds, where it is much deeper, but we shall have to use *real* (not complex) integrands. Then it will become Stokes' theorem!

In consistent fashion, we call collections of points Σm_iP_i "0-chains" and we call 0-chains "homologous" if their difference is a boundary of a 1-chain. For example, any two points P and Q in $\mathcal{R}$ can be connected (filled in) by a curve $\mathcal{K}$, hence they are homologous, $Q - P = \partial\mathcal{K}$. Thus, we have a rather trivial homology theory of 0-chains, but we have just examined a more interesting theory with 1-chains. (This matter also will be explored more systematically in Chap. 8.)

EXERCISES

Since the standards of rigor are somewhat ambiguous at this point, instead of the usual exercises we merely make remarks for the reader to verify "by intuition." Some

† This idea of filling in is often confused with the cognate idea of *homotopy:* that $\mathcal{L}$ can be "deformed into" $\mathcal{M}$. This matter is discussed again in Appendix B.

sketch of a rigorous approach is made in Chaps. 7 and 8, when the reader is better prepared.

1 By extending the concept of homology to the torus $\mathfrak{J}$ (visualized as the surface of revolution), explain why an arbitrary closed curve (visualized as an elastic string) is homologous to a chain of type $m_1\mathcal{C} + m_2\mathcal{C}^*$. (Here we use the result that if two curves are deformable into one another or are homotopic, they are homologous. This will be discussed further in Appendix B.)

2 Let us cut out a small circular disk $\mathfrak{D}$ from $\mathfrak{J}$ and obtain a "deleted" torus $\mathfrak{J}^* \subset \mathfrak{J}$. Then if $\mathcal{C}_0$ is a curve in $\mathfrak{J}^*$ concentric with the disk $\mathfrak{D}$, we can show $\mathcal{C}_0$ is homologous with zero (or with the boundary of the rectangular model $\mathcal{C} + \mathcal{C}^* - \mathcal{C} - \mathcal{C}^*$). Note that $\mathcal{C}_0$ is not homotopic to a point (although we must leave the justification of this statement to advanced texts on topology).

3 Let us refer to Fig. 1-3*a* again. Show how to draw a closed curve $\mathcal{C} \sim m_1\mathcal{C}_1 + m_2\mathcal{C}_2$ for arbitrary integers m_1, m_2.

4 For an n-tuply-connected region $\mathcal{R}$, show that a closed curve is generally homologous to $m_1\mathcal{C}_1 + \cdots + m_{n-1}\mathcal{C}_{n-1}$ by generalizing the diagram in Fig. 1-3*a*.

SINGULAR BEHAVIOR

1-7 Cauchy Kernel

The completion of the proof of Theorem 1-1 is based on the following result:

THEOREM 1-6 (*Cauchy's representation theorem*) *If a simple closed curve $\mathcal{C}$, positively oriented and lying in the region $\mathcal{R}$, contains only points z of $\mathcal{R}$ in its interior, then a function $f(z)$ analytic in $\mathcal{R}$ can be represented for points z interior to $\mathcal{C}$ by*

$$f(z) = \frac{1}{2\pi i}\int_{\mathcal{C}} \frac{f(Z)\,dZ}{Z - z} \tag{1-33}$$

As we recall, any operation we desire to perform on $f(z)$ becomes transferred through the integral sign to the so-called "Cauchy kernel"

$$C(Z,z) = \frac{1/2\pi i}{Z - z} \tag{1-34}$$

Formally, at least, we find that by taking the nth derivative, we obtain

$$\left(\frac{d}{dz}\right)^n \left(\frac{1}{Z - z}\right) = \frac{n!}{(Z - z)^{n+1}} \tag{1-35}$$

and hence differentiating (1-33) under the integral sign, we find

$$f^{(n)}(z) = \frac{n!}{2\pi i}\int_{\mathcal{C}} \frac{f(Z)\,dZ}{(Z - z)^{n+1}} \tag{1-36}$$

Expanding the Cauchy kernel (1-33) into a power series about a, we use the identity $1/(1 - \varepsilon) = 1 + \varepsilon + \varepsilon^2 + \cdots + \varepsilon^{n-1} + \varepsilon^n/(1 - \varepsilon)$ to obtain the following identity:

(1-37)
$$\frac{1}{Z-z} = \frac{1}{(Z-a)-(z-a)} = \frac{1}{Z-a}\left(1 - \frac{z-a}{Z-a}\right)^{-1}$$
$$= \frac{1}{Z-a} + \frac{z-a}{(Z-a)^2} + \cdots + \frac{(z-a)^{n-1}}{(Z-a)^n} + \frac{(z-a)^n}{(Z-z)(Z-a)^n}$$

We substitute expression (1-37) into (1-34) and integrate term by term to obtain the usual power series with remainder

(1-38) $$f(z) = f(a) + (z-a)f'(a) + \cdots + \frac{(z-a)^{n-1}f^{(n-1)}(a)}{(n-1)!} + R_n$$

The elementary texts supply the remaining details on deriving the Taylor series expansion from (1-38). More specifically, if we take $\mathcal{C}$ to contain (in its interior) the circular disk

(1-39) $$|z-a| \leq r$$

we find that the remainder $R_n \to 0$ and the power series (1-38) converges uniformly inside (see Exercise 1). Thus, Theorem 1-1 is finally proved.

The interchange of integral with derivative or with sum (of infinite series) can be justified amply on the basis of the fact that the denominator $Z - z$ is bounded away from zero as Z goes around $\mathcal{C}$. An interesting way of interpreting the Cauchy representation formula is that the singularity (nonanalytic character) of the kernel at $Z = z$ "generates" $f(z)$. In other words, $f(z)$ is constructed in some sense out of singularities. This idea, developed largely by Riemann, is a philosophical leitmotiv of the whole subject of complex variables.

By a modification of the argument it is further proved that a power series expansion can be generalized as follows:

THEOREM 1-7 (*Laurent*) *If a function $f(z)$ is analytic in an annulus and on its boundary:*

(1-40) $$r_1 \leq |z-a| \leq r_2$$

then it is representable there by a uniformly convergent (Laurent) series

(1-41) $$f(z) = \sum_{-\infty}^{\infty} c_n(z-a)^n$$

COROLLARY 1 *Both the Taylor series and the Laurent series of the preceding theorems are uniquely determined by the function (and the disk or annulus).*

We have now seen that an arbitrary convergent power series determines the most general analytic function. Actually, Lagrange (1797) had used convergent power series as the basic concept (rather than analytic functions). Although the logical equivalence of these concepts is now established, the power series is conceptually more cumbersome to use.† For example, consider the result that "the substitution of a convergent power series into a convergent power series produces a convergent power series (after collection of like powers, etc.)." This result would be unnecessarily difficult to prove directly, but in terms of analytic functions, it means that the substitution of an analytic function into an analytic function produces an analytic function—which is a result of the chain rule for differentiation: If $w = f(z)$ and $q = g(w)$ are two analytic functions of z and w, respectively, then if z changes by an *arbitrary* Δz, w will change by a *dependent*‡ Δw, which will make q change by a *dependent* Δq corresponding to this Δz. Hence, naïvely, we see

$$\lim_{\Delta z \to 0} \frac{\Delta q}{\Delta z} = \frac{\Delta q}{\Delta w} \cdot \frac{\Delta w}{\Delta z} \to g'(w)f'(z)$$

EXERCISES

1 Complete the proof of Taylor's expansion from (1-33) by inserting a circle of radius r' between the curve $\mathcal{C}$ and disk (1-39). Note that $|z - a|/|Z - a| < r/r' < 1$.

2 (*a*) Show that if $f(z)$ is analytic at a, then $f(z) \neq f(a)$ in some "punctured" disk $0 < |z - a| < r$ unless $f(z)$ is constant. (*b*) Show that if two analytic functions $f(z)$ and $g(z)$ defined in the same region $\mathcal{R}$, and that if they agree in value on a point set $\mathcal{S}$ which has a limit point a in $\mathcal{R}$, then $f(z) = g(z)$ in $\mathcal{R}$. [*Hint:* Consider the power series for $f(z) - g(z)$ around $z = a$.]

3 Demonstrate a function $f(z)$ for which different Laurent series occur in different annuli (1-40) centered at $z = a$. [*Hint:* Consider two annuli separated by a pole of $f(z)$.]

4 Set up the *integrand* for Laurent series analogously with (1-33) and expand formally into series to write the expression for c_n in (1-41). (Do not find remainders.) Show the *uniqueness* required in Corollary 1 of Theorem 1-7.

5 Let $f(z)$ be analytic in a region containing the disk $|z - a| \leq r'$ and take a smaller radius $r < r'$. Then if $M = \max |f(z)|$ for $|z - a| = r'$, the following inequality holds on $|z - a| = r$:

$$\max |f^{(n)}(z)| \leq \frac{n!Mr'}{(r' - r)^{n+1}} \tag{1-42}$$

Also show that if $f(z) = \sum_0^\infty a_n(z - a)^n$, then $|a_n| \leq M/r^n$.

6 Let the analytic functions $f_n(z) \to f(z)$ uniformly in every subregion $\mathcal{R}^*$ whose closure lies in a simply-connected region $\mathcal{R}$. (Let $\mathcal{R}^*$ be bounded by a simple closed

† A more compelling reason for avoiding Lagrange's definition appears in the applications to physical problems in Part III.

‡ The fact that Δw could be chosen arbitrarily (by determining Δz) is irrelevant (although it is correct by the existence of an analytic inverse as is established under certain conditions in Chap. 3).

curve $\mathcal{C}^*$.) Then $f_n'(z) \to f'(z)$ uniformly in every subregion $\mathcal{R}^0$ whose closure lies in $\mathcal{R}^*$. [*Hint:* Use formula (1-36) to obtain the estimate

$$|f_n'(z) - f'(z)| \leq \frac{l(\mathcal{C}^*) \sup (|f_n(z) - f(z)| \text{ on } \mathcal{R}^*)}{2\pi d^2}$$

where d is the minimum separation of $|\partial\mathcal{R}^0|$ and $|\partial\mathcal{R}^*|$.]

7 Assume uniform convergence of $f_n(z)$ to $f(z)$ in a region $\mathcal{R}$ containing a compact set $\mathcal{S}$. Let $f(z) \neq a$ on $\mathcal{S}$, then $f_n(z) \neq a$ on $\mathcal{S}$ for n large enough. [*Hint:* Suppose there were counterexamples $f_n(z_n) = a$ as $n \to \infty$ for z_n on $\mathcal{S}$; consider z_0 a limit point of the z_n.]

1-8 Poles and Residues

After having restricted our interest to functions which are analytic, we find that these functions have a restricted type of nonanalytic behavior and this behavior can also determine the function.

The following theorem, by Riemann and Weierstrass, concisely restricts nonanalytic behavior at isolated singularities:

THEOREM 1-8 *If a function $f(z)$ is analytic in the region consisting of the punctured disk $\mathfrak{D}_r'$,*

$$0 < |z - a| < r \tag{1-43}$$

then just one of the following is true:

(*a*) *$f(z)$ is bounded in $\mathfrak{D}_r'$, whence $f(a)$ can be defined so that $f(z)$ is analytic in the whole region $|z - a| < r$.*

(*b*) *$|f(z)| \to +\infty$ as $z \to a$, in which case $f(z)$ has a so-called "principal part" with pole of order n for some integer n:*

$$f_a(z) = \frac{c_{-n}}{(z-a)^n} + \frac{c_{-n+1}}{(z-a)^{n-1}} + \cdots + \frac{c_{-1}}{z-a} \qquad c_{-n} \neq 0$$

with the difference function $f(z) - f_a(z)$ bounded in $\mathfrak{D}_r'$, leading to case (*a*).

(*c*) *$f(z)$ is unbounded yet does not approach $+\infty$ as $z \to a$. Then $f(z)$ oscillates in such a fashion that for any prescribed complex b and real positive ε_1 and ε_2, the inequality $|f(z) - b| < \varepsilon_1$ can be satisfied in $\mathfrak{D}_{\varepsilon_2}'$.*

In case (*a*) we say $f(z)$ is *regular* at a or has a *removable singularity;* in case (*b*) we say $f(z)$ has a *pole* as singularity at a; in case (*c*) we say $f(z)$ has an *essential singularity* at a.

COROLLARY 1 (*Liouville's theorem*) *If a function $f(z)$ is analytic for all z and bounded as $|z| \to \infty$, then $f(z)$ is a constant.*

Proofs are reserved for Exercises 1 and 2.

This corollary has a famous consequence that every polynomial $P(z)$ has at least one complex root [as proved by setting $f(z) = 1/P(z)$].

In *all cases* of Theorem 1-8 we have a legitimate Laurent expansion in the disk $\mathfrak{D}_r'$

(1-44) $$f(z) = \sum_{-\infty}^{\infty} c_n(z - a)^n$$

[We call c_{-1} the *residue of* $f(z)\,dz$ at $z = a$ (written $\operatorname{res} f(a)\,dz$ for reasons to be clarified by the next result).]

THEOREM 1-9 (*Cauchy's residue theorem*) *If a given function $f(z)$ is analytic in a region $\mathfrak{R}$ containing a simple closed curve $\mathfrak{C}$ and is analytic in the interior of $\mathfrak{C}$ except for a finite number of isolated singularities z_j (inside $\mathfrak{C}$), then*

(1-45) $$\int_{\mathfrak{C}} f(z)\,dz = 2\pi i \sum \operatorname{res} f(z_j)\,dz$$

where Σ is the sum of the residues at these singularities.

The proof of this result is referred to elementary texts.

Here we must remark that the most practical method of finding residues of $f(z)$ is to use a succession of known Taylor series expansions in the case where $f(z)$ is built up by known functions. (This matter is pursued in elementary texts.) We shall find Theorem 1-9 to be of more theoretical value later on.

If we consider a function $f(z)$ with a zero of order n, then

$$f(z) = c_n(z - a)^n + c_{n+1}(z - a)^{n+1} + \cdots$$

with $c_n \neq 0$. Then $f'(z)/f(z)$ has residue n at $z = a$. Likewise, if $f(z)$ has a pole of order m at $z = a$, then

$$f(z) = c_{-m}(z - a)^{-m} + c_{-m+1}(z - a)^{-m+1} + \cdots$$

and $f'(z)/f(z)$ has residue $-m$ at $z = a$. Thus we recall the following result (often called the "argument principle"):

COROLLARY 1 *Let an analytic nonconstant function $f(z)$ be given in a region containing a simple closed curve $\mathfrak{C}$, let $f(z)$ be analytic in the interior of $\mathfrak{C}$ except for isolated singularities consisting of poles, and finally, let $f(z)$ be nonvanishing on $\mathfrak{C}$. Then*

(1-46) $$\int_{\mathfrak{C}} \frac{f'(z)\,dz}{f(z)} = 2\pi i(Z - P)$$

where Z is the total (finite) number of zeros inside $\mathfrak{C}$ and P is the total (finite) number of poles inside $\mathfrak{C}$ (always counted with their multiplicity).

The proof is immediate except for the fact that the *total number of zeros Z and the total number of poles P must be shown to be finite.* To see that the total number of *zeros* in $\mathfrak{C}$ is finite, suppose this set, $\mathfrak{S}$, were infinite. Then there would be a limit point z_0 for $\mathcal{S}$ at which $f(z_0) = 0$, by continuity. Now z_0 cannot lie on $\mathcal{C}$ but must be in its interior. This is a contradiction to Exercise 2*a* of Sec. 1-7. Likewise, the number of poles of $f(z)$ is the number of zeros of $1/f(z)$ (which is again finite, completing the proof).

EXERCISES

1 Prove Theorem 1-8 by showing in (*a*) that if $g(z) = f(z)(z - a)^2$ and $g(z) = 0$, then $g'(a)$ exists (and $= 0$); and in (*b*) by using $g(z) = 1/f(z)$, $g(a) = 0$; and finally in (*c*) by using $g(z) = 1/[f(z) - b]$.

2 Prove Liouville's theorem by noting that both $f(z)$ and $f(1/z)$ have Taylor series about $z = 0$ and comparing terms.

3 Invent a *real* function $u(x)$ representable around each point by a power series with $u(z)$ bounded, say <1, for $-\infty < x < \infty$. (Thus, Liouville's theorem has no *real* analog.)

4 Extend Theorem 1-9, Corollary 1 to the evaluation of

$$\int_{\mathcal{C}} \frac{\phi(z)f'(z)\,dz}{f(z)}$$

stating the hypothesis carefully. Consider just the case that $\phi(z)$ has no singularities, and show that where $f(z)$ has a simple root a, a residue of $2\pi i\phi(a)$ results.

5 Prove that if $f(z)$ is analytic in the z plane and if for some $R > 0$ and for all $|z| > R$, $|f(z)| \le M|z|^k$ (for some positive constants M, R, k), then $f(z)$ is a polynomial.

6 Let $f_n(z)$ be a set of analytic functions uniformly convergent to $f(z)$ in $\mathcal{R}^*$, any simply-connected subregion of $\mathcal{R}$. Let each equation $f_n(z) = a$ have only one (or no) solution for each n for $z \in \mathcal{R}$. Prove that the same is true for the equation $f(z) = a$. [*Hint:* If $f(z) = a$ for z_1 and z_2 ($\in \mathcal{R}$), then let $\mathcal{R}^*$ be a subregion of $\mathcal{R}$ which contains z_1 and z_2 is simply-connected, and for which $\mathcal{C}^* = \partial\mathcal{R}^*$ has no roots of $f(z) = a$.] Consider $N = 1/2\pi i \int_{\mathcal{C}^*} f_n'(z)/[f_n(z) - a]\,dz$. Show for n large enough that the denominator never vanishes on $\mathcal{C}^*$, and find the limit of N as $n \to \infty$ (using Exercise 6 of Sec. 1-7).

7 Let $f(z)$ be analytic in $\mathcal{R}$ except possibly at a finite number of isolated singularities $Z_1, Z_2, \cdots, Z_S$. Let $\mathcal{C}$ be a closed curve in $\mathcal{R}$ which contains no singularity. Then show

$$\int_{\mathcal{C}} f(z)\,dz = 2\pi i \sum_{i=1}^{S} N(\mathcal{C};Z_i)\{\text{res } f(Z_i)\,dz\}$$

Hint: First consider $F(z) = f(z) - \Sigma\{\text{res } f(Z_i)\,dz\}/(z - Z_i)$.

CHAPTER 2
RIEMANN SPHERE

The symbol ∞ is next introduced† into complex analysis. By analogy with real analysis we would set $1/0 = \infty$, but in complex analysis, $+\infty$ and $-\infty$ must have the same meaning [since $-\infty = 1/(-0) = 1/0$]; indeed, $i\infty$ or $k\infty$ ($k \neq 0$) must also mean the same thing. (Thus the use here of the same symbol ∞ as in real analysis might be considered to be an abuse of language.)

Yet the *algebraic* convenience of a symbol for $1/0$ is not enough justification. (Why not use another symbol for $0/0$ or $\infty + \infty$, etc.?) For the complete reason we must also turn to geometry.

TREATMENT OF INFINITY

2-1 Ideal Point

Let us consider the fact that the z plane is not compact. This means that certain sequences $z_1, z_2, \ldots, z_n$ for which $|z_n| \to \infty$ in the real sense [or $|z_n| > 1/\varepsilon$ for $n > N(\varepsilon)$] have no limit point in the z plane. Let us introduce an *ideal* point ∞ which is somehow to have the property that any sequence z_n (with $|z_n| \to \infty$) will have ∞ as a limit point, or $\{z_n\}$ will have a point in an arbitrary neighborhood of this "ideal point" at ∞. It can be seen, upon reflection, that the only way to accomplish this objective is to first define *neighborhoods of* ∞ as those open sets $\mathcal{O}$ which contain the exterior of some circle $|z| > k$, for a positive k depending only on $\mathcal{O}$. Clearly, if $|z_n| \to \infty$, then every such set $\mathcal{O}$ will contain at least one term of the sequence z_n. The open sets $\mathcal{O}$ actually exist in the conventional sense, so we now define ∞ (*as a fiction*) in order to call the sets $\mathcal{O}$ neighborhoods of ∞. In brief, we first define neighborhoods of ∞, and then we define ∞. This process is called "compactification."

† In Chaps. 2 and 3, the review material is not always presented in detail, but the order is considered suitable for learning the material anew.

This procedure, although it may seem artificial, is very deep. For example, let us start once again with the open disk $\mathfrak{D}$ ($|z| < 1$). Certainly $\mathfrak{D}$ is not compact; indeed, if for a sequence $\{z_n\}$ we have $|z_n| \to 1$, then no point of $\mathfrak{D}$ serves as a limit point of $\{z_n\}$. We then introduce an ideal boundary point j (not called ∞ this time!) by saying neighborhoods of j shall mean open sets (in $\mathfrak{D}$) which contain annuli of the type $1 - \varepsilon < |z| < 1$ for some ε between 0 and 1. In effect, the ideal point j is now interpreted geometrically as the circle $|z| = 1$. In some contexts, we must therefore talk of the ideal point as an "ideal boundary."† It is no trivial problem to decide on the geometric merits of the ideal point versus the ideal boundary. We shall decide that ∞ is an ideal *point* because it "transforms algebraically like a point" (see Sec. 2-2). Thus the idea of a point at ∞ is both algebraic and geometric.

EXERCISES

1 Consider the transformation of $\mathfrak{D}$ onto $\mathcal{P}$, the finite w plane, by $w = z/(1 - |z|)$. How does this affect j and ∞? [Is $w(z)$ analytic?] (*Remark:* This transformation asserts that $\mathfrak{D}$ is topologically equivalent to $\mathcal{P}$.)

2 Consider the real (euclidean) x axis. How can it be made to have one ideal point? Two ideal points?

2-2 Stereographic Projection

The (usual) finite plane together with $z = \infty$ is called the (Riemann) "sphere." An actual (euclidean) sphere is related to it by *stereographic projection* in which $z = 0$ projects on S (the south pole), $|z| = 1$ on the equator, and $z = \infty$ on N (the north pole) of the sphere. (See Fig. 2-1.)

Let $Q = (\xi,\eta,\zeta)$ be a point of the euclidean sphere‡

$$\xi^2 + \eta^2 + \zeta^2 = 1 \tag{2-1}$$

and let Q correspond to P, the point $z = x + iy$, in that xy plane which passes through the equatorial circle. Then

$$\xi = \frac{2x}{x^2 + y^2 + 1} \qquad \eta = \frac{2y}{x^2 + y^2 + 1} \qquad \zeta = \frac{x^2 + y^2 - 1}{x^2 + y^2 + 1} \tag{2-2a}$$

$$x + iy = \frac{\xi + i\eta}{1 - \zeta} \tag{2-2b}$$

† Compare the discussion of infinite Riemann manifolds in Chap. 5 and Appendix B.

‡ No direct use is made of (ξ,η,ζ) as coordinates to help use the z plane as a sphere (since this nomenclature is figurative).

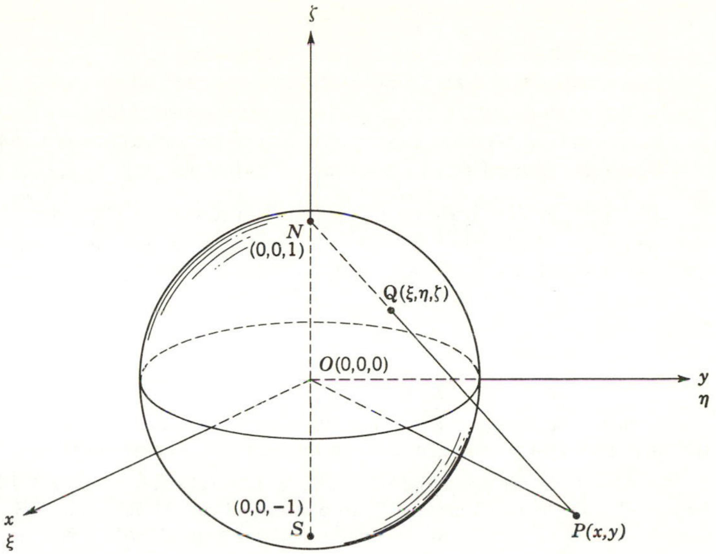

Fig. 2-1 Stereographic projection of the (ξ,η,ζ) sphere onto the z sphere.

The sphere is given above mostly because of its classical interest. Indeed, in Lemma 2-12 we shall even see that a certain class of linear transformations corresponds to rotations of the sphere. This correspondence historically was used to support the intuitive idea that the point at ∞ is just like any other point of the z sphere (to the extent that the north pole is just like any other point of the euclidean sphere).

Yet this artifice goes unnecessarily beyond the demands of intuition. What we must know is that the algebraic substitution $1/0 = \infty$ satisfies the requirements that a neighborhood of 0 be transformed topologically by $z' = 1/z$ into a neighborhood of ∞ and conversely.

Clearly (for $k \neq 0$) $\infty = \pm\infty = i\infty = k\infty = \overline{\infty}$ (conjugate), etc. Also, ∞ is a limit point of any unbounded set on the z sphere. Furthermore, the *exterior* of any circle in the z plane is a neighborhood of $z = \infty$ on the z sphere, etc. With ∞ treated like an ordinary point (on the sphere), the *interior of a closed curve on the z sphere* is determined wholly by orientation (as the left-hand region with respect to the curve).

We correspondingly permit $w = \infty$ *to be a value for the function* $w(z)$, *thus enlarging our notion of an analytic function* to include *poles* as values (but not essential singularities). Thus the w plane is enlarged to be a w sphere, as a range for $w(z)$.

Now we have the glossary:

Finite values of z	z plane
Finite values of z and $z = \infty$	z sphere
$w = f(z)$ is analytic	$f(z)$ takes values on the w plane
$w = f(z)$ is analytic or has a pole	$f(z)$ takes values on the w sphere
Principal part of $f(z)$ at $z = \infty$	principal part of $f(1/z)$ at $z = 0$

The phrase "analytic or has a pole" is rendered for convenience by *meromorphic*. Note that if $f(z)$ is defined for z on the z sphere, then $f(1/z)$ is defined for z on the z sphere. If $w = f(z)$ takes values on the w sphere, then $1/f(z)$ takes values on the w sphere. (All we really say is that $1/0 = \infty$ and $1/\infty = 0$.)

EXERCISES

1 Show that in Fig. 2-1 the line NQP makes the same angle with the xy plane at P as it does with the tangent plane to the (euclidean) sphere at Q.

2 Show from the preceding result that a dihedral angle through PQ intercepts equal angles on the two aforementioned planes. (*Note:* A consequence of this statement is the fact that angles are preserved in stereographic projection.)

3 Show that if $Q(\xi,\eta,\zeta)$ corresponds to $P(z)$, then $Q(-\xi,-\eta,-\zeta)$ corresponds to $P(z')$ where $z\overline{z'} = -1$. [*Hint:* Use (2-2*b*) only.]

2-3 Rational Functions

We next define a *rational function* of z to be a function $f(z)$ which is either identically zero or is the ratio of two nonzero polynomials $P_1(z)/P_2(z) = f(z)$ where $P_1(z)$ is of degree n and $P_2(z)$ is of degree m. Specifically,

$$\begin{aligned} P_1(z) &= a_0 z^n + a_1 z^{n-1} + \cdots + a_n \\ &= a_0(z-\alpha_1)^{t_1}\cdots(z-\alpha_r)^{t_r} \qquad a_0 \neq 0 \\ P_2(z) &= b_0 z^m + b_1 z^{m-1} + \cdots + b_m \\ &= b_0(z-\beta_1)^{u_1}\cdots(z-\beta_s)^{u_s} \qquad b_0 \neq 0 \end{aligned} \tag{2-3}$$

We assume for convenience that the complex numbers α_i and β_j are all different. Here $n = \sum_{j=1}^{r} t_j$ and $m = \sum_{j=1}^{s} u_j$, and max (n,m) is called the "order" (or "degree") of $f(z)$. When $f(z) \equiv 0$, the degree is not defined.

The rational function $f(z)$ has these properties:

(*a*) $f(z)$ is analytic or else it has poles, when z is finite. (This is obvious because the only singularities are the β_j.)

(*b*) $f(z)$ is analytic or else it has a pole, at $z = \infty$ [since then $f(z)$ behaves like z^{n-m}; for example, $f(z)/z^{n-m} \to a_0/b_0$ as $z \to \infty$].

According to our glossary, these properties mean $w = f(z)$ *is defined on the z sphere with values on the w sphere.*

These properties actually characterize rational functions:

THEOREM 2-1 *A function* $w = f(z)$ *which is rational is meromorphic on the z sphere, and conversely, any meromorphic function defined on the z sphere is rational.*

PROOF The direct part is obvious, but the converse is a bit deeper. We first show the following:

A nonconstant function $f(z)$ *defined on the z sphere with values on the w sphere can assume any value a on the w sphere only a finite number of times.*

To see this, note that we can take $a = 0$ by replacing $f(z)$ by a function of equal degree, $F(z) = f(z) - a$ (if a is finite) or $1/f(z)$ (if a is infinite). We then ask: Are there possibly an infinitude of points z_i for which $F(z_i) = 0$? Since $f(z)$ and $F(z)$ have only poles as singularities, there cannot be an infinitude of such z_i within a finite domain, such as any disk $|z| < R$ (by the method of Theorem 1-9, Corollary 1). Now, if $z_i \to \infty$, then $F(1/z)$ would have an infinitude of zeros near $z = 0$, which is likewise impossible. [Note that our ability to use $1/z$ as a variable is tantamount to knowledge that $F(z)$ is defined on the z sphere, and our ability to write $1/f(z) = F(z)$ is tantamount to knowledge that the values of $f(z)$ are on the w sphere.]

To get to the main part of the proof, let $f(z)$ have the properties (a) and (b) with the additional information that $f(z) = \infty$ for only a finite number of points β_j with multiplicity u_j ($j = 1, 2, \ldots, S$).

Then consider

$$g(z) = c_0(z - \beta_1)^{u_1} \cdots (z - \beta_S)^{u_S} f(z) \tag{2-4}$$

This function has no poles for finite z, hence the following expansion into power series is valid:

$$g(z) = b_0 + b_1 z + \cdots + b_k z^k + \cdots \tag{2-5}$$

On the other hand, $g(z)$ has a value (finite or ∞) at $z = \infty$, whence $g(1/z)$ has an expansion around $z = 0$, showing possibly a pole as follows:

$$g\left(\frac{1}{z}\right) = \frac{q_{-m}}{z^m} + \frac{q_{-m+1}}{z^{m-1}} + \cdots + q_0 + q_1 z + q_2 z^2 + \cdots \tag{2-6}$$

Now both (2-5) and (2-6) are simultaneously valid ($z \neq 0$); hence, they are identical when z is changed to $1/z$. Since (2-5) is lacking in negative powers, $q_1 = q_2 = \cdots = 0$ and $q_0 = b_0$, $q_{-1} = b_1$, $q_{-2} = b_2$, . . . , $q_{-m} = b_m$; likewise $b_{m+1} = b_{m+2} = \cdots = 0$ by the absence of additional negative powers in (2-6). Thus $g(z)$ is a polynomial and $f(z)$ is rational.

Q.E.D.

COROLLARY 1 *A nonconstant rational function $w = f(z)$ of degree n assumes every value a on the w sphere for exactly n values of z on the z sphere (using conventions of multiplicity of root and order of pole).*

PROOF We start with formulas (2-3) for $f(z) = P_1(z)/P_2(z)$, and we make the simplifying assumption that $f(\infty) = \infty$. To do this, replace $f(z)$ by $F(z)$ as follows:

$$F(z) = \begin{cases} f(z) & \text{if } n > m \\ \dfrac{1}{f(z)} - \dfrac{a_0}{b_0} & \text{if } n = m \\ \dfrac{1}{f(z)} & \text{if } n < m \end{cases} \tag{2-7a}$$

To simplify the notation, let us still write

$$F(z) = \frac{P_1(z)}{P_2(z)} \qquad \begin{cases} P_1(z) \text{ of degree } n \\ P_2(z) \text{ of degree } m \end{cases} \qquad n > m \tag{2-7b}$$

We first see that for every finite a, $F(z) = a$ for n values of z. For $F(z) = \infty$, we note that $F(z) = (a_0/b_0)z^{n-m}$ + lower powers as $z \to \infty$, hence there are $n - m$ values of $z = \infty$ while $P_2(z)$ is of degree m. Therefore, there are m finite points z where $P_2(z) = 0$, making a total of n points for which $F(z) = \infty$ as well. Q.E.D.

The cases where $n = 1$ in Corollary 1 are of special interest.

COROLLARY 2 *A meromorphic function provides a unique correspondence between the z sphere and the w sphere if and only if it is a linear-fractional function; that is,*

$$w = \frac{Az + B}{Cz + D} \qquad AD - BC \neq 0 \tag{2-8}$$

PROOF First note that the relation (2-8) may be inverted by solving for z as follows:

$$z = \frac{Dw - B}{-Cw + A} \qquad AD - BC \neq 0 \tag{2-9}$$

Hence the biunique correspondence holds. Conversely, if $f(z)$ is defined from z sphere to w sphere (regardless of uniqueness) by Theorem 2-1, $f(z)$ is rational. The uniqueness means $n = 1$ in Corollary 1 (see Exercise 2), whence $f(z)$ is linear-fractional. Q.E.D.

(Usually the word "linear" is used synonymously with "linear-fractional.")

EXERCISES

1 What would happen to the proof of Corollary 1 (Theorem 2-1) if $f = P_1(z)/P_2(z)$ where $P_1(z)$ and $P_2(z)$ are polynomials with a common factor?

2 Show that if $f = P_1(z)/P_2(z)$ is of degree $n > 1$, then it is impossible that $f = a$ have a single root (necessarily of multiplicity n) for all a. In fact, show the absurdity of the factorization (for all a)

$$P_1(z) - aP_2(z) = b(z - c)^n \tag{2-10}$$

(where b and c are functions of a). [*Hint:* Expand any *three* powers of z in (2-10) (this is why $n \geq 2$), and show the contradiction by eliminating a and b.]

2-4 Unique Specification Theorems

We now see two natural ways in which a function is specified by its singularities.

THEOREM 2-2 (*The partial-fraction decomposition*) *A rational function $f(z)$ can be specified as follows: Given the principal parts of $f(z)$ at each of s finite poles β_j of order u_j ($j = 1, 2, \ldots, s$); for example,*

$$\phi_j(z) = \frac{c_{-u_j}^{(j)}}{(z - \beta_j)^{u_j}} + \cdots + \frac{c_{-1}{}^{(j)}}{(z - \beta_j)} \qquad c_{-u_j}^{(j)} \neq 0 \tag{2-11}$$

in addition to the so-called "principal part" of $f(z)$ at ∞ (if $z = \infty$ is a pole),

$$\phi_{s+1}(z) = c_{-u_{s+1}}^{(s+1)} z^{u_{s+1}} + \cdots + c_{-1}{}^{(s+1)} z \qquad c_{-u_{s+1}}^{(s+1)} \neq 0 \tag{2-12}$$

Call $\phi_{s+1}(z) = 0$ if ∞ is not a pole. Then $f(z)$ is determined by

$$f(z) = \sum_{j=1}^{s+1} \phi_j(z) + const \tag{2-13}$$

PROOF This theorem follows from Liouville's theorem, since for each j, the difference $f(z) - \phi_j(z)$ is bounded in some neighborhood of β_j (or ∞) while it is bounded (trivially) in the complement of each neighborhood [and so is every other $\phi_j(z)$ bounded there]. Q.E.D.

For certain algebraic purposes, it develops that the *location* of the poles together with the zeros (instead of the principal parts) is the preferred mode of representation of a rational function, as follows:

THEOREM 2-3 (*Divisor structure*) *A rational function $f(z)$ is determined to within a constant factor by its finite zeros α_i and poles β_i of indicated multiplicity:*

$$\textit{Zeros:}\ \overset{t_1}{\alpha_1}, \ldots, \overset{t_r}{\alpha_r} \qquad \textit{poles:}\ \overset{u_1}{\beta_1}, \ldots, \overset{u_s}{\beta_s} \tag{2-14}$$

This means that the nature of the point at ∞ (whether zero or pole) is also determined by the finite zeros or poles.

PROOF We need only refer to Sec. 2-3. Let $\phi(z) = P_1(z)/P_2(z)$ where P_1 has the zeros as roots and P_2 has the poles as roots (a_0 and b_0 are arbitrary). Then f/ϕ is finite in the (finite) z plane, but so is ϕ/f. One (actually both) of these two fractions must be bounded at ∞, hence constant. Q.E.D.

As a matter of notation we write a zero at $z = \alpha_1$ of multiplicity t_1 as $P_{\alpha_1}{}^{t_1}$, and we write the pole of multiplicity u_1 at β_1 as $P_{\beta_1}{}^{-u_1}$. A simple zero (or pole) at ∞ is denoted by P_∞ (or $P_\infty{}^{-1}$).

Thus the so-called "divisor structure" of $f(z)$ is written as the symbolic product

$$f(z) \doteq P_{\alpha_1}{}^{t_1} \cdots P_{\alpha_r}{}^{t_r} P_{\beta_1}{}^{-u_1} \cdots P_{\beta_s}{}^{-u_s} P_\infty{}^{m-n} \qquad m = \sum_{j=1}^{s} u_j,\ n = \sum_{j=1}^{r} t_j$$

We see this describes $f(z) = P_1(z)/P_2(z)$ in Theorem 2-3. Note that *one* factor, say $P_\infty{}^{m-n}$ (conveniently at ∞) is restricted by all the others, but otherwise, factors and multiplicities of rational functions can be arbitrarily chosen. For convenience, we can also use the notation $P_\alpha{}^0$ to mean $f(\alpha)$ is neither 0 nor ∞. The total sum of positive orders over the z sphere is equal numerically to the sum of the negative orders.

A similar "conservation" law holds for residues. We first define the residue of a function $f(z)$ at $z = \infty$, where $f(z)$ has (at worst) an isolated singularity at $z = \infty$, by a method compatible with (the Cauchy residue) Theorem 1-9. Let $\mathcal{C}$ be, say, the positively oriented circle $|z| = r$, containing all other singularities in its interior. Then $-\mathcal{C}$ surrounds ∞. We therefore define *the residue of* $f(z)\,dz$ *at* ∞ as

$$\text{(2-15)} \qquad \operatorname{res} f(\infty)\,dz = \frac{1}{2\pi i}\int_{-\mathcal{C}} f(z)\,dz$$

The definition (2-15) is also used for any analytic $f(z)$ with a pole at ∞.

THEOREM 2-4 *If $f(z)$ is a function of the z sphere with only a finite number of singularities, the sum of the residues at all singularities is zero.*

PROOF In the preceding discussion the curve $\mathcal{C}$ encloses all singularities except (possibly) one at ∞. Then the result follows from Theorem 1-9 (Sec. 1-8)

$$\int_{\mathcal{C}} f(z)\,dz + \int_{-\mathcal{C}} f(z)\,dz = 0$$

Q.E.D.

Now (2-15) may seem strange because of the following example: The function $f(z) = 1/z$ is analytic at $z = \infty$ [since $f(1/z) = z$ is analytic at $z = 0$], but by (2-15), $f(z)$ has a residue -1 at $z = \infty$. *The residue, therefore, was not defined as a property of $f(z)$ but a property of the differential $f(z)\,dz$.*

Let the following be the Laurent expansion for large z:

$$\text{(2-16)} \qquad f(z) = \sum_{-\infty}^{\infty} c_n z^n = \cdots + \frac{c_{-1}}{z} + \cdots$$

Then to find the residue at ∞ for $f(z)$, we could consider

$$\Omega = f\left(\frac{1}{z}\right) d\left(\frac{1}{z}\right) = -\frac{f(1/z)\,dz}{z^2} \tag{2-17}$$

Then in (2-16) we obtain "differential" Ω, formally written as

$$\begin{aligned}\Omega &= \left(\cdots - c_2 - \frac{c_{-1}}{z} - \frac{c_0}{z^2} - \frac{c_1}{z^3} - \cdots\right) dz \\ &= \frac{-c_{-1}\,dz}{z} + \text{irrelevant terms}\end{aligned} \tag{2-18}$$

Thus, indeed, the residue at ∞ of $f(z)$ in (2-16) is $-c_{-1}$.

EXERCISES

1 From Theorem 2-3, derive

$$\frac{1+z+z^3}{z(z^2+1)^2} = \frac{1}{z} - \frac{1+i}{2(z-i)} + \frac{i}{4(z-i)^2} - \frac{1-i}{2(z+i)} - \frac{i}{4(z+i)^2}$$

2 Show how the partial-fraction decomposition of Theorem 2-2 can always be carried out in terms of *real* coefficients for $f(z)$, the ratio of *real* polynomials, by pairing the principal parts of complex conjugate roots of the denominator. For example, derive from the above (complex) partial-fraction expansion the real result.

$$\frac{1+x+x^3}{x(x^2+1)^2} = \frac{1}{x} + \frac{-x+1}{x^2+1} + \frac{-x}{(x^2+1)^2}$$

Note: This is a legitimate use of complex variables to prove a real variable result!

3 Show that an infinite set $\mathcal{S}$, of points on the z sphere, has a limit point on the z sphere (compactness). (*Hint:* Show that either all finite points in $\mathcal{S}$ lie in the disk $|z| < M$ for some positive M or else a sequence of points of $\mathcal{S}$ approaches ∞.)

4 Prove Theorem 2-1, Corollary 1, by applying Theorem 2-4 to

$$F(z) = f'(z)/[f(z) - a]$$

with $f(z) = P_1(z)/P_2(z)$. [*Hint:* Show that every g-tuple root of $f(z) = a$ at $z = z_0$ (finite) contributes the amount $g = \text{res } F(z_0)\,dz$.] Next, consider the residue contributions where $z = \infty$ and where $f(z) = \infty$ under some convenient simplifying assumption, say $n \geq m$ (P_1 of no lower degree than P_2).

5 Prove that if $f(z)$ is rational, $\int f(z)\,dz$ can be expressed as

$$R(z) + \sum_{i=1}^{t} \alpha_i \log (z - \beta_i)$$

where $R(z)$ is rational and α_i and β_i are complex constants. Express α_i in terms of $\text{res } f(\beta_i)\,dz$. What about $\text{res } f(\infty)\,dz$?

TRANSFORMATION OF THE SPHERE

2-5 Invariant Properties

We have previously seen that the only one-to-one rational transformation of the Riemann sphere or z sphere into itself is

(2-19) $$z' = \phi(z) = \frac{\alpha z + \beta}{\gamma z + \delta} \qquad \Delta = \alpha\delta - \beta\gamma \neq 0$$

We can regard (2-19) as a change in variables which only "relabels" points of the sphere. With this point of view we can regard the rational function

(2-20a) $$w = f(z)$$

in a new light in two different ways:

(*a*) If (2-20*a*) is regarded as a transformation of the z sphere into *another* z sphere, it is merely a representative of the class of transformations

(2-20b) $$\frac{\alpha' w + \beta'}{\gamma' w + \delta'} = f\left(\frac{\alpha z + \beta}{\gamma z + \delta}\right) \qquad \begin{cases} \alpha\delta - \beta\gamma = \Delta \neq 0 \\ \alpha'\delta' - \beta'\gamma' = \Delta' \neq 0 \end{cases}$$

(*b*) If (2-20*a*) is regarded as a transformation of the z sphere into *itself*, it is merely a representative of the class of transformations

(2-20c) $$\frac{\alpha w + \beta}{\gamma w + \delta} = f\left(\frac{\alpha z + \beta}{\gamma z + \delta}\right) \qquad \alpha\delta - \beta\gamma = \Delta \neq 0$$

Actually, under either interpretation we have a very difficult problem in characterizing rational functions (2-20*a*) in so-called "invariant" fashion (invariant under linear transformations).

For example, if $f(z) = P_1/P_2$ is a rational function of degree n, the polynomial P_1 or P_2 each has $n + 1$ coefficients, so $f(z)$ has $2n + 1$ independent parameters (considering ratios). Now the transformation (2-20*b*) has six independent variables (three for each linear transformation, that is, α/δ, β/δ, γ/δ). Thus we are not surprised to know that if $n \leq 2$; that is, if $f(z)$ is linear (fractional) or quadratic (fractional), a choice of six coefficients in (2-20*b*) can be made, so that (2-20*a*) becomes *free of parameters* in the new variables, say

(2-21) $$w = z \qquad \text{or} \qquad w = z^2$$

(This is done in Exercise 1.) If $n \geq 3$, however, $2n + 1 > 6$, and there is at least one *essential* invariant in a rational function $w = f(z)$ of degree n (which *cannot* be removed by transforming the w or z spheres). This matter is handled in advanced texts on invariant theory.

Let us next note that the linear transformation can be expressed as the superposition of several simple types, which have generic names:

(2-22*a*)	$z' = z + \beta$		(translation)
(2-22*b*)	$z' = \alpha z$	$\lvert\alpha\rvert = 1$	(rotation)
(2-22*c*)	$z' = \alpha z$	$\alpha > 0$ (real)	(magnification)
(2-22*d*)	$z' = \dfrac{1}{z}$		(inversion)

Actually, the term "affine" transformation is used to denote a transformation which preserves $z' = z = \infty$. The most general such linear transformation, of course, is found by setting $\gamma = 0$ in (2-19) so that $z = \infty$ becomes $z' = \infty$. (Note if $\gamma = 0$, then $\alpha \neq 0$ since $\Delta \neq 0$.) This yields

(2-22*e*) $$z' = \alpha z + \beta \qquad \alpha \neq 0 \text{ (affine)}$$

To see that (2-19) is a combination of these operations, note (if $\gamma \neq 0$)

(2-23) $$z' = \frac{\alpha}{\gamma} - \frac{\Delta}{\gamma(\gamma z + \delta)}$$

Incidentally, for further reference, if $a' = (\alpha a + \beta)/(\gamma a + \delta)$ [the image of a under (2-19)], then

(2-24) $$z' - a' = (z - a)\frac{\Delta}{(\gamma z + \delta)(\gamma a + \delta)}$$

THEOREM 2-5 *Let the rational function $f(z)$ be transformed to $g(z')$ by linear transformation* (2-19) *so that $g([\alpha z + \beta]/[\gamma z + \delta]) = f(z)$. Then the divisor structure of poles and zeros is invariant in the sense that if a is transformed into a' under the linear transformation* (2-19), *then whenever $f(z)$ has factor $P_a{}^r$, it will follow that $g(z')$ has factor $P_{a'}{}^r$ (with the same order r).*

PROOF By (2-24) we note that if a and a' are finite, then as $z \to a$, $(z' - a')/(z - a) \to$ limit $(\neq 0, \infty)$ so that the theorem is evident. It remains to consider only what happens if one (or both) are ∞. By the fact that linear transformations are generated by operations (2-22*a*) to (2-22*d*), it suffices to reduce our considerations to the two cases

(2-25) $$\begin{aligned} z' &= z \qquad & a &= \infty \qquad & a' &= \infty \\ z' &= \frac{1}{z} \qquad & a &= 0 \qquad & a' &= \infty \end{aligned}$$

These are trivial by definition of behavior at ∞ (in terms of the change to the variable $1/z$). Q.E.D.

This theorem, in effect, vindicates our procedure in discussing the behavior at $z = \infty$ by use of a *definite* transformation $1/z$ instead of, say, $1/(z + 1)$. We would get the same orders for the divisors in either case. (Compare Exercise 2.)

EXERCISES

1 Prove that by a change of variables (2-20*b*) the second degree function $w = f(z) = (a_0 + a_1 z + a_2 z^2)/(b_0 + b_1 z + b_2 z^2)$ can be reduced to the form $w = z^2$. [*Hint:* Find an α such that $f(z) - \alpha$ has a double root at, say, $z = \zeta$.] Then $1/[f(z) - \alpha] = q(z - \zeta)/(z - \zeta)^2$ where $q(z - \zeta)$ is a polynomial, at most quadratic in $z - \zeta$. Use the new variable $z' = 1/(z - \zeta)$, and complete the square.

2 Suppose the residue of $f(\infty)\,dz$ were defined not by the contour integral (2-15) but by the following mechanism: Let $z = g(w)$ be a linear transformation which maps

$z = \infty$ onto $w = a$(finite). Then change variables to $f(g(w))g'(w)\,dw$ and calculate $\operatorname{res} f(g(a))g'(a)\,dw$. Show the result will be independent of the choice of $g(w)$. [Observe that $\operatorname{res} f(g(a))\,dw$ is not independent of $g(w)$.]

2-6 Möbius Geometry

We now consider the *Möbius geometry* of the z sphere, or the geometry involved in the linear transformation (2-19) in the sense that euclidean geometry is the geometry involved in the affine transformations (2-22*e*) (see Exercise 1). There is an overly profuse literature because of the significance of such geometries as noneuclidean. We shall leave most of the steps as exercises and merely list some results in the form of 12 lemmas.

LEMMA 2-1 *If four distinct points of the z sphere are ordered, then the cross ratio*

$$\lambda = \{z_1,z_2;z_3,z_4\} = \frac{(z_1 - z_2)(z_3 - z_4)}{(z_1 - z_4)(z_3 - z_2)} \tag{2-26a}$$

is invariant under linear-fractional transformation $\phi(z) = (\alpha z + \beta)/(\gamma z + \delta)$ *if* $\alpha\delta - \beta\gamma = \Delta \neq 0$; that is,

$$\{z_1,z_2;z_3,z_4\} = \{\phi(z_1),\phi(z_2);\phi(z_3),\phi(z_4)\} \tag{2-26b}$$

(See Exercise 2.)

LEMMA 2-2 *Two ordered quadruples of distinct points* z_1, z_2, z_3, z_4, *and* z_1', z_2', z_3', z_4', *can be transformed linearly into one another if and only if they determine the same cross ratio.*

Indeed, the equation $\{z,z_2;z_3,z_4\} = \{z',z_2';z_3',z_4'\}$ expresses z' linearly in z and transforms z_2, z_3, z_4 into z_2', z_3', z_4'.

LEMMA 2-3 *Two unordered quadruples of distinct points can be transformed linearly into one another if and only if the cross ratios* λ *and* λ' *(of each quadruple in any order) satisfy*

$$\phi(\lambda) = \phi(\lambda') \qquad \textit{where} \qquad \phi(\lambda) = \frac{4(\lambda^2 - \lambda + 1)^3}{27\lambda^2(\lambda - 1)^2} \tag{2-27}$$

(See Exercises 3 to 5.)

MÖBIUS CIRCLES

We define a *Möbius circle* as a general concept embracing (ordinary) circles and (ordinary) straight lines in the z sphere (with the point at ∞ adjoined to such lines).

LEMMA 2-4 *A Möbius circle is transformed into another Möbius circle under any linear-fractional transformation.*

(See Exercises 6 and 7.)

LEMMA 2-5 *A Möbius circle $\mathfrak{M}$ with general point z is determined by any triple of distinct points z_1, z_2, z_3 in the z sphere lying on $\mathfrak{M}$ by the condition that $\{z,z_1;z_2,z_3\}$ is real. (Here ∞ is a real quantity.)*

(See Exercise 8.)

IMAGES

We now define a generalization of *conjugate* where the role of the *real axis is replaced by the Möbius circle* $\mathfrak{M}$ through the distinct points z_1, z_2, z_3. We say two points z, z^* are *images* with respect to $\mathfrak{M}$ when

$$\{z^*,z_1;z_2,z_3\} = \overline{\{z,z_1;z_2,z_3\}} \tag{2-28}$$

Note that $\overline{\{z,z_1;z_2,z_3\}} = \{\bar{z},\bar{z}_1;\bar{z}_2,\bar{z}_3\}$, hence if $\mathfrak{M}$ is the real axis, the concept of image reverts to conjugate. We shall soon see that the image of z is determined by $\mathfrak{M}$ independently of the choice of z_1, z_2, z_3.

LEMMA 2-6 *Images are invariant under linear-fractional transformations; i.e., if z and z^* are images with respect to $\mathfrak{M}$ and if $\phi(z)$ is a linear-fractional transformation, then $\phi(z^*)$ is the image of $\phi(z)$ with regard to the Möbius circle through $\phi(z_1)$, $\phi(z_2)$, $\phi(z_3)$.*

(See Exercise 8.)

LEMMA 2-7 *(a) A point z on $\mathfrak{M}$ is its own image. Moreover the following cases occur:*

(b) If $\mathfrak{M}$ is a straight line, the images are pairs of points which have the straight line as a perpendicular bisector.

(c) If $\mathfrak{M}$ is a circle, the images are pairs consisting of an interior and an exterior point on a ray through the center, situated so that the radial distances have as their product the square of the radius of the circle.

(See Exercise 9.)

We note the formula for images α, β with regard to the circle $|z| = R$, namely

$$\alpha\bar{\beta} = R^2 \qquad \text{or} \qquad \bar{\alpha}\beta = R^2 \tag{2-29}$$

so that images are, for example,

$$\alpha = |\alpha| \exp i\lambda \qquad \beta = \frac{R^2}{|\alpha|} \exp i\lambda \tag{2-30}$$

where $\arg \beta = \arg \alpha = \lambda$. Thus images agree with the cognate concept in optics.

EXERCISES

1 Usually euclidean geometry is defined as the study of affine transformations together with *symmetry*. Show that the most general symmetry operation, namely a reflection about the (axis) arbitrary cartesian line $y = Ax + B$ (A, B real), can be expressed in terms of affine transformations toegether with the *one* conjugate operation $z' = \bar{z}$. (Also take care of x = constant as axis.)

2 Show the cross ratio (2-26*a*) is invariant under the linear-fractional transformation which replaces z_i by $\phi(z_i) = (\alpha z_i + \beta)/(\gamma z_i + \delta)$. Allow for any one of the z_i or $\phi(z_i)$ to be ∞. [Symbolically, $\{z_1,z_2;z_3,z_4\} = \{\phi(z_1),\phi(z_2);\phi(z_3),\phi(z_4)\}$.]

3 Suppose four distinct points a, b, c, d are given in the z sphere in no particular order. Verify that the number of possible cross ratios which can be formed out of 4! permutations is *six* as follows:

$$\lambda_1 = \{a,b;c,d\} = \{b,a;d,c\} = \{c,d;a,b\} = \{d,c;b,a\} = \lambda$$

$$\lambda_2 = \{a,d;c,b\} = \{d,a;b,c\} = \{c,b;a,d\} = \{b,c;d,a\} = \frac{1}{\lambda}$$

$$\lambda_3 = \{a,c;b,d\} = \{c,a;d,b\} = \{b,d;a,c\} = \{d,b;c,a\} = 1 - \lambda$$

$$\lambda_4 = \{a,d;b,c\} = \{d,a;c,b\} = \{b,c;a,d\} = \{c,b;d,a\} = \frac{1}{1-\lambda}$$

$$\lambda_5 = \{a,b;d,c\} = \{b,a;c,d\} = \{d,c;a,b\} = \{c,d;b,a\} = \frac{\lambda}{\lambda-1}$$

$$\lambda_6 = \{a,c;d,b\} = \{c,a;b,d\} = \{d,b;a,c\} = \{b,d;c,a\} = \frac{\lambda-1}{\lambda}$$

Hint: Write $a = \lambda$, $b = 0$, $c = 1$, $d = \infty$. (Does this reduce the generality?)

4 (*a*) Show that if a, b, c, d are distinct points of the z sphere, the cross ratios λ_j in Exercise 3 can have all complex values except 0, 1, ∞. (*b*) Show that for distinct points, no two of the cross ratios λ_j in Exercise 3 can be equal unless $\lambda = \pm 1$, $\lambda = 2$, $\lambda = \frac{1}{2}$, $\lambda = (+1 \pm \sqrt{-3})/2$.

5 (*a*) Verify that $\phi(\lambda) = 6 + 2\sum_1^6 \lambda_i^2/27$. (*b*) Prove Lemma 2-3 by noticing that $\phi(\lambda) = \phi(1/\lambda) = \phi(1-\lambda) = \phi(1/(1-\lambda)) = \phi(\lambda/(\lambda-1)) = \phi((\lambda-1)/\lambda)$. Then observe that (2-27) has at most six roots as an equation in λ. Consider separately the special cases of Exercise 4*b*.

6 Show that all Möbius circles are expressed as

$$Lz\bar{z} + M\bar{z} + \bar{M}z + N = 0 \tag{2-31a}$$

with L and N real and M complex, and

$$|M|^2 > LN \tag{2-31b}$$

7 Prove Lemma 2-4 by showing that equation (2-31*a*) and condition (2-31*b*) are formally unchanged by the substitutions $z = 1/z'$, $z = \alpha z'$, $z = z' + \beta$.

8 (*a*) Prove Lemma 2-5 showing first that it holds when $z_1 = 0$, $z_2 = 1$, $z_3 = \infty$ (the real axis). Why does this suffice? (*b*) Prove Lemma 2-6 by repeated use of (2-26*b*).

9 Prove Lemma 2-7 [including (2-29) and (2-30)]. [*Hint:* Do (*b*) by rotation and do (*c*) by transforming z_1, z_2, z_3 to points on the unit circle and verifying that $z_1 = 1/\bar{z}_1$, etc., thus $\{\bar{z},\bar{z}_1;\bar{z}_2,\bar{z}_3\} = \{\bar{z},1/z_1;1/z_2,1/z_3\} = \{1/\bar{z},z_1;z_2,z_3\}$.]

10 Show that the great circles on the euclidean sphere project stereographically onto the Möbius circles (2-31*a*) specialized by $L = N$ and $M = -\bar{M}$. (*Hint:* Compare Exercise 3 of Sec. 2-2.)

11 Show that the most general linear transformation leaving the real axis invariant is $w = (Az + B)/(Cz + D)$ where A, B, C, D are real and $AD - BC \neq 0$. (*Hint:* Three points of the real axis map onto three preassigned points of the real axis.) (See Lemma 2-11.)

2-7 Fixed-point Classification

A very general procedure, regardless of the branch of mathematics, is to classify a transformation $w = f(z)$ by its fixed points, or the roots of $f(z) = z$. Let us consider the linear transformation

$$w = f(z) = \frac{Az + B}{Cz + D} \qquad AD - BC \neq 0 \tag{2-32}$$

in terms of the roots of $z = f(z)$. This last equation becomes

$$Cz^2 + (D - A)z - B = 0 \tag{2-33}$$

which has one or two roots [unless $C = B = D - A = 0$ or $f(z)$ is identically z, which we now exclude].

One root of (2-33) is ∞ if $C = 0$. If so, we take $D = 1$ and obtain for (2-32)

$$w = Az + B \qquad A \neq 0 \tag{2-34a}$$

The other root of (2-33) is also ∞ if $D = A$ $(= 1)$ or

$$w = z + B \tag{2-34b}$$

Otherwise, if $C \neq 0$, the roots of (2-33) are finite. They are distinct unless [referring to (2-33) again]

$$(D - A)^2 + 4BC = 0 \tag{2-35}$$

but they may be real or complex (as indeed are A, B, C, D). In particular the fixed points are ∞ and 0 precisely when $B = 0$ in (2-34*a*) or $w = Az$ $(A \neq 0, 1)$.

Next, let us consider the case where there are *two* distinct roots of (2-33) ξ_1, ξ_2 (possibly *one* may be infinite). We set

$$W = g(w) = \frac{w - \xi_1}{w - \xi_2} \qquad Z = g(z) = \frac{z - \xi_1}{z - \xi_2} \tag{2-36a}$$

(with modifications to $W = w - \xi_1$, $Z = z - \xi_1$ if $\xi_2 = \infty$, etc.). Then substituting the inverses of (2-36*a*),

$$w = \frac{W\xi_2 - \xi_1}{W - 1} \qquad z = \frac{Z\xi_2 - \xi_1}{Z - 1} \tag{2-36b}$$

we find that (2-32) reduces to a linear relation between W and Z, which we need not write down here. All we need note is that the fixed points $w = z = \xi_1$, $w = z = \xi_2$ must become

$$W = Z = 0 \qquad \text{and} \qquad W = Z = \infty$$

Hence, as we saw earlier,

(2-37) $$W = KZ \qquad K \neq 1$$

For reasons developed in texts on noneuclidean geometry, we classify transformations (2-32) as

(2-38*a*) $\begin{cases} \textit{Hyperbolic} \text{ if } K \text{ is real and } >0, \\ \text{or } f(z) \text{ is called a ``homothetic'' transformation or a} \\ \text{``magnification''} \end{cases}$

(2-38*b*) $\begin{cases} \textit{Elliptic} \text{ if } K = \exp i\theta \quad \theta \text{ real,} \\ \text{or } f(z) \text{ is called a ``rotation''} \end{cases}$

We shall soon consider the fact that these classifications, indeed the very values of K and $1/K$, are intrinsic in (2-32), not in our particular method of deriving (2-37) by equation (2-36*a*).

We finally introduce the concept of angle between two arcs $\mathfrak{M}_1$, $\mathfrak{M}_2$ of two Möbius circles intersecting in, say, ξ_1, ξ_2 with z_1 on $\mathfrak{M}_1$ and z_2 on $\mathfrak{M}_2$.

We can conveniently express the angle independently of tangents to $\mathfrak{M}_1$ and $\mathfrak{M}_2$ as follows: If $\xi_1 = 0$ and $\xi_2 = \infty$, the angle θ will be given at the origin just by the direction of z_1 and the direction of z_2. Thus for straight lines, the following definition would apply to ordinary euclidean geometry independently of the choice of z_1 on $\mathfrak{M}_1$ and z_2 on $\mathfrak{M}_2$:

$$\theta = \arg \frac{z_2}{z_1} = \arg \{z_2, 0; z_1, \infty\}$$

Hence, more generally, we define *the angle from* $\mathfrak{M}_1$ *to* $\mathfrak{M}_2$ by

(2-39) $$\theta = \arg \{z_2, \xi_1; z_1, \xi_2\}$$

We define *orthogonal* circles to be those for which $\theta = \pm\pi/2$, as usual. For *tangent* circles we set $\xi_1 = \xi_2$ and $\theta = 0$ or π.

LEMMA 2-8*a* *Let* ξ_1, ξ_2 *be two (distinct) fixed points of transformation* (2-32). *Consider the two families of curves*

(2-40) $$\arg W \textit{ or } \arg \frac{w - \xi_1}{w - \xi_2} = \text{const}$$

and

(2-41) $$|W| \textit{ or } \left| \frac{w - \xi_1}{w - \xi_2} \right| = \text{const}$$

[*with modifications for* $\xi_2 = \infty$ *as in* (2-36*a*)]. *Then these two families are mutually orthogonal Möbius circles. The hyperbolic* (*and elliptic*) *transformations preserve* (2-40) [*and,* (2-41) *respectively*] *curve by curve, and they change each curve of* (2-41) [*and* (2-40), *respectively*] *into another such curve.*

(See Exercise 7.)

This is illustrated in Fig. 2-2*a* and *b*. Note the motion of magnification for the hyperbolic case and rotation for the elliptic case (as viewed from the fixed points).

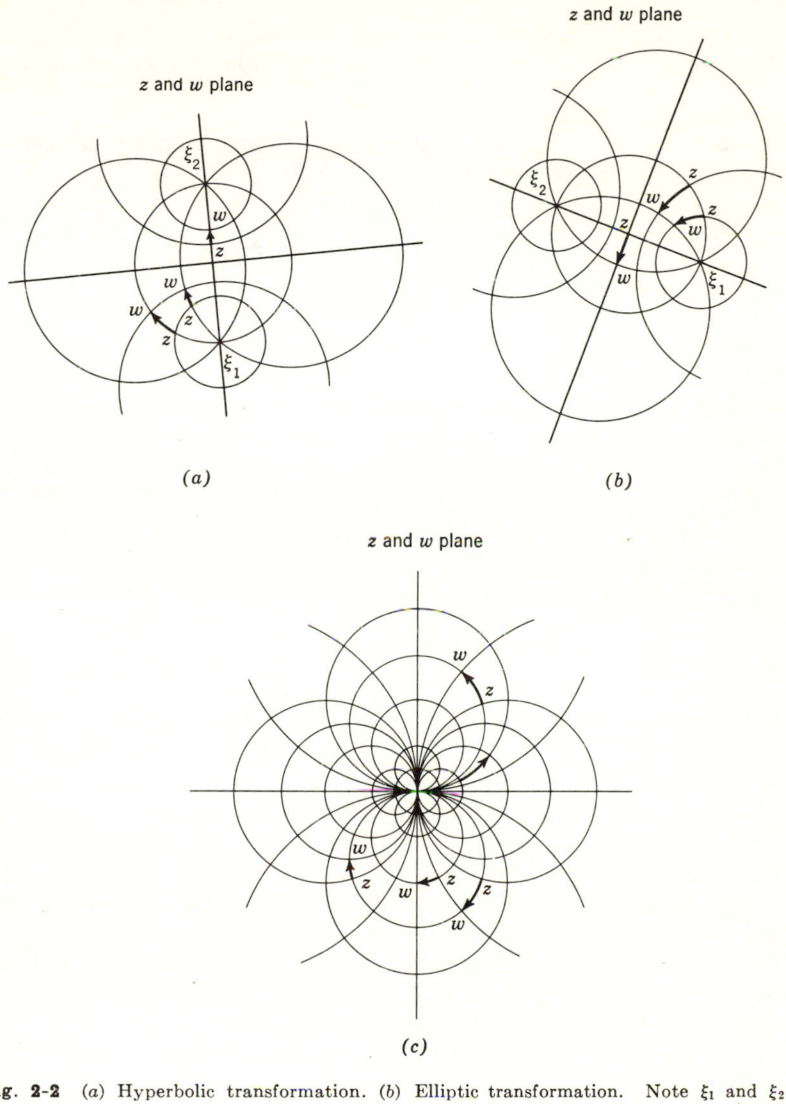

Fig. 2-2 (*a*) Hyperbolic transformation. (*b*) Elliptic transformation. Note ξ_1 and ξ_2 are images with respect to all circles shown here as enclosing them. (*c*) Parabolic transformations.

LEMMA 2-9 *Two points are images with respect to a Möbius circle $\mathfrak{M}$ if and only if all Möbius circles joining these points are orthogonal to $\mathfrak{M}$.*

(See Exercise 7.)

We now consider transformations (2-32) with a double-fixed point ξ. Assuming $\xi \neq \infty$, we write instead of (2-36a)

(2-42a) $$W = \frac{1}{w - \xi} \qquad Z = \frac{1}{z - \xi}$$

Then substituting the variables in (2-32),

(2-42b) $$w = \xi + \frac{1}{W} \qquad z = \xi + \frac{1}{Z}$$

We have a new transformation (with double-fixed point at ∞), necessarily of the form (2-34b)

(2-43) $$W = Z + B \qquad B \neq 0$$

This is called a "parabolic" transformation.

LEMMA 2-8b *In the terminology which precedes, the families of curves*

(2-44a) $$\operatorname{Re}\frac{W}{B} = \operatorname{Re}\frac{1}{B(w - \xi)} = \text{const}$$

(2-44b) $$\operatorname{Im}\frac{W}{B} = \operatorname{Im}\frac{1}{B(w - \xi)} = \text{const}$$

are mutually orthogonal Möbius circles. Under transformation (2-43), *the curves of* (2-44b) *are each preserved, and the curves of* (2-44a) *are each transformed into another such curve.*

(In Exercise 8, this case is derived as the limit of the case with distinct but confluent fixed points.)

Note that the loci (2-44a) and (2-44b) are orthogonal families of circles through $z = \xi$ at angles $-\arg B$ from the horizontal and vertical. Sometimes this is called "dipole flow." (See Fig. 2-2c.)

Summarizing Lemmas 2-8a and 2-8b we have the following result.

LEMMA 2-8 (bis) *A nonidentical linear transformation $w = f(z)$ can be put in one of the forms*

(2-37) $$W = KZ \qquad K \neq 1$$
(2-43) $$W = Z + B \qquad B \neq 0$$

by an appropriate linear transformation $W = g(w)$, $Z = g(z)$ which puts one fixed point at ∞ and the other fixed point (if there are two) at 0.

(Compare Exercise 2, where the matter is referred to matrix theory.)

Much material of historical interest is relegated to the exercises and more is referred to the literature. One particular result (Exercise 9) upon which we shall constantly draw is the following:

LEMMA 2-10 *The most general linear-fractional transformation which preserves the unit circle and which transforms* $z = \alpha$ *into* $w = 0$ *is given by*

(2-45) $$f(z) = \frac{z - \alpha}{1 - \bar{\alpha}z} \exp i\theta \qquad \theta \text{ real}, |\alpha| \neq 1$$

This last result is an application of the principle that images are preserved under linear mappings, so that if $f(z)$ preserves the unit circle, it must preserve images, namely z and $1/\bar{z}$:

(2-46a) $$f\left(\frac{1}{\bar{z}}\right) = \frac{1}{\overline{f(z)}}$$

Likewise, a linear transformation which preserves the real axis must preserve images z and $\bar{z}$ or

(2-46b) $$f(\bar{z}) = \overline{f(z)}$$

Hence, the following analog of Lemma 2-10 can be established (see Exercise 10).

LEMMA 2-11 *The most general linear-fractional transformation which preserves the real axis is given by*

(2-47) $$f(z) = \frac{Az + B}{Cz + D} \qquad AD - BC \neq 0$$

where A, B, C, D *are real.*

We can go even further and observe that a special type of linear transformation of the z sphere corresponds to a rotation of the euclidean sphere (referring to stereographic projection). Such a rotation of the sphere must preserve diametrically opposite points (z and $-1/\bar{z}$ by Exercise 3 of Sec. 2-2). Therefore,

(2-46c) $$f\left(\frac{-1}{\bar{z}}\right) = -\frac{1}{\overline{f(z)}}$$

Hence, we finally establish the following† further result (see Exercise 11):

† It can be seen by a more elegant argument that *all* rotations of the euclidean sphere correspond to linear transformations of the z sphere. For, by Exercise 2 of Sec. 2-2, such a rotation induces a conformal (angle-preserving) mapping of the z sphere onto the w sphere, and such mapping, by Sec. 3-4, corresponds to an *analytic* function $w = f(z)$ (which is one-to-one, hence linear).

LEMMA 2-12 *The most general linear-fractional transformation of the z sphere which corresponds to the rotation of a euclidean sphere is given by*

$$f(z) = \frac{\beta z - \alpha}{\bar{\alpha} z + \bar{\beta}} \qquad \alpha\bar{\alpha} + \beta\bar{\beta} = 1 \tag{2-48}$$†

EXERCISES

1 Show that if the transformations (2-32) are normalized so that $AD - BC = 1$ (by dividing A, B, C, D by $\sqrt{\Delta}$), then those transformations correspond uniquely to the matrices $\pm\begin{pmatrix} A & B \\ C & D \end{pmatrix}$ so that functional substitution corresponds to multiplication. For instance, if $\phi(z)$ corresponds to the matrix $\pm\Phi$ and $\psi(z)$ corresponds to the matrix $\pm\Psi$, then $\phi(\psi(z))$ corresponds to $\pm\Phi\Psi$.

2 Show how the process of finding *distinct* fixed points ξ_1, ξ_2 and determining the value of K in (2-37) is equivalent to the reduction of the matrix $\begin{pmatrix} A & B \\ C & D \end{pmatrix}$ for $f(z)$ to diagonal form under similarity transformations. [*Hint:* From (2-36*a*) and (2-37), $g(w) = Kg(z)$ and from (2-32) $w = f(z)$; hence, $g[f(z)] = Kg(z)$. But multiplication by K corresponds to the matrix $\pm\begin{pmatrix} \sqrt{K} & 0 \\ 0 & 1/\sqrt{K} \end{pmatrix}$, hence $g'f(g^{-1}(g(z)))) = Kg(z)$, or the matrix for gfg^{-1}, has the indicated diagonal form.] What is the significance of $K = 1$ (eigenvalue of ± 1 in the matrix for f)?

3 Show that $w = a^2/z$ is elliptic with $K = -1$.

4 Show that the transformation (2-48) (the rotation of the euclidean sphere) is always elliptic. Also show that transformations (2-45) or (2-47) might be elliptic, hyperbolic, or parabolic.

5 Show that the parabolic transformations correspond to matrices which are reducible to $\begin{pmatrix} 1 & 0 \\ B & 1 \end{pmatrix}$ with $B \neq 0$, hence not diagonalizable, according to well-known theorems on matrices. [*Hint:* Try $1/w = B + 1/z$, and compare (2-43).]

6 Justify the definition in (2-39) by euclidean intuition, by showing θ is equivalent to the angle between tangents at ξ_1 (by letting $z_1 \to \xi_1$, $z_2 \to \xi_1$) and showing (2-39) becomes $\theta = \lim \arg [(z_2 - \xi_1)/(z_1 - \xi_1)]$.

7 Prove Lemma 2-8*a* by transferring the case to the Z and W planes (with fixed points at 0 and ∞). Do likewise with Lemma 2-9.

8 Prove Lemma 2-8*b* by taking the limit of the hyperbolic transformation as $\xi_1 \to \xi_2$ in the following fashion: $\xi_2 = \xi$, $\xi_1 = \xi - \varepsilon/B$ where ε is real. Then show that in (2-40)

$$\frac{1}{\varepsilon} \arg \frac{w - \xi_1}{w - \xi_2} \to \operatorname{Im} \frac{1/B}{w - \xi}$$

† If we were to write out $\beta = p + iq$, $\alpha = r + is$, then we would be able to write the matrix of (2-48) (see Exercise 2) as

$$\begin{pmatrix} \beta & -\alpha \\ \bar{\alpha} & \bar{\beta} \end{pmatrix} = p\begin{pmatrix} 1 & 0 \\ 0 & 1 \end{pmatrix} + q\begin{pmatrix} i & 0 \\ 0 & -i \end{pmatrix} + r\begin{pmatrix} 0 & -1 \\ 1 & 0 \end{pmatrix} + s\begin{pmatrix} 0 & -i \\ -i & 0 \end{pmatrix}$$

where $p^2 + q^2 + r^2 + s^2 = 1$. This gives the classical representation of the rotation group of the sphere in terms of quaternions $\pm(p + qI + rJ + sK)$.

and that in (2-41)

$$\frac{1}{\varepsilon}\left(\left|\frac{w-\xi_1}{w-\xi_2}\right|-1\right)\to \operatorname{Re}\frac{1/B}{w-\xi}$$

9 (*a*) Deduce Lemma 2-10 from (2-46*a*) by using the fact that if $f(\alpha) = 0$, then $f(1/\bar{\alpha}) = \infty$ (the image of 0). This determines numerator and denominator to within a constant factor. This factor is determined by $|f(z)| = 1$ when $|z| = 1$. Note that (2-45) embraces "in the limit" $f(z) = \exp i\theta/z$ (with $\alpha = \infty$) but a more useful form would be

(2-49)
$$f(z) = \frac{\beta z - \alpha}{-\bar{\alpha}z + \bar{\beta}} \qquad \beta\bar{\beta} - \alpha\bar{\alpha} = 1$$

(*b*) Show (2-49) covers all cases of (2-45) except that β and α can be replaced by $-\beta$ and $-\alpha$ in (2-49) producing the same transformation. Note the only such mapping which preserves the origin is $f(z) = z \exp i\theta$.

10 Prove Lemma 2-11 from (2-46*b*). [The roots of $f(z) = 0$ or ∞ must be real since 0 and ∞ are real!]

11 Prove Lemma 2-12 from (2-46*c*). (Compare Exercise 9.)

12 In Exercise 1 a relationship is developed between the linear transformation $w = f(z)$ and the unimodular matrix $\pm S$. Show the $\pm$ sign *cannot be fixed so as to produce a biunique correspondence* between f and S. This can be shown by noticing that the transformations

$$f_0(z) = z \qquad f_1(z) = -z \qquad f_2(z) = \frac{1}{z} \qquad f_3(z) = -\frac{1}{z}$$

correspond to the matrices $\pm S_j$ where

$$S_0 = \begin{pmatrix}1 & 0\\ 0 & 1\end{pmatrix} \qquad S_1 = \begin{pmatrix}i & 0\\ 0 & -i\end{pmatrix} \qquad S_2 = \begin{pmatrix}0 & i\\ i & 0\end{pmatrix} \qquad S_3 = \begin{pmatrix}0 & 1\\ -1 & 0\end{pmatrix}$$

Draw the required conclusion from the fact that

$$f_2(f_1(z)) = f_1(f_2(z)) = f_3(z)$$

whereas the corresponding matrices produce the result

$$S_2S_1 = -S_1S_2 = S_3$$

(The reader will notice the relevance of the preceding footnote on quaternion representations of the rotation group of a sphere.)

13 (*a*) Show that in Lemma 2-10, the interior of the unit circle in the z plane is mapped onto the interior of the unit circle in the w plane [$w = f(z)$] exactly when $|\alpha| < 1$. (*b*) Show that in Lemma 2-11, the upper half z plane is mapped onto the upper half w plane exactly when $AD - BC > 0$. [*Hint:* Consider $f(0)$ and $f(i)$, respectively.]

14 Show that hyperbolic, elliptic, and parabolic transformations are characterized by the fact that a circle remains invariant.

CHAPTER 3
GEOMETRIC CONSTRUCTIONS

The meromorphic functional relationship $w = f(z)$ is sometimes called a "mapping" of the z sphere into the w sphere because the understanding of this relationship is so deeply geometric.† The relationship $w = f(z)$ indeed produces a natural geometrical construction on the domain z called "analytic continuation" and also natural consequences for the range w known variously as "conformal" or "analytic" mappings.

ANALYTIC CONTINUATION

3-1 Multivalued Functions

As is well known, $w = \sqrt{z}$ is multivalued in the sense that the equation $w^2 - z = 0$ has *two* roots in w (for each $z \neq 0, \infty$). If we were concerned with real variables, we should be able to write the positive root as, say,

$$\sqrt{x} = \exp\left(\frac{1}{2}\int_1^x \frac{dx}{x}\right) = \sum_{\nu=0}^{\infty} \frac{1}{\nu!}\left(\frac{1}{2}\int_1^x \frac{dx}{x}\right)^\nu \tag{3-1}$$

and thereby disguise any connection with multiple roots.

In complex variable theory things are more difficult; e.g., if $z = |z| \exp i(\arg z)$, then although the quantity $|z|^{1/2}$ (>0) is well defined, still $\arg z$ is not, and therefore the values of $\sqrt{z}$ are found by fixing $\arg z$ and writing

$$\sqrt{z} = |z|^{1/2} \exp \frac{i}{2}(\arg z + 2\pi m) \qquad m = 0, \pm 1, \pm 2, \ldots \tag{3-2}$$

Then for $m = 0$ and $m = 1$ we get the two values $\pm\sqrt{z}$. This is an intuitive type of argument which was used with such skill that almost until the end of the nineteenth century its basic soundness remained unquestioned. Ultimately, ideas of Riemann (1851) and Weierstrass (1842) were invoked to explain what had been done by intuition. The explanations amounted to the idea that we really do not *use* multi-

† Nowadays the term "mapping" is used so broadly as to lose its original geometric connotations; for example, the real sequence a_n is called a "mapping of the positive integers into the real numbers."

valued functions even if we appear to do so by writing $\pm\sqrt{z}$. First we consider Weierstrass' interpretation (leaving Riemann's to Sec. 3-3).

A general situation is that we begin with a function $f(z)$ analytic in a region $\mathcal{A}$ and defined by some property *valid initially in region* $\mathcal{A}$ (perhaps by convergence of a power series, perhaps by the solution to an equation, etc.).

We now consider two analytic functions in the z plane, $f(z)$ defined in region $\mathcal{A}$ and $g(z)$ defined in region $\mathcal{B}$, where the regions have a nonvoid intersection $\mathcal{A} \cap \mathcal{B}$ in which $f(z) = g(z)$. Then we say $f(z)$ is an "analytic continuation" (or extension) of $g(z)$ and vice versa. Indeed, if $\mathcal{D}$ is the union of $\mathcal{A}$ and $\mathcal{B}$, we can speak of a function $h(z)$ which represents the individual (and common) values in $\mathcal{D}$ and which is an analytic continuation in $\mathcal{A}$ and $\mathcal{B}$ of $f(z)$. As a consequence of Exercise 2 of Sec. 1-7, we have the following results:

THEOREM 3-1 *If a function $f(z)$ can be extended analytically from given region $\mathcal{A}$ to given region $\mathcal{B}$, such a continuation of $f(z)$ is unique for the given regions.*

COROLLARY 1 *Let a function $f(z)$ be defined in the neighborhood $\mathcal{A}$ of the point a, and let $\mathcal{C}$ be a curve joining a to point b. Let $\mathcal{B}$ be any region containing $\mathcal{C}$ and such that $f(z)$ can be extended analytically to $\mathcal{B}$. Then the extension on $\mathcal{C}$ is independent of the particular choice of the surrounding region $\mathcal{B}$.*

This is called the "extension" (or continuation) along a curve $\mathcal{C}$ from a to b.

Naturally, there is nothing in the last theorem or corollary to contradict the possibility that $f(z)$ has different continuations for different regions or different curves we might construct. It is also possible that $f(z)$ might have a singularity on curve $\mathcal{C}$ so that the definition of an extension along curve $\mathcal{C}$ is invalid [then the continuation of $f(z)$ from $\mathcal{A}$ into $\mathcal{B}$, or $f(z)$ itself, is called *singular*].

Here it is convenient to introduce the terminology that a function $f_{\mathcal{A}}(z)$ defined in $\mathcal{A}$ is a *function element* for any point in $\mathcal{A}$ and that two functions $f_{\mathcal{A}}(z)$ and $f_{\mathcal{A}'}(z)$ represent the same function elements at $z = z_0$ when they agree in a small neighborhood of z_0. For example, Weierstrass thought of a function as being represented "in embryo"† as a power series about z_0. If the power series has an infinite radius of convergence, it (numerically) *represents* the function. Otherwise it is a function element $f_{\mathcal{A}}(z_0)$, say, valid in a circle $\mathcal{A}$ (within the circle of convergence) about z_0.

† In modern terminology, the aggregate of functions which are continuations of a certain function element at z is called (essentially) a "germ." Characteristic of the set-theoretic revolution, an aggregate $\{f(z)\}$ is the *primary* entity and the element of the aggregate $f(z)$ is the *derived* entity!

At some point in $\mathcal{A}$, say z_1, we construct another circle of convergence $\mathcal{A}'$ (centered at z_1) such that $\mathcal{A} \cup \mathcal{A}' \supset \mathcal{A}$, thus extending the domain of definition. For a point z_2 in $\mathcal{A}'$ we can construct another circle of convergence $\mathcal{A}''$ (centered at z_2) such that $\mathcal{A} \cup \mathcal{A}' \cup \mathcal{A}''$ is a still bigger domain of definition. If the regions $\mathcal{A} \cup \mathcal{A}' \cup \mathcal{A}'' \cup \cdots$ are a sequence following a curve $\mathcal{C}$ from z_0 to z^*, we call the function value thus derived $f_{\mathcal{A}}(z^*;\mathcal{C})$, "the function element $f_{\mathcal{A}}(z)$ evaluated at z^* through the path $\mathcal{C}$."

Thus, as shown in Fig. 3-1, we can start by defining $\sqrt{z}$ near $z = 1$ by, say, its principal value

$$(3\text{-}3)\qquad f_{\mathcal{A}}(z) = \sqrt{z} = \{1 + (z-1)\}^{1/2} = 1 + \frac{z-1}{2} - \frac{(z-1)^2}{8} + \cdots$$

Looking at power series (3-3) we might not suspect anything. Nevertheless, as indicated in the diagram, the values of f at -1 differ; for example,

$$(3\text{-}4)\qquad f_{\mathcal{A}}(-1;\mathcal{C}_1) = +i \qquad f_{\mathcal{A}}(-1;\mathcal{C}_2) = -i$$

Thus we are not dealing with multivalued functions but multivalued *extensions.*

Let us suppose $f_{\mathcal{A}}(z)$ is a function element defined in a region $\mathcal{A}$ which lies in a larger region $\mathcal{R}$ (of the z sphere). Let us suppose that the process of analytic continuation along curves from $\mathcal{A}$ to $\mathcal{R}$ *never produces singularities.* We say $f_{\mathcal{A}}(z)$ is *monodromic* in $\mathcal{R}$ if for each $z \in \mathcal{R}$, $f_{\mathcal{A}}(z;\mathcal{C})$ is defined uniquely by extension, regardless of choice of curve $\mathcal{C}$ lying in $\mathcal{R}$. Thus

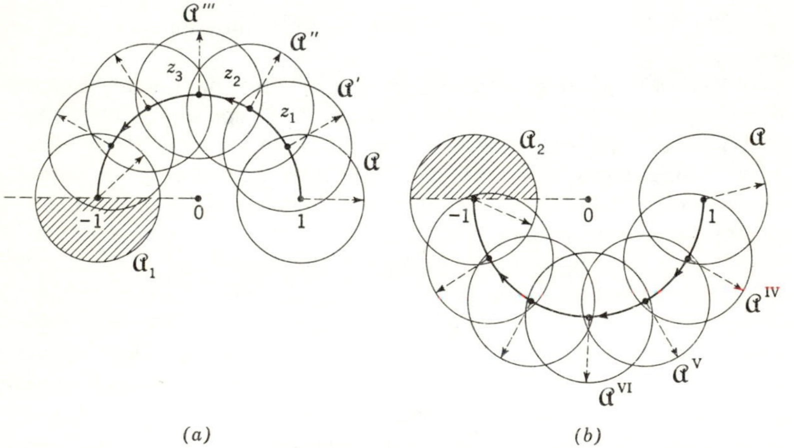

Fig. 3-1 Two different analytic continuations of the function element around $z = 1$, $f_{\mathcal{A}}(z) = \sqrt{z} = 1 + (z-1)/2 + \cdots$ are given for a region surrounding $z = -1$. The shaded portions of each figure will produce values of $f_{\mathcal{A}_1}(z)$ and $f_{\mathcal{A}_2}(z)$, respectively, which contradict the values in the other figure.

(*a*) Continuation along $\mathcal{C}_1$;

(*b*) continuation along $\mathcal{C}_2$.

$f_\alpha(z;\mathcal{C}) = f_\alpha(z) = f(z)$ symbolically. (This also means that no singularities are encountered on any curve $\mathcal{C}$ lying in $\mathcal{R}$.) Otherwise, we say that the nonsingular function $f_\alpha(z)$ is *polydromic* in $\mathcal{R}$.

THEOREM 3-2 (*Weierstrass monodromy theorem*) *If $f_\alpha(z)$ is a function element on $\alpha \subseteq \mathcal{R}$, a simply-connected region, and if $f_\alpha(z;\mathcal{C})$ is nonsingular for any $\mathcal{C}$ in $\mathcal{R}$, then $f_\alpha(z)$ is monodromic in $\mathcal{R}$; that is, all extensions $f_\alpha(z;\mathcal{C})$ are embraced by a single analytic function $f_{\mathcal{R}}(z)$.*

It is assumed that the reader has encountered the proof in an elementary course, so it will be relegated to Exercises 4 to 9 as a review.

Meanwhile we check the hypothesis of Theorem 3-2 critically.

If $\mathcal{R}$ is *not* simply-connected, say, if $\mathcal{R}$ is the annulus $1 < |z| < 2$, there are obviously many functions such as $f_\alpha(z) = \sqrt{z}$ or $\log z$ definable, say, near $z_0 = \frac{3}{2}$, for which every curve $\mathcal{C}$ in $\mathcal{R}$ drawn from $z_0 = \frac{3}{2}$ to an arbitrary z determines a value of $f_\alpha(z;\mathcal{C})$, but there is *no unique* continuation of $f_\alpha(z)$ in $\mathcal{R}$.

A more difficult situation is this: Suppose $\mathcal{R} \supset \alpha$ is simply-connected and for every $z \in \mathcal{R}$ there exists *at least one* path $\mathcal{C}$ (in $\mathcal{R}$) for which $f_\alpha(z;\mathcal{C})$ is definable, but *not every* path $\mathcal{C}$ from z_0 to z has that property (since the extension from α to $\mathcal{R}$ is possibly singular). It then does *not* follow that all definable values of $f_\alpha(z;\mathcal{C})$ agree at each z (regardless of $\mathcal{C}$). The counterexamples are not easy, and the matter is relegated to Exercises 2, 4, and 8 in Sec. 5-4 where the easiest counterexample seems to be

$$w^3 + z^3 - 3azw = 0 \tag{3-5}$$

Here we can define at least one power series solution $w(z)$ centered at z_0 and analytic in z for every point z_0 on the whole z sphere, but the functions thus obtained are not single-valued in the disk $|z| < r$ if $r > \sqrt[3]{4}\, a$.

The terms "uniform" and "multiform" are sometimes used for monodromic and polydromic.

Finally, another important counterexample should be borne in mind, namely

$$\phi_0(z) = \sum_{n=0}^{\infty} z^{n!} = 1 + z + z^2 + z^6 + \cdots \tag{3-6}$$

It can be verified that $\phi_0(z)$ is definable only in the unit disk $|z| < 1$ and cannot be extended to any region α' which contains a point (hence interval) of the unit circle. The proof of this last result involves the use of the points

$$z_{p/q} = \exp \frac{2\pi i p}{q} \qquad p, q \text{ integers}$$

which are everywhere dense on the unit circle. Indeed, $z_{p/q}^{n!} = 1$ if $n \geq q$, hence the series for $\phi_0(z_{p/q})$ contains ultimately all 1s, and by Exercise 10 $\phi_0(z)$ is not analytic at $|z| = 1$.

EXERCISES

1 Show how the definition (3-1) of $\sqrt{z}$ leads to multiple roots if z is complex, by analytic continuation.

2 *The most general continuation by function elements can be done using circles of convergence (Weierstrass).* [*Hint:* Let $f_{\mathcal{A}}(z)$ be any function element. Show that any two points in $\mathcal{A}$ can be connected by an arc inside a subregion of $\mathcal{A}$ lying within a nonzero distance δ of the boundary of $\mathcal{A}$. Then circles of convergence on $\mathcal{A}$ are at least of radius δ, and we can use the Heine-Borel theorem on such circular disks to reduce them to a finite number.]

3 Show that a function element $f_{\mathcal{A}}(z)$ can be built by analytic continuation solely (*a*) along polygonal arcs and (*b*) along polygonal arcs with *rational* complex coordinates at each vertex, for example, $p/q + ir/s$. (*c*) Show that the number of circular regions $\mathcal{A}_j$ required for the construction of all the elements of an analytic function is denumerable.

4 Let $f_{\mathcal{A}}(z)$ be a function element and let $f_{\mathcal{B}}(z)$ be an arbitrary element derived from $f_{\mathcal{A}}(z)$ by analytic continuation to $b \in \mathcal{B} \subset \mathcal{R}$ along some path from $z_0 \in \mathcal{A}$ to b. Let $\mathcal{K}$ be a closed path from b back to b ($\mathcal{K} \subset \mathcal{R}$). Then call $\mathcal{K}$ a "path of polydromy" if the process of analytic continuation of $f_{\mathcal{B}}(z)$ along $\mathcal{K}$ *either encounters a singularity or else produces a (different) value* $f_{\mathcal{B}}(b;\mathcal{K}) \not\equiv f_{\mathcal{B}}(b)$. (Here $\mathcal{R}$ need not be simply-connected.) Show that if $f_{\mathcal{A}}(z)$ is not monodromic in $\mathcal{R}$ it has a path of polydromy.

5 (*a*) Show Exercise 4 is still valid if we restrict paths to polygons. (*b*) Show that if $\mathcal{K}$ is a path of polydromy and if $\mathcal{K} = \mathcal{K}_1 + \mathcal{K}_2$ for closed paths $\mathcal{K}_1$ and $\mathcal{K}_2$ (in $\mathcal{R}$), then $\mathcal{K}_1$ or $\mathcal{K}_2$ is a path of polydromy. (Here we think of adding $\mathcal{K}_1$ as chains.)

6 Show that if $f_{\mathcal{A}}(z)$ is not monodromic in a *simply-connected* region $\mathcal{R}$, it has *triangular* paths of polydromy (three actual straight sides) which are smaller in diameter than any prescribed positive quantity δ. (Recall the triangulation procedure used for proving Cauchy's integral theorem of Sec. 1-3.)

7 Show that if $f_{\mathcal{A}}(z)$ is not monodromic in $\mathcal{R}$, it is not monodromic in some closed subregion $\mathcal{R}^*$ of $\mathcal{R}$ separated from $|\partial\mathcal{R}|$ by some minimum distance $\delta = \text{dist}\,(|\partial\mathcal{R}|,|\partial\mathcal{R}^*|) > 0$.

8 Show that if $f_{\mathcal{A}}(z)$ is nonsingular in $\mathcal{R}$, then for any closed subregion $\mathcal{R}^*$ of $\mathcal{R}$ a positive δ exists such that any function element $f_{\mathcal{B}}(z)$ derived from $f_{\mathcal{A}}(z)$ and centered at b ($\in\mathcal{R}^*$) has a radius of convergence at least δ, and hence all $f_{\mathcal{B}}(z)$ are monodromic in disks of radius δ about b.

9 (*Weierstrass*) Show that if $f_{\mathcal{A}}(z)$ is not monodromic in a simply-connected region $\mathcal{R} \supset \mathcal{A}$, it has a singular analytic continuation in $\mathcal{R}$.

10 Show that if $a_n > 0$ and $\sum_0^\infty a_n = \infty$, then $\sum_0^\infty a_n z^n = \psi(z)$ cannot be extended analytically beyond $z = 1$ (assuming the power series converges for $|z| < 1$). [*Hint:* Show $\psi(z) \to \infty$ as $z \to 1$ along the real axis (with $z < 1$).]

11 Referring to equation (3-6), show that $\phi_1(z) = \phi_0(z/2) + \phi_0(1/z)$ can be defined by analytic continuation only in the annulus $\mathcal{A}(1 < |z| < 2)$. Is it monodromic in $\mathcal{A}$? What about $\phi_1(\sqrt{z})$?

3-2 Implicit Functions

The natural way we arrived at function elements was the consideration of analytic functions with nonanalytic inverses (such as z^2). More generally we might consider

$$z - \phi(w) = 0 \tag{3-7a}$$

where $\phi(w)$ is analytic at w_0 and $\phi'(w_0) \neq 0$. Then we would want to assert that an inverse $w = w(z)$ is analytic near $z = z_0$ $[= \phi(w_0)]$. For later purposes (in Chap. 5) it is desirable to generalize (3-7a) slightly further to

$$\Phi(z,w) = 0 \tag{3-7b}$$

where $\Phi(z,w)$ is assumed to be expandable into a convergent power series.

THEOREM 3-3 *Let $\Phi(z,w)$ be a function of z and w which vanishes at $z = z_0$, $w = w_0$, and let $\partial\Phi/\partial w \neq 0$ at that point. Furthermore, let $\Phi(z,w)$ be expandable as a power series*

$$\Phi(z,w) = \Sigma a_{ij}(z - z_0)^i(w - w_0)^j \tag{3-8}$$

which converges for some values $z_1 \neq z_0$, $w_1 \neq w_0$. Then it is possible to find a disk (simply-connected region) $\mathfrak{D}_z$ about $z = z_0$ and a disk $\mathfrak{D}_w$ about $w = w_0$ such that every point in $\mathfrak{D}_z$ corresponds to precisely one point in $\mathfrak{D}_w$ under the relationship (3-7b). *Also, $w = w(z)$ is an analytic root function [for $\Phi(z,w(z)) \equiv 0$ in z] near $z = z_0$, $w = w_0$. The disks can be chosen to lie in any preassigned neighborhood of z_0 and w_0.*

Note that we cannot expect to imply that the correspondence between $\mathfrak{D}_z$ and $\mathfrak{D}_w$ is one-to-one since $z^2 - w = \Phi$ would satisfy the conditions at $z_0 = 0$, $w_0 = 0$. The proof will be outlined, and the more manipulative steps will be left to the reader. For convenience let us translate the planes so that $z_0 = 0$, $w_0 = 0$.

First we consider $\Phi(0,w) = a_{01}w + a_{02}w^2 + \cdots$. Since

$$\left.\frac{\partial\Phi}{\partial w}\right|_0 = a_{01} \neq 0$$

$\Phi(0,w)$ is an analytic function with only a simple root $w = 0$. We let γ_1 be such a quantity that no other root occurs when $|w| < \gamma_1$. Now consider

$$J(z) = \frac{1}{2\pi i}\int_{\mathcal{C}_w} \frac{\partial\Phi(z,w)/\partial w}{\Phi(z,w)}\,dw \tag{3-9}$$

where $\mathcal{C}_w$ is any simple closed curve surrounding the origin in the w plane and lying within $|w| < \gamma_1/2$ (or any smaller prescribed neighborhood). Clearly, we have just shown $J(0) = 1$. Furthermore, $\Phi(z,w) \neq 0$ on any such $\mathcal{C}_w$ if z varies in a small enough neighborhood $\mathfrak{D}_z$ of the origin by (uniform) *continuity* of Φ in z. But, of course, $J(z)$ can take on only integral values [always $J(z) = 1$], hence a unique function $w(z)$ is defined. Actually we can write (subject to the same restrictions),

$$w(z) = \frac{1}{2\pi i} \int_{\mathcal{C}_w} w \frac{\partial\Phi(z,w)/\partial w}{\Phi(z,w)}\, dw \tag{3-10}$$

by Exercise 4 of Sec. 1-8. Here the right-hand side employs w as a variable of integration, with z used as a constant. Hence we can even differentiate (3-10) with respect to z under the integral sign. Formally, of course, we know how to differentiate the fraction $(\partial\Phi/\partial w)/\Phi$ with regard to z, but to guarantee the interchangeability of $\int$ and $\partial/\partial z$ as operations, we must take some special precautions. What we actually do is to show that the "delta process" gives

$$\frac{1}{z_1 - z}\left(\frac{\partial\Phi(z_1,w)/\partial w}{\Phi(z_1,w)} - \frac{\partial\Phi(z,w)/\partial z}{\Phi(z,w)}\right) \to \frac{\partial}{\partial z}\left(\frac{\partial\Phi/\partial w}{\Phi}\right) \tag{3-11}$$

as $z_1 \to z$ with w on $\mathcal{C}$ and z in $\mathfrak{D}_z$. (See Exercise 4.)

The differentiation establishes $w(z)$ as analytic, completing the proof.

If we add the condition $\partial\Phi/\partial z \neq 0$ at (z_0,w_0), we have a symmetric version of Theorem 3-3.

We use the symbol

$$\mathcal{S} = f(\mathcal{R})$$

to denote $\mathcal{S}$ the aggregate of values of w satisfying $w = f(z)$ for $z \in \mathcal{R}$. We say the mapping $w = f(z)$ is *interior at* $z = z_0$ if for every small neighborhood $\mathfrak{N}$ of z_0, $f(\mathfrak{N})$ is a neighborhood of w_0. We say the mapping is *interior in a neighborhood of* (*or near*) z_0 if the property holds for all points in a particular neighborhood of z_0.

THEOREM 3-4 *Let* $\Phi(z,w)$ *be, as before, the convergent series in* z *and* w *near* z_0 *and* w_0 *and let* (z_0,w_0) *be a point at which* $\Phi = 0$, $\partial\Phi/\partial w \neq 0$ *and* (*now*) $\partial\Phi/\partial z \neq 0$. *Then there is a function* $\begin{Bmatrix} w = w(z) \\ z = z(w) \end{Bmatrix}$ *satisfying* $\Phi = 0$ *which provides interior mappings at* $\begin{Bmatrix} z = z_0 \\ w = w_0 \end{Bmatrix}$. *Furthermore, there exist two positive radii (which can be taken as small as we please)* $\begin{Bmatrix} r_z \text{ and } R_z \\ r_w \text{ and } R_w \end{Bmatrix}$ *such that the circular disk* $\begin{Bmatrix} \mathfrak{D}_z\colon & |z - z_0| < r_z \\ \mathfrak{D}_w\colon & |w - w_0| < r_w \end{Bmatrix}$ *has a one-to-one image* $\begin{Bmatrix} w(\mathfrak{D}_z) \\ z(\mathfrak{D}_w) \end{Bmatrix}$ *lying in the disk* $\begin{Bmatrix} \mathfrak{D}'_w\colon & |w - w_0| < R_w \\ \mathfrak{D}'_z\colon & |z - z_0| < R_z \end{Bmatrix}$, *and the function* $\begin{Bmatrix} w(z) \\ z(w) \end{Bmatrix}$ *provides the unique solution to the relation* $\Phi = 0$ *for* $\begin{Bmatrix} z \in \mathfrak{D}_z,\ w \in \mathfrak{D}'_w \\ w \in \mathfrak{D}_w,\ z \in \mathfrak{D}'_z \end{Bmatrix}$.

PROOF Actually, the last statement (taken with z or w) is a direct result of the symmetric version of Theorem 3-3. What is important is the fact that the interior property also follows! The trick is to consider an arbitrary (sufficiently small) $\mathfrak{D}_z$ and ask for a $\mathfrak{D}_w$ small enough so that $z(\mathfrak{D}_w) \subseteq \mathfrak{D}_z$. Then if $w(\mathfrak{D}_z)$ does not provide a covering of the $\mathfrak{D}_w$, there exists a w_* such that $w_* \not\in w(\mathfrak{D}_z)$, but $w_* \in \mathfrak{D}_w$. Then $z(w_*) = z_* \in \mathfrak{D}_z$ and $w(z_*) = w_{**}$ [$\neq w_*$ since $w_* \not\in w(\mathfrak{D}_*)$]. Thus

$$\Phi(z_*,w_*) = \Phi(z_*,w_{**}) = 0$$

contradicting uniqueness! Q.E.D.

As a matter of fact, for $\mathfrak{D}_z$ close enough to z_0, the one-to-one image $w(\mathfrak{D}_z)$ is also necessarily a disk† (simply-connected).

COROLLARY 1 *Under the hypotheses of Theorem* 3-4, *for some neighborhoods of* z_0 *and* w_0 *all points which correspond under* $\Phi(z,w) = 0$ *have interior mappings under* $w(z)$ *and* $z(w)$ (*or these functions are interior in neighborhoods of* z_0 *and* w_0).

For proof we use considerations of continuity.

EXERCISES

1 (*a*) Show that under the conditions of Theorem 3-3 the coefficients satisfy $|a_{ij}| < M/r_1^i r_2^j$ where $r_1 = |z_1 - z_0|$, $r_2 = |w_1 - w_0|$, and M is a constant. (*Hint:* The general term of the series approaches 0 and is therefore bounded.) (*b*) Show Φ is analytic in w ($|w - w_0| < r_2$) for each fixed z ($|z - z_0| < r_1$). Here we might note that the series is dominated term by term ($\ll$) as follows:

$$\Phi + M \ll M\left(1 - \frac{|z - z_0|}{r_1}\right)^{-1}\left(1 - \frac{|w - w_0|}{r_2}\right)^{-1}$$

(*c*) Show the same is true for partial derivatives of all orders in sufficiently small neighborhoods of the origin.

2 Use (3-10) to prove just the continuity of $w(z)$.

3 Consider $\{\Phi(z_1,w) - \Phi(z_2,w)\}/(z_1 - z_2)$ as a power series in z_1, z_2, w (near $z_0 = 0$, $w_0 = 0$). Show this series is dominated term by term by $M(1 - |w|/r_2)^{-1}(1 - |z_1|/r_1)^{-1}(1 - |z_2|/r_2)^{-1}r_1^{-1}$ by using Exercise 1*a*.

4 From Exercise 3, as applied to Φ and to $\partial\Phi/\partial w$, establish the limit (3-11) by rewriting the left-hand terms using the substitutions

$$\Phi(z_1,w) = \Phi(z,w) + \Delta\Phi$$
$$\frac{\partial\Phi(z_1,w)}{\partial w} = \frac{\partial\Phi(z,w)}{\partial w} + \Delta\,\frac{\partial\Phi}{\partial w}$$

Then show the limits $\Delta\Phi/\Delta z \to \partial\Phi/\partial z$ ($\Delta z = z_1 - z$), etc., are uniform in w.

† This result will be rediscovered in Theorem 3-7 from the point of view that the mapping of the boundary $w(|\partial\mathfrak{D}_z|)$ can often determine the mapping inside $\mathfrak{D}_z$, regardless of how *large* $\mathfrak{D}_z$ is taken to be.

3-3 Cyclic Neighborhoods

Our attention inevitably turns to the case $w = \Phi(z)$ where $\Phi'(z_0) = 0$. Let us begin with the following example (with variables interchanged):

$$z - w^n = 0 \tag{3-12a}$$

near $w = z = 0$. This mapping becomes one-to-one again if a new geometrical artifact is introduced, a so-called "cyclic region" (or neighborhood). The object is to find a way of "attaching n values to each z" so as to have a one-to-one correspondence with the n values of w which come from the use of roots of unity,

$$w = z^{1/n} \exp\left(\frac{2\pi ik}{n}\right) \qquad k = 1, 2, \ldots, n \tag{3-12b}$$

if the symbol $z^{1/n}$ denotes one of the n roots of (3-12a).

To visualize this process, consider the relationship, as shown in Fig. 3-2, as a collection of n (= 3) disks $\mathfrak{D}_1, \ldots, \mathfrak{D}_n$ on the z plane and one disk $\mathfrak{R}_1$ on the w plane with the provision that all n disks on the z plane *share the origin* ($z = 0$). They are joined on cuts as shown, so that a revolution of 2π about the origin in the z plane brings z ($\neq 0$) from a point on disk $\mathfrak{D}_k$ to the point z on disk $\mathfrak{D}_{k+1}$ (here $\mathfrak{D}_{n+1}$ is regarded as $\mathfrak{D}_1$ again by the cyclic structure). This sequence is shown in Fig. 3-2 as z_{I}, z_{II}, z_{III}. To show that these disks cover the same z plane, we often "paste them together" on the slits to form an object resembling a "screw-like" structure. To connect $\mathfrak{D}_n$ with $\mathfrak{D}_1$, however, requires the "fourth dimension" since they do not actually intersect any intermediate $\mathfrak{D}_k$ ($1 < k < n$).

A *cyclic neighborhood* of the origin means a set of points consisting of neighborhoods in each disk $\mathfrak{D}_1, \ldots, \mathfrak{D}_n$ (overlapping in $z = 0$). We define a cyclic neighborhood of $z = a$ by translation and for $z = \infty$ by inversion (i.e., as a cyclic neighborhood of 0 in the $1/z$ plane). The points in question, 0, a, or ∞, are called "branch points." Either Fig. 3-2b or c represents the cyclic neighborhood of 0. We call n the "order of the neighborhood" (and $n - 1$ the "order of branching" or "ramification").

A *cyclic disk z neighborhood* is one whose w image is a disk (simply-connected region surrounded by a simple closed curve).

In general, geometric devices to make multivalued mappings single-valued are called "Riemann surfaces," but a broader definition of them will be given in Chap. 5. For purposes of reference, however, we shall still use the w plane; e.g., we shall speak of a simple closed curve surrounding the origin in a cyclic neighborhood if this is a property of the image in the w plane. We call ζ (= w here) a "local uniformizing parameter" (for the branch point) if there is a local one-to-one mapping into the w plane of a neighborhood of $\zeta = 0$ by means of a suitable analytic

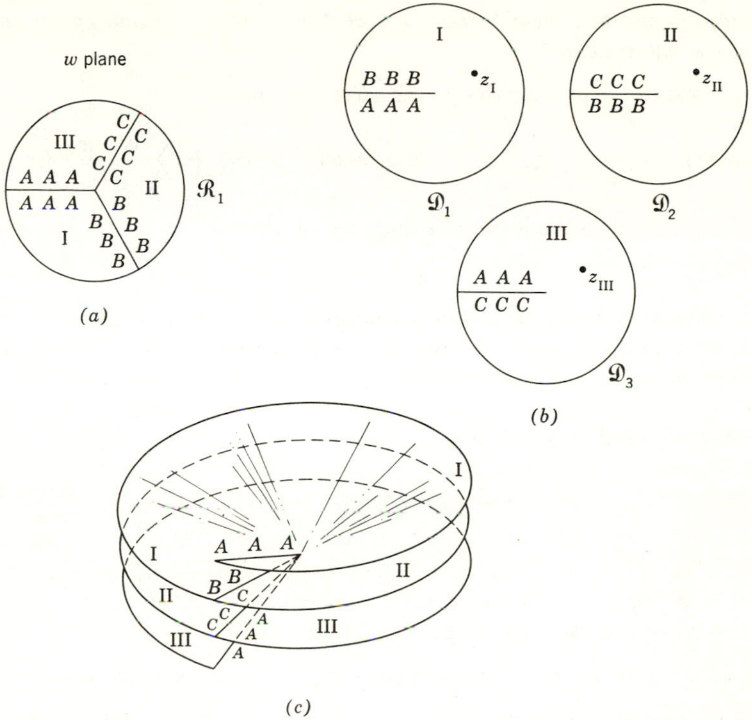

Fig. 3-2 The transformation of the w plane to the z plane by analytic continuation for $w = z^{1/3}$, ($z = \zeta^3$, $w = \zeta$). The disks $\mathfrak{D}_1$, $\mathfrak{D}_2$, $\mathfrak{D}_3$ (sharing the origin) constitute the image of $\mathfrak{R}_1$ (in the w plane). The location of the cut (where the disks merge) is unimportant.

function; for example, $\zeta = f(w)$ or $\zeta = f(z^{1/n})$. Most often, naturally, we try to use

$$\zeta = (z - a)^{1/n} \tag{3-13a}$$

for branch point a and

$$\zeta = z^{-1/n} \tag{3-13b}$$

for branch point ∞.

THEOREM 3-5 *Let $z = \phi(w)$ be an analytic function at $z = z_0$, $w = w_0$, and let*

$$\phi'(w_0) = \phi''(w_0) = \cdots = \phi^{(n-1)}(w_0) = 0$$

with $\phi^{(n)}(w_0) \neq 0$. Then we can restrict w to the disk $\mathfrak{D}_w(|w - w_0| < r_w)$ in such a fashion that the relation $z = \phi(w)$ produces a one-to-one interior *map of some cyclic neighborhood of z_0 of order n onto an ordinary neighborhood of w_0 contained in $\mathfrak{D}_w$.*

PROOF Consider the power series expansion

$$\phi(w) = \phi(w_0) + \sum_{t=n}^{\infty} \frac{\phi^{(t)}(w_0)}{t!}(w - w_0)^t \tag{3-14}$$

If we introduce the local uniformizing parameter

$$z - z_0 = \zeta^n \tag{3-15a}$$

it remains only to show that a single-valued analytic function $w = g(\zeta)$ has the property that all pairs (z,w) satisfying $z = \phi(w)$ and sufficiently close to (z_0,w_0) are parametrized by

$$z = z_0 + \zeta^n \qquad w = g(\zeta)$$

Thus we need only solve

$$\zeta^n = z - z_0 = \phi(w) - \phi(w_0) = \sum_{t=0}^{\infty} \frac{\phi^{(n+t)}(w_0)(w - w_0)^{n+t}}{(n+t)!} \tag{3-15b}$$

In effect, $\zeta^n = (w - w_0)^n[a_0 + a_1(w - w_0) + a_2(w - w_0)^2 + \cdots]$, where $a_0 = \phi^{(n)}(w_0)/n! \neq 0$ and generally $a_t = \phi^{(n+t)}(w_0)/(n+t)!$. By Exercise 2, we can take nth roots and obtain

$$\zeta = a_0^{1/n}(w - w_0) + b_2(w - w_0)^2 + \cdots + b_n(w - w_0)^n + \cdots \tag{3-15c}$$

Here we apply Theorem 3-3 and express w as $g(\zeta)$ by analyticity.

Q.E.D.

Note there are n values of ζ (hence of w)† to each z near z_0, and they are distinct when $z \neq z_0$.

COROLLARY 1 (*Maximum-modulus principle*) *Let $w = f(z)$ be analytic and nonconstant for z in a region $\mathfrak{R}$. Then $|f(z)|$ does not achieve a maximum for any point of $\mathfrak{R}$.*

COROLLARY 2 (*Maximum-minimum principle*) *Under the hypothesis of Corollary 1, neither* Re $f(z)$ *nor* Im $f(z)$ *can achieve a maximum or minimum for z in $\mathfrak{R}$.*

Both these corollaries follow from Theorem 3-5, according to the interior mapping property. We recall that a region is open, by definition. If $z_0 \in \mathfrak{R}$, then some neighborhood of z_0, namely $\mathfrak{D}$, has the property that

† Here the reader's attention should be called to the disenchanting fact that the stimulating formalism of the cyclic neighborhood has a rather unimaginative counterpart in analysis in the form of the mere substitution of ζ for z as the independent variable. Thus, as much as it was a tremendous achievement for Riemann to build a surface with the cyclic neighborhood, it was also a tremendous achievement for his successors to eventually abandon it in favor of uniformizing parameters (see Chap. 5)!

$f(\mathfrak{D})$ contains $w_0 = f(z_0)$ as an interior point [regardless of whether or not $f'(z_0) \neq 0$]. Thus neither $|w_0|$, Re w_0, nor Im w_0 is greater than some of the "surrounding" values for $w \in f(\mathfrak{D})$.

In effect this means that sup $|f(z)|$, for instance, is $|f(Z_*)|$ for $Z_* \in |\partial\mathfrak{R}|$, or else it is lim $|f(z_n)|$ where z_n approaches the ideal boundary $\partial\mathfrak{R}$ (see Sec. 2-1).

EXERCISES

1 Write the counterpart of Theorem 3-5 for the case where z_0 or w_0 is ∞ (or both).

2 Consider the power series $\mu = 1 + \sum_1^\infty a_t\lambda^t$ convergent for $|\lambda| = r$. Show that some series $\rho = 1 + \sum_1^\infty b_t\lambda^t$ has the property $\rho^n = \mu$. [*Hint:* Try $\rho = 1 + \sum_{s=1}^\infty \left\{\sum_1^\infty a_t\lambda^t\right\}^s \binom{1/n}{s}$.] To prove the convergence (under rearrangement of powers of λ), use the majorizing coefficients $a_t = -M/r^t$ $(M > 0)$ so $1 - \rho$ is majorized by $1 - \left\{1 - \dfrac{M\lambda}{r - \lambda}\right\}^{1/n}$ (which has all *positive* terms).

CONFORMAL MAPPING

3-4 Local and Global Results

The term "conformal" was introduced by Gauss (1822) to describe a transformation in which the image of a figure is "similar to the original figure in regard to the smallest parts." Here the term "similar" is used in the euclidean sense of similar triangles. Thus, it should be an equivalent intuitive idea to define conformal as "angle-preserving," since two triangles with three pairs of equal angles are similar (proportional).

To arrive at a formal definition, we write an arbitrary transformation

$$u = u(x,y) \qquad v = v(x,y) \tag{3-16}$$

as a pair of continuously differentiable functions, which map a region $\mathfrak{R}$ of the xy plane in biunique fashion on a region $\mathfrak{R}'$ of the uv plane. Such a transformation is called conformal if every pair of curves $\mathfrak{C}$, $\mathfrak{C}'$ intersecting at (x,y) in $\mathfrak{R}$ has images $\mathfrak{K}$, $\mathfrak{K}'$ which make the same angle as that of $\mathfrak{C}$, $\mathfrak{C}'$, preserving the sense of rotation. We use the term "isogonal" to mean "equal or negatively equal" so that the conjugation transformation (or reflection $u = x$, $v = -y$) is isogonal, but not conformal as angles are reversed.

THEOREM 3-6 *Let the transformation $u = u(x,y)$, $v = v(x,y)$ perform a one-to-one bicontinuous mapping of the region $\mathfrak{R}$ of the xy plane onto the* *region* $\mathcal{R}'$ *of the uv plane such that the (first and) second derivatives are continuous. Then the transformation is conformal exactly when* $w = u + iv$ *is an analytic function of* $z = x + iy$ *on* $\mathcal{R}$, *with* $dw/dz \neq 0$ *on* $\mathcal{R}$.

This theorem, in effect, tells us that conformal mapping is just the relation of image to original figure under biunique analytic functions. The attention given to this one special property (conformality) is historically due to concern with map-making, where a flat map could be used by navigators to "set a course" (by reading an angle) provided it was a conformal map of the globe. We saw in Exercise 2 of Sec. 2-2 that the stereographic projection is conformal. A more significant motivation by Riemann (1851) is one derived in Chap. 9 from hydrodynamics, hence we shall not pursue the intuitive import of conformality any further.

One half of Theorem 3-6 is easy. Let $w(z)$ be analytic at z_0 with $w'(z_0) \neq 0$; we want to show it is conformal. We let $z(s)$ be some curve through z_0 (at, say $s = 0$). Then the angle of $z(s)$ with the horizontal (real) axis is $\arg z'(0)$. Consider $w(z(s))$ the image. Then the angle of $w(z(s))$ with the horizontal axis is (by differentiability)

$$\arg \left[\frac{dw(z(s))}{ds}\right]_0 = \arg\,[w'(z_0)z'(0)] = \arg z'(0) + \theta_0$$

where $\theta_0 = \arg w'(z_0)$ is a constant (to within multiples of 2π) as $z'(0)$ takes on values for arbitrary curves. More crudely, we could have written

$$dw = \left(\frac{dw}{dz}\right)_0 dz$$

so that dz is related to dw by a magnification $|dw/dz|_0$ ($\neq 0$), and (again) the rotation by $\theta_0 = \arg w'(z_0)$. (This bit of intuition is closer to Gauss' original description; see Exercise 2.)

The converse half of Theorem 3-6 (essentially the work of Riemann) is relegated to Exercises 5 and 6 in Sec. 8-3 because it involves the separate study of the real and imaginary parts of an analytic function (contrary to the spirit of Part I).

The most important single consequence of Theorem 3-6 (i.e., the "easy" part) is the following:

Corollary 1 *The analytic function* $w(z)$ [*with* $w'(z) \neq 0$ *in* $\mathcal{R}$] *maps orthogonal curves in the z plane onto orthogonal images in the w plane to the extent that it is a one-to-one transformation.*

The above property of Theorem 3-6 and Corollary 1 is a *local* property because it prevails in a neighborhood (of biuniqueness) which usually cannot be determined in advance. Actually, in principle, it is often no harder to establish *global* properties (valid in an entire region assigned in advance). We consider, for this purpose, how an analytic function $f(z)$ maps a curve $\mathcal{C}$: $z = z(t)$ onto an image curve $\mathcal{C}'$: $f(z(t))$; we presently show how uniqueness on $\mathcal{C}$ extends to a region bounded by $\mathcal{C}$ without any restriction that $\mathcal{C}$ must lie near enough to some point z.

THEOREM 3-7 *Let $w = f(z)$ be analytic in a region $\mathcal{R}^*$ which contains a simple closed, positively oriented curve $\mathcal{C}$ and its interior $\mathcal{R}$. Let the image of $\mathcal{C}$ under $f(z)$ be a one-to-one image which consists of a simple closed curve $\mathcal{C}'$ in the w plane taking its orientation from $f(z)$ as z traverses $\mathcal{C}$. Then*

(*a*) *$\mathcal{C}'$ is positively oriented and encloses a region $\mathcal{R}'$.*

(*b*) *The equation $w = f(z)$ defines a one-to-one correspondence between $\mathcal{R}$ and $\mathcal{R}'$.*

PROOF By the results of Sec. 1-8 the number of solutions to $f(z) = w_0$ (for $z \in \mathcal{R}$) is

$$N = \frac{1}{2\pi i} \int_{\mathcal{C}} \frac{f'(z)\,dz}{f(z) - w_0} \tag{3-17}$$

Now the parametrization of $\mathcal{C}$ displays positive orientation. The function $w = f(z)$ imposes a parametrization of $\mathcal{C}'$ which we first suppose to be positive. Then (3-17) becomes

$$N = \frac{1}{2\pi i} \int_{\mathcal{C}'} \frac{dw}{w - w_0} \tag{3-18}$$

and $N = 1$ if $w \in \mathcal{R}'$. [If, however, $\mathcal{C}'$ were negatively oriented, we should get $N = -1$, which would not be possible in (3-17) unless $f(z)$ had a pole.]

Thus $f(z) = w_0$ has one and only one solution for $z \in \mathcal{R}$ when $w_0 \in \mathcal{R}'$. Q.E.D.

COROLLARY 1 *In Theorem 3-7, let $f(z)$ be known to be analytic on $\mathcal{C}$ but otherwise meromorphic with possible poles inside $\mathcal{C}$ as long as $f(\mathcal{R}^*)$ does not contain $\mathcal{C}'$. (The hypotheses are otherwise unchanged.) Then one of the following is true:*

(*a*) *$\mathcal{C}'$ is positively oriented, $f(z)$ is analytic, and the theorem follows.*

(*b*) *$\mathcal{C}'$ is negatively oriented, and $w = f(z)$ provides a one-to-one correspondence between the interior of $\mathcal{C}$ and the exterior of $\mathcal{C}'$ [and in particular $f(z)$ has precisely one pole for $z = z_0 \in \mathcal{R}$].* (See Exercise 3.)

A *local* version of Theorem 3-7 now emerges as a corollary of a global result!

COROLLARY 2 *If $w = f(z)$ is analytic at $z = z_0$ and $f'(z_0) \neq 0$, then a sufficiently small circular disk $\mathcal{D}_0$ exists, centered at z_0 and having the property that every disk $\mathcal{D}$ in $\mathcal{D}_0$ has a biunique image $\mathcal{D}' = f(\mathcal{D})$ which is surrounded by an oriented curve $\partial\mathcal{D}'$ obtaining its orientation from $\partial\mathcal{D}$ by $|\partial\mathcal{D}'| = f(|\partial\mathcal{D}|)$.*

This follows with the help of Theorem 3-4. The next result, by contrast, is a *global-local* result:

THEOREM 3-8 *Let* $w = f(z)$ *be analytic in a region* $\mathcal{R}_z$ *containing a closed segment of arc* $\mathcal{C}$*; and furthermore, let* $f'(z) \neq 0$ *on* $\mathcal{C}$*, and let the mapping* $f(|\mathcal{C}|)$ *be biunique on* $\mathcal{C}$*. Then some region* $\mathcal{R}$ *exists such that* $\mathcal{R}_z \supseteq \mathcal{R} \supset \mathcal{C}$ *and* $f(z)$ *is a biunique map on* $\mathcal{R}$ *as well as* $\mathcal{C}$*.*

The proof is easier than the earlier results and is relegated to Exercise 6.

Here we introduce a useful term "schlicht" (or simple) to describe a meromorphic mapping $w = f(z)$ which is biunique. Thus, more concisely, a local condition for *schlichtness* near z_0 is $f'(z_0) \neq 0$ (Theorem 3-5 above); a global condition for schlichtness is schlichtness of boundary (Theorem 3-7); and a global-local condition for schlichtness near $\mathcal{C}$ is schlichtness on $\mathcal{C}$ with $f'(z_0) \neq 0$ for z_0 on $\mathcal{C}$.

EXERCISES

1 Deduce from Theorem 3-6 a consequence which covers isogonal (rather than conformal) transformations.

2 Show that the function $w = w(z)$, analytic at z_0 with $w_0 = w(z_0)$ and $w'(z_0) \neq 0$, magnifies and rotates the limiting neighborhoods. [*Hint:* Consider $z_1(t)$ a curve $(t_0 \leq t \leq t_1)$ such as a closed triangle, and let $\lambda > 0$. Consider the family of *similar* curves (in the *euclidean* sense) $\mathfrak{F}_\lambda$: $z = z_0 + \lambda z_1(t)$ and the image curves $\mathfrak{F}'_\lambda$: $w_1(t) = w(z)$. Then show that as $\lambda \to 0$ the "magnified" image curves $\mathfrak{F}''_\lambda$: $w_2(t) = [w_1(t) - w_0]/[\lambda w'(z_0)]$ are arbitrarily close (point by point) to $z_1(t)$ as $\lambda \to 0$.]

3 Prove Corollary 1 of Theorem 3-7.

4 From the hypotheses of Theorem 3-7 draw some conclusions on the *nonvanishing* of $f'(z)$. Do the same for Corollary 1.

5 Rewrite Theorem 3-7 when the mapping of $\mathcal{C}$ onto $\mathcal{C}'$ is 1 to n (for example, consider $w = z^n$ as z traces $\mathcal{C}$ the unit circle).

6 Under the hypotheses of Theorem 3-8, let $\mathcal{R}^{(\varepsilon)}$ be a region satisfying $\mathcal{R}_z \supseteq \mathcal{R}^{(\varepsilon)} \supset \mathcal{C}$ and such that $\mathcal{R}^{(\varepsilon)}$ lies within positive distance ε of curve segment $\mathcal{C}$. If $f(\mathcal{R}^{(\varepsilon)})$ is not a biunique mapping, let z_ε and z'_ε be points of $\mathcal{R}^{(\varepsilon)}$ such that $f(z_\varepsilon) = f(z'_\varepsilon)$. If we could make $\varepsilon \to 0$ and always obtain mappings which are not biunique, we would obtain an infinite aggregate of z_ε and z'_ε with limit points z_0 and z'_0 on $\mathcal{C}$. Show that a contradiction would be obtained, regardless of whether $z_0 = z'_0$ or $z_0 \neq z'_0$. (*Hint:* Use Theorem 3-4 in the first case.)

3-5 Special Elementary Mappings

The study of *schlicht* conformal mapping of regions of the z sphere onto regions of the w sphere has reached a very high state of art. There are atlases of "known" results and there is a well-developed "approximation theory." We leave most of the finer aspects of the art to the Bibliography since our objective is essentially to consider mappings which take into consideration the nonschlicht regions such as cyclic neighborhoods. We do note a few simple results in passing.

First of all, we might summarize some elementary mappings already used in some form or other:

(*a*) *Linear-fractional mappings* (see Sec. 2-7).

(*b*) *Powers* $w = z^n$ where n is a fixed real positive number. This function maps the sector of the z plane

(3-19) $$\theta_1 < \arg z < \theta_2 \qquad (0<)r_1 < |z| < r_2$$

into the sector of the w plane

(3-20*a*) $$n\theta_1 < \arg w < n\theta_2 \qquad (0<)r_1^n < |w| < r_2^n$$

provided

(3-20*b*) $$\theta_2 - \theta_1 < 2\pi \qquad n(\theta_2 - \theta_1) < 2\pi$$

(*c*) *Logarithmic mapping* $w = \log z$. This maps the angular sector

(3-21*a*) $$\theta_1 < \arg z < \theta_2 \qquad 0 < r_1 < |z| < r_2 \qquad (2\pi > \theta_2 - \theta_1)$$

into the rectangle

(3-21*b*) $$\theta_1 < \operatorname{Im} w < \theta_2 \qquad \log r_1 < \operatorname{Re} z < \log r_2$$

(see Fig. 3-3*a* and *b*). If $r_1 = 0$ and $r_2 = \infty$, the sector makes an angle $(\theta_2 - \theta_1)$ from 0 to ∞, and the rectangle extends to a horizontal infinite strip. If, however, the rectangle extends from $\operatorname{Im} w = -\pi$ to $\operatorname{Im} w = +\pi$, the sector becomes an annulus (or disk) with the negative real axis removed (see Fig. 3-3*c* and *d*). Other situations are similar and self-explanatory.

(*d*) *Multivalued mappings on cyclic neighborhoods.* We refer to the mapping $w^n = z$ (see Sec. 3-3).

(*e*) *Slit-opening map.* Let us now see how a multivalued map $w = z^{1/2}$ can be interpreted as a single-valued map by a suitable restriction, namely, to the region $\mathcal{R}^*$ of the z plane exterior to a simple curve (slit) $\mathcal{C}$ from 0 to $-\infty$. [We can visualize $\mathcal{C}$ as, say, the negative real axis for convenience.] A value of w can be found for every point of $\mathcal{R}^*$ by analytic continuation, hence by the Weierstrass monodromy† theorem (Sec. 3-1) a single value of $\sqrt{z}$ is determined in $\mathcal{R}^*$. Thus the mapping $w = z^{1/2}$ for z on $\mathcal{R}^*$ opens the slit, since for every w in the image of $\mathcal{R}^*$ the value $-w$ lies *outside* the image of $\mathcal{R}^*$. In other words, the image of $\mathcal{R}^*$ has exterior points (e.g., it is not the complement of some other slit in the w plane). To get down to cases, when $\mathcal{R}^*$ is the complement of the negative real axis, its image is the right half w plane ($\operatorname{Re} w > 0$).

† Strictly speaking, we always have required that $\mathcal{R}^*$ be bounded by a simple closed curve. Actually $\mathcal{R}^*$ can be approximated by such a region to any degree of accuracy, or we can regard the slit as being two arcs (retraced so as to cancel from ∞ to 0 and 0 to ∞). The concept of a Riemann manifold enables us to disassociate the slit into an original and a retraced arc. (See Fig. 3-4 for an analogous case based on Exercises 3 to 5.)

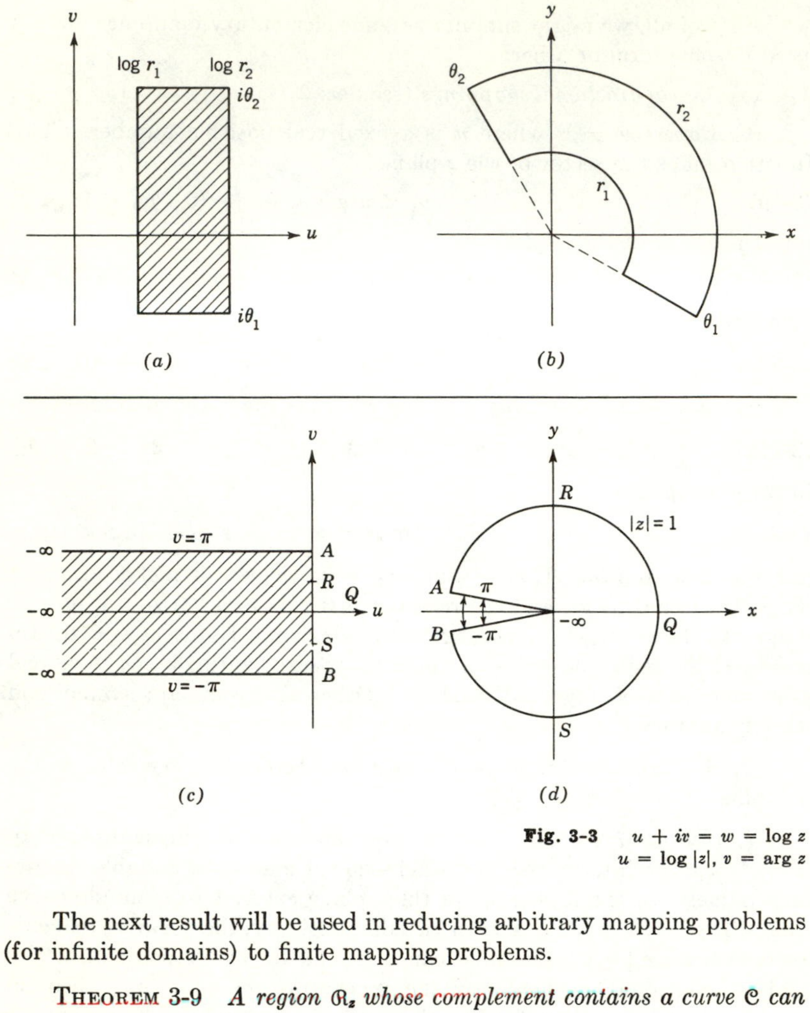

Fig. 3-3 $u + iv = w = \log z$
$u = \log |z|,\ v = \arg z$

The next result will be used in reducing arbitrary mapping problems (for infinite domains) to finite mapping problems.

THEOREM 3-9 *A region $\mathcal{R}_z$ whose complement contains a curve $\mathcal{C}$ can be mapped conformally onto a region lying inside the unit circle.*

PROOF Let the complement of $\mathcal{R}_z$ contain an arc of $\mathcal{C}$ joining $z = \alpha$ with $z = \beta$. If we let $\zeta = (z - \alpha)/(z - \beta)$, then $\mathcal{R}_z$ is mapped into $\mathcal{R}_\zeta$, a subregion of $\mathcal{R}^*$ [given in (*e*) above]. The mapping $\omega = \zeta^{1/2}$ opens the slit so that $\mathcal{R}_\omega$, the image of $\mathcal{R}_\zeta$, has an exterior point ω_0, from which $\mathcal{R}_\omega$ remains at least, say, distance r. Finally, $w = r/(\omega - \omega_0)$, as a function of z, maps the region $\mathcal{R}_\omega$ onto a subregion of $|w| < 1$ (since $|\omega - \omega_0| > r$). All steps are biunique in the regions concerned. Q.E.D.

Another elementary result will be used soon, particularly in Chap. 6. It extends Lemma 2-1 (of Sec. 2-6) by stating that a mapping does not have to be linear to preserve images.

THEOREM 3-10 *Let $w = f(z)$ be any schlicht conformal mapping of a region $\mathcal{R}_z$, surrounding an arc $\mathcal{C}_z$ of a Möbius circle, onto another region $\mathcal{R}_w$, surrounding an arc $\mathcal{C}_w$ of a Möbius circle $\mathcal{C}_w$, such that $\mathcal{C}_z$ is mapped onto $\mathcal{C}_w$. Then if z_1, z_2 are images (in $\mathcal{R}_z$) with respect to $\mathcal{C}_z$, it follows that $f(z_1)$, $f(z_2)$ are images (in $\mathcal{R}_w$) with respect to $\mathcal{C}_w$.*

PROOF First consider the case where $\mathcal{C}_z$ is a segment of the real z axis and $\mathcal{C}_w$ is a segment of the real w axis. Then $f(z)$ is real for z real and images z_1, z_2 with respect to $\mathcal{C}_z$ are merely conjugates. Since $f(z)$ is real on $\mathcal{C}_z$, so are all its derivatives (and coefficients a_n) in the expansion

$$f(z) = \sum_0^\infty a_n(z - x_0)^n \tag{3-22}$$

for x_0 any (real) point of $\mathcal{C}_z$. Thus if $\bar{z}_1 = z_2$ and *if the series* (3-22) *converges* for both z_1 and z_2, we see (since $\bar{a}_n = a_n$, $\bar{x}_0 = x_0$)

$$\overline{f(z_1)} = f(\bar{z}_1) = f(z_2) \tag{3-23}$$

This would prove the result, but we need not require that (3-22) converge as far as z_1 and z_2. Indeed, consider the analytic function $g(z) = \overline{f(\bar{z})}$ (see Exercise 3, Sec. 1-1). Clearly, as we just saw, $g(z) = f(z)$ for z near x_0 and the same result must obtain at $z = z_1(\bar{z} = z_2)$ by analytic continuation.

We must now show that the theorem holds for $\mathcal{C}_z$ and $\mathcal{C}_w$ taken as arbitrary Möbius arcs. To see this, we note that for some linear-fractional functions, $\zeta = L_1(z)$ and $\omega = L_2(w)$. Then the images $\mathcal{C}_\zeta$ and $\mathcal{C}_\omega$ of the Möbius arcs $\mathcal{C}_z$ and $\mathcal{C}_w$ are on the real axis. If images with respect to $\mathcal{C}_\zeta$ transform into images with respect to $\mathcal{C}_\omega$ (as we just saw), then the image property is assuredly preserved for $\mathcal{C}_z$ and $\mathcal{C}_w$ by the functions L_1 and L_2 (as we saw by Lemma 2-7 of Sec. 2-6). Q.E.D.

Theorem 3-10 is often called the "weak reflection principle" [weak because it requires knowledge of $f(z)$ on both sides of the arc $\mathcal{C}_z$]. In Chap. 11 we shall consider "stronger" principles. The formulation of these principles was mainly the work of Schwarz (1870).

In concluding these elementary results we shall state a result which dominates the subject of conformal mapping. It will be proved in Chap. 10, but it will be used for incidental illustrative purposes at various times in advance. It is the *Riemann*† *mapping theorem:*

An arbitrary region $\mathcal{R}$ bounded by a simple closed curve can be mapped in one-to-one fashion onto the interior of the unit disk $|w| < 1$ by a function $f(z)$ analytic for $z \in \mathcal{R}$.

† This theorem was conjectured by Riemann (1852) on the basis of physical intuition and unaccountably remained unproved for almost fifty years. At the time it was a focal point and unifying force of both pure and applied mathematics, as we shall attempt to recount.

It is clear that linear-fractional functions $f(z)$ cannot suffice for this result (because they preserve Möbius circles and therefore cannot map a noncircular domain onto a circle).

In the proof which is given ultimately (in Chap. 10), we show the $f(z)$ required for the Riemann mapping theorem is expressible as the *limit* of functions expressible in terms of linear transformations *and* the square root (compare Exercise 2).

EXERCISES

1 *The interior of any region of the z sphere bounded by two straight lines (or circular arcs) can be mapped onto the upper half w plane.* Note that this category includes a region bounded by parallel straight lines as well as segments of a circle and lunes. (*Hint:* If the boundary of this "biangle" has two vertices, transform them into 0 and ∞ by a linear-fractional transformation.) If the boundary has only one (double) vertex (parallel straight lines or tangent circles), then transform the double vertex to ∞. Then use $w = A(x - a)^r$ (for r real) or $w = A(\exp bz)$ (as required) to transform the region bounded by two straight lines (or rays) into the half plane.

2 Prove that the (not necessarily uniform) limit $f(z)$ of a sequence of linear-fractional functions $f_n(z)$ on a region $\mathfrak{R}$ is also linear-fractional. (*Hint:* Consider the cross ratio $\{z,z_1;z_2,z_3\}$ for z_1, z_2, z_3 in $\mathfrak{R}$.)

3 Prove that the function $2w = z + 1/z$ maps $\mathfrak{R}_z$, the interior of the unit circle $|z| < 1$, biuniquely onto $\mathfrak{R}_w$, the exterior of the slit $-1 \leq w \leq +1$ (relative to the Riemann sphere). (*Hint:* Consider $z = \exp i\theta$ and $w = \cos\theta$.) Show the same holds for the exterior of the unit circle $|z| > 1$.

4 Do Exercise 3 by applying the procedure of Theorem 3-9 to $\mathfrak{R}_w$, the exterior of the slit $-1 \leq w \leq +1$, by the following sequence of steps: Take $\omega = (w - 1)/(w + 1)$ (so that $\mathfrak{R}_\omega$ is the exterior to slit $-\infty < \omega < 0$). Take $\zeta = \sqrt{\omega}$ (so that $\mathfrak{R}_\zeta$ is the half plane $\operatorname{Re}\zeta > 0$), and finally, $\zeta = (1 - z)/(1 + z)$ maps $\mathfrak{R}_\zeta$ onto $\mathfrak{R}_z$.

5 For the mapping $2w = z + 1/z$ show the images of the lines $|z|$ = constant (and $\arg z$ = constant) are ellipses (and hyperbolas) with foci at $w = \pm 1$. (See Fig. 3-4.)

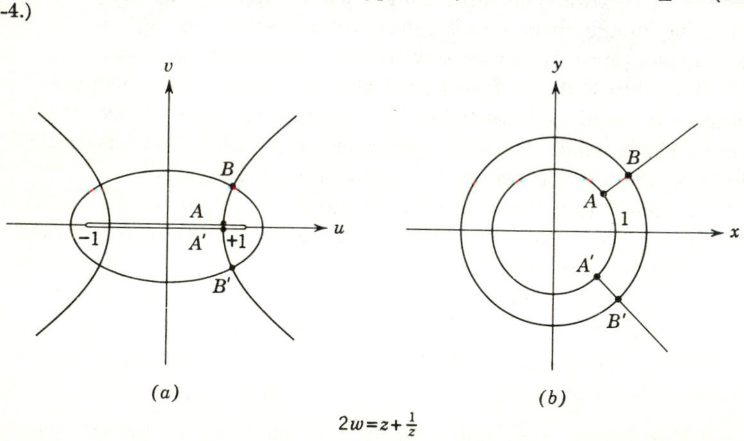

Fig. 3-4 Mapping of *exterior* of slit onto *exterior* of circle. Note that the slit from -1 to 1 is opened into the circle.

PART II
RIEMANN MANIFOLDS

DEFINITION OF RIEMANN MANIFOLD THROUGH GENERALIZATION

In the next three chapters we shall consider the process of abstraction which gave rise to the Riemann manifold. In principle, it is not hard to explain intuitively the concept of Riemann manifold. If a manifold is a point set which "looks" two-dimensional at every point, then an analytic (or Riemann) manifold is one with an analytic coordinate system.

We shall approach this concept in stages. The easiest illustration is the Riemann sphere (or the finite plane, or any subregion). The second easiest illustration is the torus. (Both these illustrations have been discussed earlier.) It is remarkable that the concept of the Riemann manifold had to be introduced *before the easiest illustration*, the Riemann sphere, could be appreciated (1851).

A period of abstraction followed, culminating in 1913 in Weyl's topological description of a Riemann manifold by justifying a "scissors and paste" version which then characterized topology itself. We content ourselves in Part II with this plateau of development, and we present a minimal amount of material here to make the reader appreciative of it.

CHAPTER 4
ELLIPTIC FUNCTIONS

In this chapter, we present an abridged treatment of the theory of elliptic functions. We consider it necessary to show the historical roots in order to place the later concept of Riemann manifolds in better perspective. Most importantly, we want to eventually justify a statement often made somewhat glibly but which reflects the spirit of the subject: "Elliptic functions are functions defined on a torus, just as rational functions are functions defined on a sphere."

ABEL'S DOUBLE-PERIOD STRUCTURE

4-1 Trigonometric Uniformization

It is a well-known result that if $R(z)$ is a rational function of z, then the indefinite integral $\int R(z)\,dz$ can be expressed explicitly in terms of a rational function of z possibly with "logarithmic terms" [for example, $\alpha \log (z - \beta)$] added on. (See Exercise 5, Sec. 2-4.)

If $q^2(z)$ is a quadratic polynomial, say

$$q^2(z) = (z - \alpha_1)(z - \alpha_2)$$

then it is also an elementary result that the indefinite integral $\int R(z,q)\,dz$ (for R rational in z and q) can be found explicitly. The calculus student usually does this by use of trigonometric functions. For instance, if

$$z = \alpha_2 + (\alpha_1 - \alpha_2) \cos^2 \theta$$

then $q(z) = \sqrt{\alpha_2 - \alpha_1} \sin \theta$, and $\int R(z,q)\,dz$ becomes $\int R_0 (\cos \theta, \sin \theta)\,d\theta$ where R_0 is also rational. This latter integral is then evaluated by the half-angle substitution $\tan \theta/2 = \zeta$ to yield $\int R(\zeta)\,d\zeta$ again (and explicit integration of a *rational* function).

Note that we used $\int R_0(\cos \theta, \sin \theta)\,d\theta$ as an intermediate step from $\int F(z,q)\,dz$ to $\int R(\zeta)\,d\zeta$. This use of trigonometry for purposes of uniformization was completely relevant, as we shall now explain.

The term "uniformization" means in a generic sense "the process of making something single-valued." We had occasion in Sec. 3-3 to introduce $w = z^{1/n}$ as a uniformizing parameter to deal with $z^{1/n}$ (merely by transferring our attention to the w plane!). Here, in dealing with $\int R(z,q)\,dz$ we seem to be saved by the use of a *trigonometric* uniformizing parameter. We must ask the following question: *How would we discover the use of trigonometric functions if we had not known them before being confronted with $\int R(z,q)\,dz$?* In answering this question we shall discover how elliptic integrals lead to a special (doubly periodic) manifold.

Take, for simplicity, $q(z) = (1 - z^2)^{1/2}$. This function involves a $\pm$ square root, but $q(z)$ can be defined uniquely in the region of the z sphere $\mathcal{R}_z$ *exterior to the slit* $-1 \leq z \leq 1$. This can be seen many ways, in particular by noting that the region $\mathcal{R}_z$ (containing the point at ∞) is for practical purposes bounded by $\mathcal{C}^*$, a simple closed curve, namely, one enclosing the slit and arbitrarily close (although oriented as in Fig. 4-1 so as to surround $z = \infty$). Hence by the monodromy theorem $q(z)$ is single-valued in $\mathcal{R}_z$. We find, however, that if we include the slit

$$\mathcal{S}\colon\ z = x \qquad -1 \leq x \leq +1$$

then $q(z)$ cannot be single-valued on $\mathcal{S}$. If $z \to x$ on the top (along $\mathcal{K}_1$),

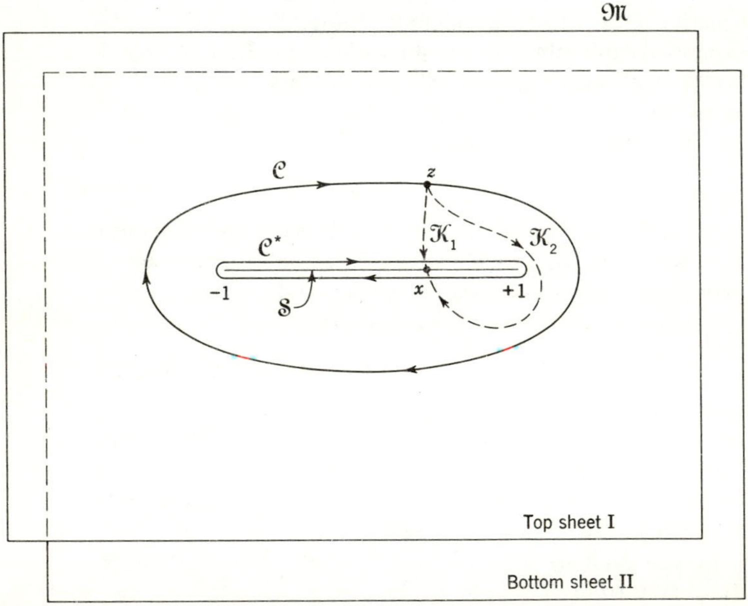

Fig. 4-1 Path showing period of sine function on Riemann surface $\mathfrak{M}$. The top sheet (I) is identified with $\mathcal{R}_z$.

then $q(z)$ approaches a value $q(x)$ which is the negative of the value $q(z)$ approaches if $z \to x$ on the bottom (along $\mathcal{K}_2$) (see Fig. 4-1 and Exercise 1).

A geometric contrivance for making $q(z)$ single-valued is to construct $\mathfrak{M}$, a two-sheeted *Riemann surface* which consists of two replicas of $\mathcal{R}_z$. We can otherwise describe $\mathfrak{M}$ as a replica of the z sphere joined to the present z sphere at the points ± 1 by cyclic neighborhoods of order 2 (see Fig. 4-1 and Sec. 3-3). Thus each of the branch points ± 1 belong to both sheets, and the top and bottom approaches to $\mathcal{S}$ (the slit $-1 < x < +1$) determine two replicas of $\mathcal{S}$ one on each sheet. The point at ∞ is actually duplicated—once on each sheet (like the general value of $z \neq \pm 1$) since $q(z)$ is definable as *single-valued* near $z = \infty$. The Riemann surface $\mathfrak{M}$ now has the property that the integrand or *differential* $\mathcal{R}(z,q)\,dz$ *is single-valued on* $\mathfrak{M}$. We next see the *integrals* are not necessarily single-valued on $\mathfrak{M}$.

Let us restrict our attention once again to $\mathcal{R}_z$ (essentially the "top sheet" of $\mathfrak{M}$), and consider

$$\theta = \int_0^z \frac{dz}{q(z)} = \int_0^z \frac{dz}{(1 - z^2)^{1/2}} \tag{4-1}$$

We can define a function element $\theta_Z(z)$ which is clearly analytically continuable in a region Z near any z (if we avoid $z = \pm 1$). If we extend z by analytic continuation in $\mathcal{R}_z$, however, we can find several different continuations of $\theta_Z(z)$, even if we remain on $\mathcal{R}_z$ (or on a single sheet of the Riemann surface).

Let us extend z analytically along a curve $\mathcal{C}$ that encloses the slit (see Fig. 4-1). Then $\theta_Z(z)$, by analytic continuation, produces $\theta_Z(z;\mathcal{C})$ such that

$$\theta_Z(z;\mathcal{C}) - \theta_Z(z) = \int_{\mathcal{C}} \frac{dz}{(1 - z^2)^{1/2}} \tag{4-2}$$

To evaluate the integral on the right, we note that the integrand is analytic (meaning single-valued here) in the annulus between $\mathcal{C}$ and $\mathcal{C}^*$, and thus we can distort $\mathcal{C}$ to the slit retraced (as shown by shrinking $\mathcal{C}^*$ in Fig. 4-1). Thus

$$\begin{aligned} \int_{\mathcal{C}} \frac{dz}{(1 - z^2)^{1/2}} &= \int_{-1}^{1} \frac{dx}{(1 - x^2)^{1/2}} + \int_{1}^{-1} \frac{dx}{-(1 - x^2)^{1/2}} \\ &= 2\int_{-1}^{1} \frac{dx}{(1 - x^2)^{1/2}} \end{aligned} \tag{4-3}$$

if we take into account the fact that the sign of $q(z)$ on the slit differs depending on whether we approach from top or bottom (Sheet I or Sheet II). Now let us define $z = \sin\theta$ as "any inverse" of $\theta(z)$ we wish, and let us define 2π as $2\int_{-1}^{1} dx/(1 - x^2)^{1/2}$ (the real integral). Then we have just shown that

$$\theta_Z(z;\mathcal{C}) = \theta_Z(z) + 2\pi \tag{4-4a}$$

hence we would be able to write

$$\sin(\theta + 2\pi) = \sin\theta \tag{4-4b}$$

provided $\sin\theta$ *is known to be analytic (single-valued) in the whole* θ *plane.* This is, for our present context, how the periodic functions of trigonometry enter.

The last condition on the single-valuedness of $\sin\theta$ is not trivial (see Exercise 2), but fortunately the analog for elliptic integrals is easier, so that there is all the more reason for not pursuing the single-valuedness of the sine. To obtain some useful manipulative practice, however, we consider not the arcsine (4-1) but (the arc cotangent)

$$\theta = -\int_\infty^k \frac{dk}{1+k^2} \tag{4-5}$$

and show (in Exercises 3 and 4) that it has a unique inverse $k(\theta)$ by a direct construction for $k(\theta)$.

EXERCISES

1 (*a*) Show how $q(z)$ is single-valued in $\mathcal{R}_z$ (exterior to the slit $-1 \leq z \leq 1$) by showing that any simple closed path in $\mathcal{R}_z$ encircles $z = +1$ and $z = -1$ so that a sign change is produced in $(1-z)^{1/2}$ and $(1+z)^{1/2}$ on analytic continuation. (*b*) Show that if $z \to x$ where x lies on the slit, then there is a difference in the sign of $q(z)$, depending on whether $z \to x$ from the top ($\mathcal{K}_1$) or from the bottom ($\mathcal{K}_2$). [*Hint:* Use the principal value of $\sqrt{Z}$ which is continuous except across the negative (real) Z axis where Z is the radicand.]

2 According to the "rigorous definition" of trigonometric functions, we set $\exp z = \sum_0^\infty z^n/n!$ ($0^0 = 1$ by definition), and we define $\sin z$ and $\cos z$ by Euler's laws, $\exp(\pm iz) = \cos z \pm i \sin z$ (two equations). Verify the relation (4-1) as a definition of $\sin z$ and verify $2\int_{-1}^1 dx/(1-x^2)^{1/2}$ as the definition of 2π where π is by definition the smallest positive root of $\sin z$.

3 Consider the function

$$\kappa(\theta) = \frac{1}{\theta} + \sum_{k=1}^\infty \left(\frac{1}{\theta+\pi k} + \frac{1}{\theta - \pi k}\right)$$

(*a*) Show that $\kappa(\theta+\pi) = \kappa(\theta)$ by writing out the first few terms of each. (*b*) Show that $|\kappa(\theta) - 1/\theta| <$ constant in the vertical strip of the complex θ plane given by $-\pi/2 < \operatorname{Re}\theta \leq \pi/2$. [*Hint:* Write $\kappa(\theta) - 1/\theta = \Sigma a_k(\theta)$ where $a_k(\theta) = 2\theta/(\theta^2 - \pi^2k^2)$, and write $\theta = s + it$ where $|s| \leq \pi/2$ and t is taken > 0 for convenience.] Then it suffices to show $|\kappa(\theta) - 1/\theta|$ is bounded for t large, say $t > \pi$. Verify the estimates: $|\theta| < t + \pi/2 < 2t$, $|-\pi^2k^2 + \theta^2| = |-\pi^2k^2 - t^2 + 2its + s^2| \geq t^2 + \pi^2k^2$

$-2t(\pi/2) - (\pi/2)^2 \geq (t/2)^2 + \pi^2k^2/2$. Hence $|a_k(\theta)| \leq 2t/(t^2/4 + \pi^2k^2/2) = \mathfrak{a}_k(t)$, and

$$\sum |a_k(\theta)| < \mathfrak{a}_0(t) + \int_0^\infty \mathfrak{a}_k(t)\,dk$$

(*c*) Verify the same "vertical strip property" for $|\kappa^2(\theta) - 1/\theta^2|$ and for $|\kappa'(\theta) + 1/\theta^2|$.

4 Expanding $\kappa(\theta)$ in the preceding exercise, show $\kappa(\theta) = 1/\theta + c_1\theta + \cdots$ where $c_1 = -2\pi^2 \sum_1^\infty 1/k^2$. (*a*) Verify the result that $-3c_1 + \kappa^2(\theta) + \kappa'(\theta) = g(\theta)$ vanishes for all θ by showing that it vanishes as $\theta \to 0$ and that $g(\theta)$ is bounded for all θ. [*Remark:* If we can prove $\kappa(\theta) = \cot\theta$, as done in supplementary texts, it will follow that $c_1 = -\frac{1}{3}$ or $\sum_1^\infty 1/k^2 = \pi^2/6$. In the meantime, however, we have a direct proof that for some $c_1 < 0$,

$$\frac{d\kappa}{d\theta} + \kappa^2(\theta) = 3c_1 \tag{4-6}$$

4-2 Periods of Elliptic Integrals

Elliptic integrals are expressions of the type $\int R(z,w)\,dz$ where R is rational and w^2 is either a cubic or fourth-degree polynomial with distinct roots. Let us take the latter case for now (see Exercise 7),

$$w^2(z) = (z - a_1)(z - a_2)(z - a_3)(z - a_4) \tag{4-7}$$

where $a_i \neq a_j$ for different subscripts. Any rational function $R(z,w)$ can be simplified (by the reduction of "surd expressions") to the form $R(z,w) = r_1(z) + r_2(z)w$ where $r_1(z)$ and $r_2(z)$ are rational. We can eliminate $\int r_1(z)\,dz$ as being the rational case and rewrite $r_2(z)w = r_0(z)/w$ where $r_0(z) = r_2(z)w^2$ (rational). Hence the difficulties resolve themselves to $\int[r_0(z)/w]\,dz$. Furthermore, $r_0(z)$, being rational, is expressible as a sum of partial fractions and polynomials (see Theorem 2-12). We are therefore led to integrals of the form (see Exercise 2)

$$I_1 = \int \frac{dz}{w} \qquad I_2 = \int \frac{dz}{(z - a_i)^t w} \qquad I_3 = \int \frac{dz}{(z - \beta)^t w} \tag{4-8}$$

where $\beta \neq a_i$ and $t \geq 1$ is an integer. These integrals are said to be of the *first, second,* and *third type* according to an adaptation of the designation of Legendre, who appreciated the fact that they are not expressible in terms of "ordinary" known functions.

The integrals (4-8) are called "abelian integrals," and they are of greater significance than evident now. For now we shall see how Abel showed I_1 leads to a double-period system. We write formally

$$u(z) = \int_a^z \frac{dz}{w} \tag{4-9}$$

(for some a) which provides an analytic function element in a region Z around any point z where $w \neq 0$. We now consider simple closed curves $\mathcal{C}$ on which $w(z)$ can be extended analytically, preserving value (i.e., sign). Thus

$$w_Z(z) = w_Z(z;\mathcal{C}) \tag{4-10}$$

On such a curve, we shall define a period as

$$u_Z(z;\mathcal{C}) - u_Z(z) = \int_{\mathcal{C}} \frac{dz}{w} \tag{4-11}$$

Clearly, by the argument used earlier we can take for $\mathcal{C}$ any simple closed curve on the z sphere surrounding an even number of the points a_1, a_2, a_3, a_4. (Actually, if $\mathcal{C}$ surrounds no points or all four points, $\mathcal{C}$ can be distorted to a point, possibly ∞.) We see the nontrivial cases are $\mathcal{C}_1$ surrounding a_1 and a_2 and $\mathcal{C}_2$ surrounding a_2 and a_3 (we consider a curve surrounding a_3 and a_4 in Exercise 6).

In Fig. 4-2 (using real a_i) we show the *paths of monodromy* satisfying (4-10) by the use of a *Riemann surface*, as in the last section. We consider again two replicas of the z sphere lying on top of one another so that "physically" they cover the same space. They are each cut on a_1a_2 and a_3a_4 and "sewn crosswise," so that the paths $\mathcal{C}_1$ and $\mathcal{C}_2$ go as shown in Fig. 4-2. (Compare the cyclic neighborhoods of Fig. 4-1.) We shall again maintain our unrigorous description of this Riemann surface for

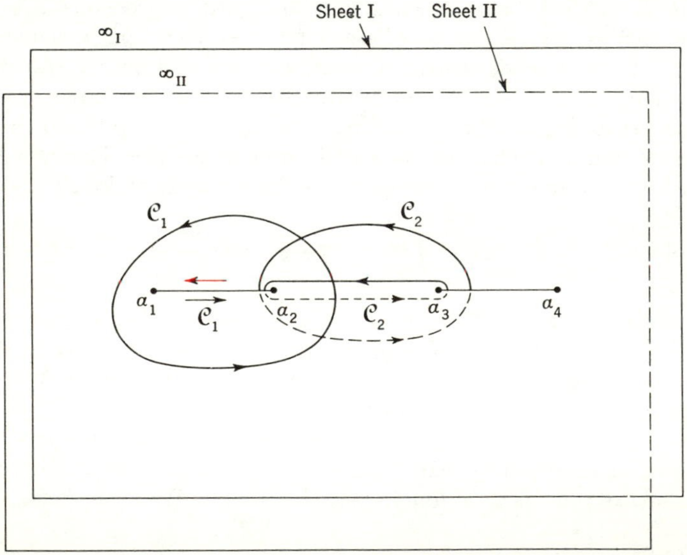

Fig. 4-2 Riemann surface of $w(z)$.

now, but let us note that if we are to think of "shrinking" these paths, $\mathcal{C}_1$ is deformable only to the cut a_1a_2 (traversed twice) and $\mathcal{C}_2$ is deformable only to a loop consisting of a_2a_3 traversed twice (as shown). The branch points a_1, a_2, a_3, a_4 appear only once for *both* replicas (or sheets), but all other points appear twice (once for each sheet), including the points on the cuts between end points and the points at ∞. The idea is, as usual, that a point appears only once (or is shared by both sheets) exactly when the two values $\pm w$ are interchanged by a cycle about such a point, thus ∞ is no branch point since $w = z^2\phi_0$ near $z = \infty$ for ϕ_0 analytic at ∞.

HYPOTHESIS 1 *The function element $u(z)$ defined by* (4-9) *has an inverse $z = z(u)$ which is (single-valued) meromorphic for every point u on the u plane (by analytic continuation).*

On the basis of this hypothesis we have a *doubly periodic function*

(4-12*a*) $$z(u + 2\omega_1) = z(u + 2\omega_2) = z(u)$$

where the periods are given by (4-11)

(4-13*a*) $$2\omega_1 = \int_{\mathcal{C}_1} \frac{dz}{w} = -2 \int_{a_1}^{a_2} \frac{dz}{w}$$

(4-13*b*) $$2\omega_2 = \int_{\mathcal{C}_2} \frac{dz}{w} = -2 \int_{a_2}^{a_3} \frac{dz}{w}$$

In a certain sense (to be explained later) these are the only periods† (see Exercise 3).

From the relation $u = \int dz/w$, it follows that $w = dz/du$; and by differentiating (4-12*a*) "straight across" we obtain

(4-12*b*) $$w(u + 2\omega_1) = w(u + 2\omega_2) = w(u)$$

so that $R(z,w)$, the integrand, is also doubly periodic in u as defined by (4-9), just as the integrand $R(z,q)$ of Sec. 4-1 was (simply) periodic in θ as defined by (4-1).

The justification of Hypothesis 1 will for our purposes be a major point of interest of elliptic function theory.

EXERCISES

1 Express the length of the ellipse in terms of integrals (4-8) using $x = a \cos \phi$, $y = b \sin \phi$ $(0 \leq \phi \leq 2\pi)$. (This, of course, is the derivation of the term "elliptic integral.")

2 Show that $\int z^t\, dz/w$ [which might arise in (4-8)] can be reduced to one of the other types by the change in variables $z = 1/\zeta$. (Assume all $a_i \neq 0$. Why is this convenient?)

† The symbol 2ω is used for periods since (in Greek) $\bar{\omega}$ is a variant letter for π (and 2π is the well-known period in trigonometry).

3 With $a_1 < a_2 < a_3 < a_4$ (real) show $\int_{a_1}^{a_2} dz/w = -\int_{a_3}^{a_4} dz/w$. (*Hint:* Let $J_1 = \int_{a_1}^{a_2}, J_2 = \int_{a_2}^{a_3}, J_3 = \int_{a_3}^{a_4}, J_4 = \int_{a_4}^{\infty} + \int_{-\infty}^{a_1}$ and show J_1, J_3 are pure imaginary while J_2, J_4 are real.)

4 Consider $\int d\theta/(\cos\theta - \cos\theta_0)^{1/2}$. Show that if $\theta = 2 \arcsin z$ and $\theta_0 = 2 \arcsin k$, the integral leads to $\int dz/[(1 - z^2)(k^2 - z^2)]^{1/2}$ (essentially Legendre's form). Show that the periods can be expressed in terms of the integrals of type $\int_{\theta_0}^{\pi}$ and $\int_0^{\theta_0}$ (in θ).

5 Show that for any distinct a_1, a_2, a_3, a_4 (not necessarily real) we can choose a k so that these cross ratios match: $\{a_1,a_2;a_3,a_4\} = \{-1,-k;k,1\}$. Here k (written $1/k$ in classical texts) $\neq 0, 1, \infty$. With this k show that a linear-fractional transformation will change w from (4-11) so that it may be replaced by W, where $W^2 = (z^2 - 1)(z^2 - k^2)$. Show the integrals I_1 (of the first kind) are transformed into the same kind in W. What can be said of I_2 and I_3?

6 Consider a unique definition of $w(z)$ in Sheet I and $-w(z)$ in Sheet II (relative to Fig. 4-2). Let $\mathcal{C}_3$ be a simple closed curve (not shown) surrounding a_3, a_4 only, and let $\mathcal{C}_0$ be a simple closed curve (not shown) surrounding all of a_1, a_2, a_3, a_4. Show that $\int_{\mathcal{C}_3} dz/w + \int_{\mathcal{C}_1} = \int_{\mathcal{C}_0} = 0$.

7 Show that by a change of variables $z - a_4 = 1/\zeta$, the integrals $\int R(z,w)\,dz$ are transformed to $\int R_0(\zeta,y)\,d\zeta$ where y^2 is a cubic polynomial in ζ [with distinct roots, assuming those of w are distinct in (4-7), of course].

4-3 Physical and Topological Models

Let us examine one of many realizations of double periodicity: Consider a circular pendulum of mass m held at length a from the center. The displacement is measured in terms of the angle θ from the lowest point, and g (>0) is the downward gravitational acceleration. The velocity $d\theta/dt$ is given by the energy equation

$$\frac{m}{2}\left(\frac{a\,d\theta}{dt}\right)^2 = mg(\cos\theta - \cos\theta_0) \tag{4-14}$$

where θ_0 represents one of two symmetric positions at which the pendulum is momentarily at rest ("high" point). A quarter period is obviously

$$\frac{T_1}{4} = \int_{\theta=0}^{\theta=\theta_0} dt = \sqrt{\frac{a^2}{2g}} \int_0^{\theta_0} \frac{d\theta}{\sqrt{\cos\theta - \cos\theta_0}} \qquad g > 0 \tag{4-15a}$$

On the other hand, if we imagine the direction of gravity reversed, then g becomes $-g$ (<0), but otherwise Eq. (4-14) still describes the motion of the pendulum with θ_0 still the position of momentary rest (but now θ_0 represents the "low" point since the arc of the pendulum is now complementary to the earlier case, oscillating "over the top under reversed gravity").

A new quarter period is obtained (after setting $-g$ for g), namely,

$$\frac{T_2}{4} = \sqrt{\frac{a^2}{-2g}} \int_{\theta_0}^{\pi} \frac{d\theta}{\sqrt{\cos\theta - \cos\theta_0}} \qquad g > 0 \tag{4-15b}$$

Actually the periods T_1 and T_2 are both real, but in terms of $\int_0^{\theta_0}$ and $\int_{\theta_0}^{\pi}$ we have two integrals of the same integrand, one real and one pure imaginary. In Exercise 4 of Sec. 4-2, we showed that T_1 and iT_2 are periods of an elliptic integral like $2\omega_1$ and $2\omega_2$ of (4-13a) and (4-13b).

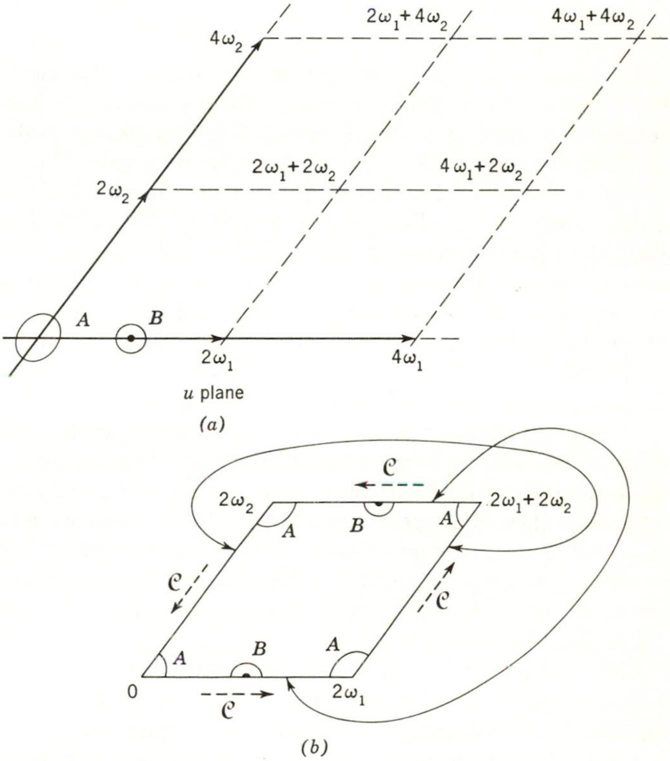

Fig. 4-3 Periodic structure of plane (a) and period parallelogram (b). Note the identification of sides by translation in the period parallelogram in (b). The neighborhood of A consists of four corners, the neighborhood of B consists of two semicircular sections. For the closed path $\mathcal{C}$ about the parallelogram the opposite sides are oriented so as to cancel in (b) under translation.

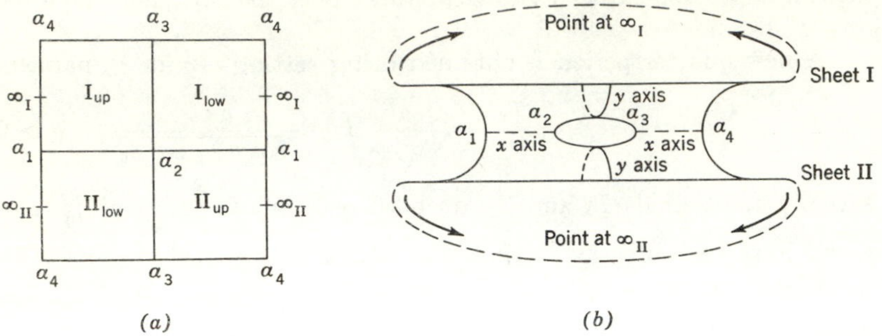

Fig. 4-4 Identification of Riemann surface of $w^2 = (z - a_1) \cdots (z - a_4)$ with the period parallelogram (*a*) and torus (*b*).

Now the double-period structure also has a geometrical significance. Under the system of double periods it means that $z(u)$ is known throughout the u plane if it is known merely in a parallelogram determined by the vectors $2\omega_1$ and $2\omega_2$. (It will be seen eventually that these vectors are never parallel, although they need not be at 90°.) Hence $z(u)$ is defined over the parallelogram in Fig. 4-3. We note that in defining $z(u)$, the opposite sides of the period parallelogram are identified in accordance with the period so that all four corners of the parallelogram are identified with one another. If we compare Fig. 1-2, we see how the period parallelogram is equivalent to a torus, so that we say that the integrand $R(z,w)$ of the elliptic integral is "defined† on a torus" (just because z and w were each seen to be doubly periodic).

Here our intuition leads us to something which is not immediately obvious, namely, to the conclusion that the Riemann surface of Fig. 4-2 is topologically equivalent to a torus. We could justify this statement from Hypothesis 1 directly if an additional hypothesis were valid, namely, that each $z = z(u)$ is achieved exactly twice (except for branch points) when u goes over the torus. Characteristically such (analytic) arguments are somewhat difficult (although valid as we shall see by Lemma 4-3 in Sec. 4-5).

An easier way exists for seeing that the Riemann surface in Fig. 4-2 is topologically equivalent to a torus. This approach consists of abandoning complex transformations for continuous transformations and thinking in terms of "rubber-sheet" geometry. Consider Fig. 4-4*a* and *b*.

In Fig. 4-4*a*, the upper and lower half planes I_{up}, I_{low}, II_{up}, II_{low} are cut out of Sheet I and Sheet II of Fig. 4-2 by cutting both sheets along the real axis. We reassemble them (rubber-sheet style) in accordance

† In Sec. 5-2, we speak generally of functions defined on a manifold.

with rules evident in Fig. 4-2:

I_{up} joins I_{low} on segments lying over a_4a_1 and a_2a_3
I_{up} joins II_{low} on segments lying over a_1a_2 and a_3a_4
I_{low} joins II_{up} on segments lying over a_1a_2 and a_3a_4
II_{low} joins II_{up} on segments lying over a_2a_3 and a_4a_1

Of course, we must not forget that the opposite sides of Fig. 4-4*a* are identified as in Fig. 4-3*b*. (Also a_4a_1 is a segment through ∞.)

In Fig. 4-4*b*, we see the Riemann surface of Fig. 4-2 as a torus by a more direct means. We look at the Fig. 4-2 "sideways" so that the two sheets are flattened out nearly into two straight lines. The Sheets I and II join across a_1a_2, a_3a_4 where they "crisscross," and the Sheets I and II are separate across a_2a_3. Now actually, if we draw the sheets so that they *close up* at ∞_I and ∞_{II}, we have a "hole in the middle" between a_2, a_3, which gives us the torus. In so doing we must "go into the fourth dimension" to undo the crisscrossing of the sheets at a_1a_2, a_3a_4.

WEIERSTRASS' DIRECT CONSTRUCTION

4-4 Elliptic Functions

We now have a plausible statement that the relation

$$w^2 = (z - a_1) \cdot \cdot \cdot (z - a_4)$$

is satisfied by the use of doubly periodic functions $z(u)$ and $dz/du = w(u)$ (based on the Hypothesis 1). Rather than prove this directly, Weierstrass constructed two functions $w(u)$ and $z(u)$ primarily as doubly periodic and satisfying the required relation almost "as by coincidence."

We defined *doubly periodic functions* over a fixed period parallelogram $(2\omega_1, 2\omega_2)$ in the u plane to be functions $f(u)$ with the property that

$$f(u + 2\omega_1) = f(u + 2\omega_2) = f(u) \tag{4-16}$$

We require for now that ω_2/ω_1 *be nonreal* (see Exercise 1). If, in addition, $f(u)$ has only a finite number of poles as singularities in the parallelogram, we call $f(u)$ an *elliptic function*. It follows then for an elliptic function, that near every value u_0 of u the behavior of $f(u)$ has an order (zero or pole or constant); that is, $f(u)/A(u - u_0)^t \to 1$ as $u \to u_0$ for t some integer (positive, negative, or zero) and some constant $A \neq 0$. Using the term from algebra, we note that elliptic functions (of a fixed period parallelogram) form a "field"† (over complex constants).

† A field is a system closed under rational operations [addition, subtraction, multiplication, and division (except by 0)]. *The concept was originally introduced for this kind of function-theoretic application* by Riemann but was preempted by algebraists who introduced the actual *term* "field" for the concept.

We also note that in addition to $2\omega_1$, $2\omega_2$, any integral combination

$$\Omega_{m,n} = 2\omega_1 m + 2\omega_2 n \qquad m,n = 0, \pm 1, \pm 2, \ldots \tag{4-17}$$

is also a period, by the repetition of property (4-16). The set $\mathbf{\Omega}$ of elements (4-17) is closed under addition and subtraction and is often called a module (over integral coefficients). The property (4-17) is said to make $2\omega_1$, $2\omega_2$ a "basis" of $\mathbf{\Omega}$. Two values of u differing by an $\Omega_{m,n}$ are written for convenience as

$$u_1 \equiv u_2 \qquad (\text{mod } \mathbf{\Omega}) \text{ (read "modulo } \mathbf{\Omega}\text{")} \tag{4-18}$$

(using a notation borrowed from number theory). For example, each $\Omega_{m,n} \equiv 0 \pmod{\mathbf{\Omega}}$.

Clearly from (4-18), $f(u_1) = f(u_2)$.

Lemma 4-1 *Let the zeros and poles of a nonconstant elliptic function $f(u)$ in a period parallelogram be listed as $\alpha_1, \ldots, \alpha_p$ and $\beta_1, \ldots, \beta_q$ repeated in accordance with multiplicities. Then the number of zeros p equals the number of poles q, and*

$$\sum_1^p \alpha_i \equiv \sum_1^p \beta_i \qquad (\text{mod } \mathbf{\Omega}) \tag{4-19}$$

For proof, let us first assume that no zeros or poles lie on the sides of the period parallelogram. (Referring to Fig. 4-3 this would mean no zeros or poles lie on ACA or ABA.) To accomplish this, we need only consider $f(u + \lambda) = g(u)$, another elliptic function whose zeros and poles are those of $f(u)$ but shifted by $-\lambda$. Then calling $f(u)$ our function again, we note the number of zeros less poles is

$$\frac{1}{2\pi i}\int_{\mathcal{C}} \frac{f'(u)\,du}{f(u)} = \int_0^{2\omega_1} + \int_{2\omega_1}^{2\omega_1+2\omega_2} + \int_{2\omega_1+2\omega_2}^{2\omega_2} + \int_{2\omega_2}^{0} \tag{4-20}$$

where $\mathcal{C}$ is the path around the periphery in Fig. 4-3 as indicated. Clearly the integrals on the right in (4-20) *cancel* [i.e., since $\dfrac{1}{2\pi i}\dfrac{f'(u)}{f(u)} = \psi(u)$ is elliptic, then

(4-21a)

$$\int_0^{2\omega_1} \psi(u)\,du + \int_{2\omega_1+2\omega_2}^{2\omega_2} \psi(u)\,du = \int_0^{2\omega_1} \psi(u)\,du + \int_{2\omega_1}^{0} \psi(u + 2\omega_2)\,du = 0$$

since $\psi(u + 2\omega_2) = \psi(u)$, etc.]. Thus the number of zeros equals the number of poles.

For the second result, consider $\dfrac{1}{2\pi i}\displaystyle\int uf'(u)\,du/f(u)$, which is certainly not even a doubly periodic integrand. Here, however, *two* sides of the parallelogram $\mathfrak{C}$ give

$$\text{(4-21}b) \quad \int_0^{2\omega_1} \psi(u)u\,du + \int_{2\omega_1+2\omega_2}^{2\omega_2} \psi(u)u\,du$$
$$= \int_0^{2\omega_1} \psi(u)u\,du + \int_{2\omega_1}^{0} \psi(u + 2\omega_1)(u + 2\omega_2)\,du$$
$$= -2\omega_2 \int_0^{2\omega_1} \psi(u)\,du$$

Nevertheless, $\int_0^{2\omega_1} \psi(u)\,du = \frac{1}{2\pi i}\int_0^{2\omega_1} d\log f(u) = n$ where n is some integer (not readily determined) since $f(0) = f(2\omega_1)$ by periodicity. Hence the logarithms can differ only by $2\pi i n$. Thus the expression in (4-21b) reduces to $-2\omega_2 n$. Taking $\mathfrak{C}$ to be the whole parallelogram, we find a like contribution from the remaining pair of sides. Thus

$$\text{(4-22)} \qquad \frac{1}{2\pi i}\int_{\mathfrak{C}} \frac{f'(u)u\,du}{f(u)} \equiv 0 \qquad (\text{mod } \mathbf{\Omega})$$

Meanwhile we can evaluate the integral on the left directly by noticing the integrand has a residue of $k\alpha_i$ for each zero α_i of order k and $l\beta_i$ for each pole β_i of order l; that is, $f'(u)/f(u)$ behaves like $k/(u - \alpha_i)$ when $f(u) = A(u - \alpha_i)^k + \cdots +$, etc. Thus (4-19) follows if we can avoid the zeros and poles on the boundary of the parallelogram.

We note that if we change u to $u + \lambda$, the number p of zeros (or poles) is certainly unchanged; and the same constant $p \cdot \lambda$ is added to both sides of (4-19). Thus we can avoid zeros or poles on the boundary, and the proof of Lemma 4-1 is now complete. Q.E.D.

The number of poles p is the same for $f(u)$ as for $f(u) - a$, hence the number p is the number of times $f(u)$ takes on any given value a [number of zeros of $f(u) - a$]. This constant p is called the *order* of $f(u)$.

LEMMA 4-2 *There can be no elliptic function with only one pole (order* 1).

PROOF For otherwise, a zero α and a pole β would satisfy $\alpha \equiv \beta$ (mod $\mathbf{\Omega}$); hence $f(\alpha) = f(\beta)$, or we would have $0 = \infty$! Q.E.D.

Thus we have a situation more complicated than that of the Riemann sphere where functions of order 1 exist (e.g., linear-fractional functions of z). But, reassuringly, *a bounded elliptic function is a constant,* by Liouville's theorem (which was originally developed in this context around 1850).

EXERCISES

1 *A doubly periodic function cannot have two periods* $(2\omega_1, 2\omega_2)$ *with a real ratio unless this ratio* ω_2/ω_1 *is rational.* The classical proof (Jacobi, 1835) goes as follows: If the ratio were real and irrational, all values of $\Omega_{n,m}$ would be distinct (only $\Omega_{0,0} = 0$) and they would lie on a straight line (through the origin) with inclination arg ω_2 (or arg ω_1). Thus if we take a positive integer N and vary m and n such that $|n| \leq N$

and $|m| \leq N$, we get a total of $(2N + 1)^2$ *different* values of $\Omega_{n,m}$ satisfying $-GN < |\Omega_{n,m}| < GN$ where $G = 2|\omega_1| + 2|\omega_2|$. Hence no matter how these $(2N + 1)^2$ values were spaced on a *segment* of length $4GN$, at least two would be as close as $4GN/(2N + 1)^2 < G/N$ ($\to 0$ as $N \to \infty$). Thus there would be periods Ω^* of arbitrarily small length; and consequently the roots of $f(u) = a$ [$f(u - \Omega^*) = a$] would not be isolated (Q.E.D.).

(*a*) Show how the above proof allows for the case where ω_2/ω_1 is rational (since in this case suitable exponential functions exist with periods $2\omega_1$, $2\omega_2$)! (*b*) Why does the proof fail when ω_2/ω_1 is nonreal?

2 Prove Liouville's theorem from Lemma 4-1 [by referring to $f(u) - f(u_0)$].

4-5 Weierstrass' $\wp$ Function

We shall now actually write down an elliptic function, the Weierstrass $\wp$ function, and then try to identify it with the doubly periodic (elliptic) function $z(u)$ defined in (4-9).

The trick needed here was used (unfortunately in more complicated form) in Exercises 3 and 4 of Sec. 4-1. Essentially, we note that $\phi(x) = \sum_{-\infty}^{\infty} f(x - n)$ is *intrinsically* periodic (if it is absolutely convergent); for example, changing x to $x + 1$ is the same as changing n to $n - 1$, thus $\phi(x) = \phi(x + 1)$. In accordance with this fact, the sum

$$S = \sum_{-\infty}^{\infty} \frac{1}{(x - n)^2}$$

is absolutely convergent on a closed interval bounded away from integral x, and indeed, S can be identified (see Bibliography) as $(\pi \csc \pi x)^2$.

In a similar vein we consider $\sum_{m,n}' \Omega_{m,n}^{-t}$ where $\sum_{m,n}'$ denotes a sum over all integral m,n, *excluding* $m = n = 0$. This sum [unlike $\sum_n' n^{-2}$ (see Exercise 5)] is absolutely convergent only for $t \geq 3$. Hence we must start with something more complicated than an analog of $\Sigma(x + n)^{-2}$. The simplest such analog is the famous Weierstrass function (1860):

$$\wp(u) = \frac{1}{u^2} + \sum_{m,n}' \left[\frac{1}{(u - \Omega_{m,n})^2} - \frac{1}{\Omega_{m,n}^2} \right] \tag{4-23}$$

which is somewhat akin to the function

$$\pi \cot \pi x = \frac{1}{x} + \sum_n' \left[\frac{1}{x - n} + \frac{1}{n} \right]$$

Here again, if we change u to $u + 2\omega_1$, we change $\Omega_{m,n}$ to $\Omega_{m+1,n}$ and if we rearrange the summation, we see $\wp(u)$ is periodic (indeed doubly periodic), or

$$\wp(u_1) = \wp(u_2) \qquad \text{if } u_1 \equiv u_2 \qquad (\text{mod } \boldsymbol{\Omega}) \tag{4-24}$$

If we differentiate (4-23), we obtain (by dropping constants)

$$\wp'(u) = \frac{-2}{u^3} - 2\sum_{m,n}{}' \frac{1}{(u - \Omega_{m,n})^3} = -2\sum_{m,n} \frac{1}{(u - \Omega_{m,n})^3} \tag{4-25}$$

By uniform convergence properties (relegated to Exercise 5), it follows that $\wp(u)$ is of order 2 and $\wp'(u)$ is of order 3. (The multiple roots are explained in Lemma 4-3.) To identify these functions, we next expand $\wp(u)$ and $\wp'(u)$ into Laurent series about the origin (their sole singularity in the period parallelogram). By the binomial theorem

$$(u - \Omega_{m,n})^{-k} = (-1)^k \Omega_{m,n}^{-k} \left[1 + k\frac{u}{\Omega_{m,n}} + \frac{k(k+1)}{2!}\frac{u^2}{\Omega_{m,n}^2} + \cdots\right] \tag{4-26}$$

and summing, we find

$$\wp(u) = u^{-2} + 3G_4u^2 + 5G_6u^4 + \cdots \tag{4-27a}$$

$$\wp'(u) = -2u^{-3} + 6G_4u + 20G_6u^3 + \cdots \tag{4-27b}$$

where we introduce the so-called "Eisenstein (1847) series"

$$G_r = G_r(2\omega_1, 2\omega_2) = \sum_{m,n}{}' \Omega_{m,n}^{-r} \tag{4-28}$$

(of course for odd r, $G_r = 0$ by cancellations of symmetry; for instance, $\Omega_{m,n}^{-r}$ cancels $\Omega_{-m,-n}^{-r}$). Now we are ready for the main result:

$$\wp'^2(u) = 4\wp^3(u) - g_2\wp(u) - g_3 \tag{4-29a}$$

where we introduce g_2, g_3 defined by

$$g_2 = 60G_4 \qquad g_3 = 140G_6 \tag{4-29b}$$

To see this, we expand both sides of (4-29a) by (4-27a) and (4-27b) and find that the terms in $u^{-6}, \ldots, u^{-1}$ and the constant r match one another on both sides. Hence the difference of right- and left-hand members is a bounded elliptic function (vanishing at $u = 0$), hence a constant (namely, 0).

We might expect that $g_2^3 \neq 27g_3^2$ (Exercise 1), for otherwise, (4-29a) factors with multiple roots, which would reduce the solution of (4-29a) for $\wp(u)$ to ordinary functions.

Now since G_4, G_6 are *dependent* functions of ω_1, ω_2 it is natural to ask if they may be arbitrarily chosen:

HYPOTHESIS 2 *It is possible to select the periods* $2\omega_1$, $2\omega_2$ *as complex numbers (with nonreal ratio* ω_2/ω_1*) such that* $G_4 = \sum_{m,n}' \Omega_{m,n}^{-4}$ *and* $G_6 = \sum_{m,n}' \Omega_{m,n}^{-6}$ *have any preassigned (finite) complex values, provided only that*

$$20G_4^3 \neq 49G_6^2 \qquad (\text{i.e., } g_2^3 \neq 27g_3^2)$$

Now, viewing matters differently, we can *start with the values of* g_2, g_3 in the equation

(4-30) $$y^2 = 4z^3 - g_2z - g_3 \qquad g_2{}^3 - 27g_3{}^2 \neq 0$$

(the *Weierstrass normal form*). Then (with $G_4 = g_2/60$ and $G_6 = g_3/140$) we select ω_1, ω_2 by Hypothesis 2. Now the identity (4-30) is a consequence of the substitution of ω_1, ω_2, and a variable u in (4-23), where

(4-31) $$z = \wp(u) \qquad y = \wp'(u)$$

From this, all integrands $R(y,z)\,dz$ will take the form $f(u)\,du$ where $f(u)$ is an elliptic function. Thus the integrals $\int R(y,z)\,dz$ all become integrals in the u plane.†

LEMMA 4-3 *Every value of z on the sphere corresponds to exactly two values of u in the period parallelogram under* (4-31) (*counting multiplicities*). *If, however, we preassign z and y* [*by assigning the sign of y or of the square root in* (4-30)], *then only one value of u is determined in the period parallelogram.*

PROOF The first part of the lemma is evident from the fact that $\wp(u)$ is of order 2. Actually, $\wp(-u) = \wp(u)$ directly from the definition (4-23) (since $-\Omega_{m,n} = \Omega_{-m,-n}$). Hence for each z the two values u and $-u$ correspond. On the other hand $\wp'(-u) = -\wp'(u)$, so the sign of y $(= \wp'(u))$ distinguishes $-u$ from u.

This would seem to prove the lemma if not for two possibilities: If $u \equiv -u \pmod{\boldsymbol{\Omega}}$, then u and $-u$ are the *same* values as far as the period parallelogram is concerned; furthermore if $y = 0$ or $y = \infty$, there is no sign choice for y. Actually, matters are explained completely by factoring (4-30) as

(4-32) $$y^2 = 4(z - e_1)(z - e_2)(z - e_3)$$

where e_1, e_2, e_3 are distinct roots. Thus $\wp'(u)$ (being of order 3) has zeros only at $\wp(u_1) = e_1$, $\wp(u_2) = e_2$, $\wp(u_3) = e_3$ for certain unknowns u_1, u_2, u_3. On the other hand, the points $U_1 = \omega_1$, $U_2 = \omega_2$, $U_3 = \omega_1 + \omega_2$ (called $-\omega_3$) all satisfy $U_i \equiv -U_i$ [or $2U_i \equiv 0 \pmod{\Omega}$], hence

$$\wp'(U_i) = \wp'(-U_i) = 0$$

Thus $U_i \equiv u_i$ in some order, and we have *completely* accounted for the correspondence between the pair (z,y) and the single variable u. Q.E.D.

The one-to-one correspondence between (z,y) and u is expressed as the statement that u is a "uniformizing parameter" of the curve (4-30), regarding the curve as consisting of the (two) complex Riemann y and z spheres in correspondence (not two real cartesian axes).

† This is more precisely formulated as Corollary 1 to Theorem 4-3 in Sec. 4-8.

In Exercise 3 we shall show that the equation (4-30) (with distinct roots) is equivalent to $w^2 = (Z - \alpha_1) \cdot \cdot \cdot (Z - \alpha_4)$ in the sense that any integral $\int R(z,y)\, dz$ can be transformed into any integral $\int R(Z,w)\, dZ$ by a linear-fractional transformation of Z into z and conversely. Thus Hypothesis 1 is replaced by Hypothesis 2, which is more directly computational (and less abstract).

EXERCISES

1 Verify that $[(e_1 - e_2)(e_2 - e_3)(e_3 - e_1)]^2 = -27(e_1e_2e_3)^2 - 4(e_1e_2 + e_2e_3 + e_3e_1)^3$ if $e_1 + e_2 + e_3 = 0$. From this verify that (4-30) or (4-32) has distinct roots if and only if $g_2^3 \neq 27g_3^2$.

2 Consider the quadruple $[e_1,e_2,e_3,\infty]$. Show that for its cross ratio λ, the invariant defined in Sec. 2-6 is $\phi(\lambda) = 4(e_1^2 + e_2^2 + e_3^2 - e_1e_2 - e_2e_3 - e_3e_1)^3/27[(e_1 - e_2)(e_2 - e_3)(e_3 - e_1)]^2$. (See Lemma 2-3.) Show that under the conditions of Exercise 1, $\phi(\lambda) = g_2^3/(-27g_3^2 + g_2^3)$.

3 Transform $\int R(z,w)\, dz$ of Sec. 4-2 into $\int R(Z,y)\, dZ$ by a linear transformation of z into Z, taking the distinct values $z = \alpha_1, \alpha_2, \alpha_3, \alpha_4$ into some values e_1, e_2, e_3, ∞ (where $e_1 + e_2 + e_3 = 0$ and the e_i are finite and distinct). (Compare Exercise 7, Sec. 4-2.)

4 Show that when we transform from w^2 = fourth degree to y^2 = cubic, the three kinds of integrals of (4-8) in Sec. 4-2 are (essentially like)

$$I_1 = \int \frac{dz}{y} \qquad I_2 = \int \frac{dz}{(z - e_1)^t y} \quad \text{or} \quad \int \frac{z^t\, dz}{y} \qquad I_3 = \int \frac{dz}{(z - a)^t y}$$

Changing to the variable u, observe that

(*a*) I_1 is finite (although unbounded) in the whole (finite) u plane.

(*b*) I_2 has only poles (no logarithmic singularities).

(*c*) I_3 has logarithmic singularities (generally).

[*Hint:* For I_2, show $z - e_1 = \wp(u) - \wp(\omega_1)$ = power series in $(u - \omega_1)^2$ near $u = \omega_1$.] [This follows from differentiating the identity $\wp(u) = \wp(-u)$ at $u = \omega_1$, a point where $u \equiv -u \pmod{\mathbf{\Omega}}$.]

5 (*a*) Show that $\sum_{m,n}' (m^2 + n^2)^{-t/2}$ is convergent by comparison with

$$\iint_{-\infty}^{\infty} \frac{dx\, dy}{(x^2 + y^2)^{t/2}}$$

(*b*) Show $|\Omega_{m,n}|^2/(m^2 + n^2)$ lies between two positive constants, say r_1^2, r_2^2, for all integral (indeed for all real!) m, n (not both zero). [*Hint:* Consider the locus $|\Omega_{m,n}|^2 = 1$ as a conic in the real variables m, n, and show it lies between two circles $m^2 + n^2 = r_1^2$ and r_2^2.] (*c*) Consider the convergence problems presented by (4-23) and (4-25). (*d*) Consider the convergence problem in going from (4-26) to (4-27*a*). [*Hint:* Cut off the series (4-26) with a remainder before summing on m and n.]

4-6 The Elliptic Modular Function

The problem of justifying Hypothesis 2 is more fundamental than the obvious convenience gained by using doubly periodic functions instead of inverting elliptic integrals. Actually, a new type of function $J(z)$ is presently introduced, called an "elliptic modular function," which was historically instrumental in the development of the concept of the Riemann surface.

We can simplify Hypothesis 2 in obvious ways: We write out the system of equations for ω_1, ω_2 in the form

$$\sum_{m,n}{}' (2m\omega_1 + 2n\omega_2)^{-4} = \frac{g_2}{60} \tag{4-33a}$$

$$\sum_{m,n}{}' (2m\omega_1 + 2n\omega_2)^{-6} = \frac{g_3}{140} \tag{4-33b}$$

where g_2, g_3 are prescribed finite constants satisfying the condition [of Exercise 1, Sec. 4-5 (also see Exercise 8)]

$$g_2^3 \neq 27g_3^2 \tag{4-34}$$

Certainly we do not expect to solve (4-33a) and (4-33b) for ω_1, ω_2 directly, but we can reduce the system to a single equation by using the obvious homogeneity. Let us set $\tau = \omega_2/\omega_1$ and define the analytic functions

$$\gamma_r(\tau) = \sum_{m,n}{}' (m + n\tau)^{-r} \qquad r = 4, 6 \tag{4-35a}$$

thus

$$\frac{g_2}{60} = (2\omega_1)^{-4}\gamma_4(\tau) \qquad \text{and} \qquad \frac{g_3}{140} = (2\omega_1)^{-6}\gamma_6(\tau) \tag{4-35b}$$

Now taking advantage of the homogeneity, we define

$$J(\tau) = \frac{g_2^3}{g_2^3 - 27g_3^2} \qquad \left(= \frac{20\gamma_4^3(\tau)}{20\gamma_4^3(\tau) - 49\gamma_6^2(\tau)} \right) \tag{4-36}$$

Clearly, by (4-34) we see $J(\tau)$ is to equal some *finite* complex constant [that is, $J(\tau)$ is not singular for finite τ].

We now see that to establish Hypothesis 2, *it suffices to find an inverse to* $J(\tau) = J$, *namely* $\tau = \tau(J)$ valid for all finite complex J. For if we are given g_2, g_3 by (4-36), we know $J(\tau)$. Hence, using the inverse $\tau(J)$ we can find τ, and from this γ_4, γ_6. Then we know $\pm\omega_1$ from the quotient of the equations in (4-35b)

$$(2\omega_1)^2 \frac{\gamma_4}{\gamma_6} = \left(\frac{140}{60}\right) \frac{g_2}{g_3} \tag{4-37}$$

and we know $2\omega_2 = 2\omega_1 \cdot \tau$.

To invert $J(\tau)$, we establish certain important properties with τ

restricted to the upper half plane ($\text{Im}\,\tau > 0$):

(4-38a) $$J(\tau) \text{ has no singularities for } \text{Im}\,\tau > 0$$

(4-38b) $$J(\tau + 1) = J(\tau)$$

(4-38c) $$J\left(-\frac{1}{\tau}\right) = J(\tau)$$

(4-38d)† $$J(\tau) = (\text{const})\, q + (\text{difference bounded as } q \to \infty)$$

where $q = \exp(-2\pi i\tau)$ so $q \to \infty$ as $\text{Im}\,\tau \to \infty$. The constant is 1/1,728, but we need only know it is nonzero.

The actual details involved in establishing these remarkable properties are of more interest to the number theorist than to the complex analyst‡ (see Exercises 2 to 4). According to (4-38b), we note that $J(\tau)$ has a single value for each value of $q = \exp(-2\pi i\tau)$; that is, if we know q, we know τ to within an integer, hence by (4-38b) we know $J(\tau)$ precisely. From this fact we can necessarily expand $J(\tau)$ in a Laurent series in q about $q = \infty$, but from (4-38d) the series has a polynomial principal part and $J(\tau)$ has the form of the series

(4-39) $$J(\tau) = c_{-1}q + c_0 + \frac{c_1}{q} + \cdots \qquad q = \exp(-2\pi i\tau)$$

which converges by (4-38a) for all q ($\neq 0$).

We also observe that in (4-38c) we chose $-1/\tau$ rather than $1/\tau$ [although actually $J(1/\tau) = J(\tau)$ if we follow (4-36) literally]. We do this because $\text{Im}\,(-1/\tau) > 0$ when $\text{Im}\,\tau > 0$. This keeps us in the upper half plane.

Now we consider the region $\mathcal{R}$ shown in Fig. 4-5, and defined by

(4-40a) $$-\tfrac{1}{2} < \text{Re}\,\tau < \tfrac{1}{2}$$

(4-40b) $$1 < |\tau|$$

The boundary of this region consists of the straight lines

(4-41a) $$\text{Re}\,\tau = \pm\tfrac{1}{2} \qquad \text{Im}\,\tau \geq \frac{\sqrt{3}}{2}$$

and the circular arc

(4-41b) $$(\text{Re}\,\tau)^2 + (\text{Im}\,\tau)^2 = 1 \qquad |\text{Re}\,\tau| \leq \tfrac{1}{2}$$

We note that the region $\mathcal{R}$ has the property that boundary points symmetric with regard to the imaginary axis ($\text{Im}\,\tau = 0$) produce the same

† Actually, in Sec. 6-3, we shall have occasion to describe a somewhat larger class of functions called "elliptic modular functions" (which will be permitted poles and which can have any degree polynomial in q, rather than (constant) q, as principal part when $\text{Im}\,\tau \to \infty$).

‡ One important analytic application is Picard's theorem. (See Appendix B.)

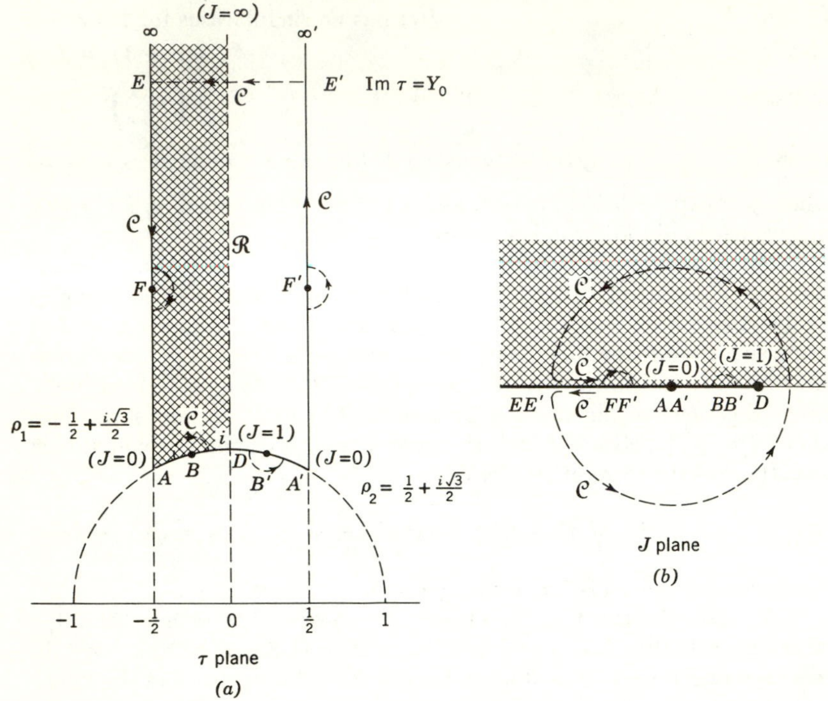

Fig. 4-5 (*a*) The domain $\mathfrak{M}$ consists of the region $\mathfrak{R}$ with boundary arcs DA, $AE\infty$ (only). The points A, A'; B, B'; etc., are identified in pairs. The semicircles near F, F'; B, B' are explained in Exercise 5. (*b*) The J plane is explained in Exercise 10. Note that the upper half J plane is crosshatched.

values of $J(\tau)$. Thus if $\tau = x + iy$, by (4-38*b*)

(4-42*a*) $$J(yi - \tfrac{1}{2}) = J(yi + \tfrac{1}{2}) \qquad y \geq \frac{\sqrt{3}}{2}$$

and by (4-38*c*)

(4-42*b*) $$J(\exp i\theta) = J(\exp i(\pi - \theta)) \qquad \frac{\pi}{3} \leq \theta \leq \frac{2\pi}{3}$$

We now consider the domain $\mathfrak{M}$ obtained by identifying symmetric boundary points of $\mathfrak{R}$, according to (4-42*a*) and (4-42*b*). This is very much like identifying boundary points† of the period parallelogram in Fig. 4-3*b*.

We call $\mathfrak{M}$ the "fundamental domain or Klein's domain of $J(\tau)$" (for

† Here we are identifying *adjacent* boundary segments as in (4-42*b*) (segments adjacent to $\exp \pi i/2$) and in (4-42*a*) (segments adjacent to ∞i). Thus we shall find $\mathfrak{M}$ to be a sphere topologically (and not a torus, where *opposite* sides were identified).

reasons developed in Sec. 6-3). *For the time being, let us note that $J(\tau)$ is defined uniquely for τ on $\mathfrak{M}$* (in the same way that $\wp(u)$ is defined uniquely for u on the period parallelogram).

LEMMA 4-4 *The function $J(\tau)$ takes on every (finite) value J_0 exactly once for some τ in the fundamental domain $\mathfrak{M}$.*

PROOF We shall consider the equation $J(\tau) = J_0$ when no solutions τ lie on the boundary of $\mathfrak{M}$. The remaining case is relegated to Exercises 5 to 7. The demonstration consists of noting that the number of solutions to $J(\tau) = J_0$ for $\tau \in \mathfrak{M}$ is

$$N = \frac{1}{2\pi i}\int_{\mathfrak{C}} d \log\,[J(\tau) - J_0] \tag{4-43}$$

where $\mathfrak{C}$ is the boundary of $\mathfrak{M}$ truncated by the dotted line (in Fig. 4-5) at such a level Y_0 that $|J(\tau)| > |J_0|$ for $\operatorname{Im}\tau > Y_0$. (We ignore the dotted semicircles about F,F' and B,B'; they refer to Exercise 5 only.)

It is clear that in evaluating N, the portion of the integral over $A'E'$ cancels the portion of EA and the portion over AD cancels the portion over DA' [by properties (4-42*a*) and (4-42*b*)]. Thus N is *precisely*

$$N = \frac{1}{2\pi i}\int_{E'E} d \log\,[J(\tau) - J_0]$$

On the other hand, as $Y_0 \to \infty$, by (4-39), $\log\,[J(\tau) - J_0]$ comes arbitrarily close to $\log\,(c_{-1}q) = -2\pi i\tau + \text{const}$. Thus N is approximated to increasingly great accuracy (as $Y_0 \to \infty$) by the *constant*

$$\frac{1}{2\pi i}\int_{E'E} d\,(-2\pi i\tau + \text{const}) = -\tau\Big|_{E'}^{E} = +1$$

Therefore $N = 1$ and the lemma is proved (subject to restrictions which are removed in Exercises 5 to 7). Q.E.D.

EXERCISES

1 Verify the identities $G_r(-\omega_2,\omega_1) = G_r(\omega_1,\omega_2)$ (by replacing the summation indices m and n by n and $-m$). Likewise, show $G_r(\omega_1, \omega_2 + \omega_1) = G_r(\omega_1,\omega_2)$ (by replacing m by $m + n$ as n remains fixed). From this, show

$$\gamma_r(\tau + 1) = \gamma_r(\tau) \qquad \text{and} \qquad \gamma_r\left(-\frac{1}{\tau}\right) = \tau^r\gamma_r(\tau) \tag{4-44}$$

Also verify (4-38*b*) and (4-38*c*).

2 There are well-known expansions for Eisenstein series (see Exercise 4)

$$\gamma_4(\tau) = [(2\pi)^4/720](1 + 240/q + \cdots) \qquad q = \exp\,(-2\pi i\tau) \tag{4-45a}$$

$$\gamma_6(\tau) = [(2\pi)^6/42 \cdot 720](1 - 504/q - \cdots) \tag{4-45b}$$

$$J(\tau) = [q/1{,}728](1 + 744/q + \cdots) \tag{4-46}$$

where the omitted terms are higher powers of $1/q$. [The omitted terms are of number-theoretic significance; our main interest is in the nonvanishing of the leading terms of $J(\tau)$.] Verify the expansion for $J(\tau)$ from those of $\gamma_r(\tau)$.

3 Verify that the constant terms of $\gamma_4(\tau)$, and $\gamma_6(\tau)$ can be found from the expansion

$$\sum_{-\infty}^{\infty} (\tau + n)^{-2} = \pi^2 \csc^2 \pi\tau \tag{4-47}$$

for example, $\gamma_{2r}(i\infty) = 2\sum_1^\infty n^{-2r}$. For instance, in (4-47) if we set $\tau = \frac{1}{2}$,

$$\sum_{-\infty}^{\infty} \frac{1}{(m + \frac{1}{2})^2} = 8\sum_0^\infty \frac{1}{(2m + 1)^2} = \pi^2 \csc^2 \frac{\pi}{2} = \pi^2$$

Furthermore $$\sum_1^\infty n^{-2r} = \sum_0^\infty \frac{1}{(2m + 1)^{2r}} \left\{1 + \frac{1}{2^{2r}} + \frac{1}{4^{2r}} + \cdots\right\} = \sum_0^\infty \frac{(2n + 1)^{-2r}}{1 - 2^{-2r}}$$

by virtue of the fact that every integer $n = (2m + 1)2^k$ for some integers $m \geq 0$, $k \geq 0$.

4 (*a*) Show how we can verify (4-45*a*) and (4-45*b*) simply from knowledge of the constant terms. For example, $\sum_{-\infty}^{\infty} (\tau + n)^{-2} = \pi^2 \csc^2 \pi\tau = -4\pi^2 q/(q - 1)^2 = -4\pi^2/q + \cdots$. Hence $\Sigma(\tau + n)^{-4} = 8\pi^4/3q + \cdots$. If we replace τ by $m\tau$ $(m > 0)$, we replace q by q^m. Thus, finally, show $\Sigma(m\tau + n)^{-4} = \Sigma' n^{-4} + 2\Sigma(\tau + n)^{-4} + \cdots = \Sigma' n^{-4} + 16\pi^4/3q + \cdots$ which is essentially (4-45*a*). (*b*) Likewise find (4-45*b*). (*c*) Show all coefficients of (4-45*a*) and (4-45*b*) are real.

5 Verify Lemma 4-4 in the case where $J(\tau) = J_0$ has a root F on the side AE of $\mathfrak{M}$ (see Fig. 4-5) and possibly other roots, of course. Show that $\mathcal{C}$ can be deformed to include F and exclude F' by the use of deformations of the contour subject to the translation $\tau' = \tau - 1$. Likewise, show how we would take care of a root B on the arc AD by deformations related by $\tau' = -1/\tau$. [This does not take care of the case where $J(\tau) = J_0$ for $\tau = i$ or $\tau = \frac{1}{2} + i\sqrt{3}/2$ (see Exercise 7).]

6 (*a*) Verify that $J(i) = 1$ and $J(\pm\frac{1}{2} + i\sqrt{3}/2) = 0$, by showing $\gamma_6(i) = 0$ and $\gamma_4(\pm\frac{1}{2} + i\sqrt{3}/2) = 0$ directly from the properties (4-44). For instance, $\gamma_6(-1/i) = i^6\gamma_6(i) = 0$. (The case of γ_4 involves $-1/\rho_1 = \rho_1 + 1$, etc.) (*b*) Verify that $[J(z) - 1]/(z - i)^2$ and $J(z)/(z - \rho_1)^3$ each approach a nonzero constant as $z \to i$ or $z \to \rho_1$, respectively. [Note where $\gamma_4 = 0$, necessarily $\gamma_6 \neq 0$ by (4-34) above.]

7 Complete Exercise 5 and Lemma 4-4, by showing $J(\tau) = 1$ or 0 has only a single root in $\mathfrak{M}$. [*Hint:* If $J(\tau) = 1$ has another root $\lambda_0(\neq i)$ in $\mathfrak{M}$, then some neighborhood of $\tau = \lambda_0$ maps under $J(\tau)$ into an image neighborhood near $J = 1$, and some values of J_0 (near 1) give rise to solutions of $J(\tau) = J_0$ for τ near λ_0 (rather than i).] What about $J(\tau) = 0$?

8 Show that if ω_2/ω_1 is nonreal, then g_2, g_3 as defined *ab initio* by (4-33*a*) and (4-33*b*) must satisfy (4-34). Show that otherwise in Sec. 4-5, $z = \wp(u)$ and $y = \wp'(u)$ will satisfy an equation $y^2 = 4z^3 - g_2 z - g_3$ with *equal* roots, and therefore, $\int du = \int dz/y$ will become infinite for some finite z (yielding a contradiction).

9 Show that $J(\tau)$ is real on the boundary of $\mathfrak{R}$, and in fact $J(E) = J(E') \to -\infty$ as $Y_0 \to \infty$. [*Hint:* By the reflection principle (Theorem 3-10), the fact that $J(\tau)$ is real (positive) on the imaginary axis implies that $J(B') = \overline{J(B)}$ for B a point on the arc $\widehat{AD}$.]

10 Show that $J(\tau)$ maps $\mathfrak{R}$ onto the exterior of the slit running from $-\infty$ to $+1$ in the J plane, as seen in Fig. 4-5*b*. Verify the behavior of the dotted lines in Fig. 4-5*b*.

EULER'S ADDITION THEOREM

4-7 Evolution of Addition Process

The quintessence of the theory of elliptic functions is a theorem known as the *addition theorem*. We say that a function $f(u)$ defined in the complex plane has an addition theorem precisely when $f(u)$, $f(v)$, $f(u+v)$ are algebraically related, i.e., when a polynomial $P(x_1,x_2,x_3)$ with complex coefficients exists such that

$$P(f(u), f(v), f(u+v)) = 0 \tag{4-48}$$

THEOREM 4-1 *All elliptic functions have an addition theorem.*

In a sense Theorem 4-1 characterizes (as well as epitomizes) elliptic functions. In fact, all functions with an addition theorem can be described in terms of rational, exponential or elliptic functions, according to a rather advanced result of Weierstrass (1860). The simplest case which is not trivial (i.e., not rational) is perhaps $f_1(u) = \exp u$. There

$$P(x_1,x_2,x_3) = x_1x_2 - x_3$$

or

$$f_1(u)f_1(v) - f_1(u+v) = 0 \tag{4-49}$$

The case $f_2(u) = \sin u$ is more labored. There the trigonometric identities yield

$$f_2(u+v) - f_2(u)\sqrt{1 - f_2{}^2(v)} - f_2(v)\sqrt{1 - f_2{}^2(u)} = 0 \tag{4-50}$$

It takes a little bit of manipulation to clear the radicals and produce a polynomial satisfying (4-48), for instance,

$$P(x_1,x_2,x_3) = [x_3{}^2 - x_1{}^2(1 - x_2{}^2) - x_2{}^2(1 - x_1{}^2)]^2 - 4x_1{}^2x_2{}^2(1 - x_1{}^2)(1 - x_2{}^2) \tag{4-51}$$

We shall confine ourselves only to special elliptic functions, referring the rest of the details to the literature. We shall observe the evolution of the addition theorem in a little detail since it presents in microcosm the direct evolution from calculus to geometric (now topological) function theory.

FIRST STAGE: DIFFERENTIAL EQUATIONS

The earliest evolutionary stage occurred when Euler (1754) showed

$$\frac{dx}{\sqrt{1-x^4}} + \frac{dy}{\sqrt{1-y^4}} = \frac{dz}{\sqrt{1-z^4}} \tag{4-52a}$$

has the integral

$$\frac{x\sqrt{1-y^4} + y\sqrt{1-x^4}}{1+x^2y^2} = z \tag{4-52b}$$

Here the mode of thinking was calculus (or differential equations). This situation is quite analogous with (4-50) and (4-51), which merely state that the following equation:

$$\frac{dx}{\sqrt{1-x^2}} + \frac{dy}{\sqrt{1-y^2}} = \frac{dz}{\sqrt{1-z^2}}$$

has the integral $x\sqrt{1-y^2} + y\sqrt{1-x^2} = z$.

SECOND STAGE: COMPLEX FUNCTION THEORY

Later on, however, Liouville showed that every elliptic function also had an addition theorem. This follows from a result of much greater depth.

THEOREM 4-2 *Every elliptic function is expressible as a rational function of $\wp(u)$ and $\wp'(u)$ (with complex coefficients).*

Deferring the proof of Theorem 4-2 (for a time), let us see how it implies Theorem 4-1. Let $f(u)$ be an elliptic function, and let λ be a complex constant. The function $f(u+\lambda)$ is also doubly periodic (as noted earlier). Thus,

$$f(u) = R_2(\wp(u), \wp'(u)) \tag{4-53a}$$
$$f(u+\lambda) = R_1(\wp(u), \wp'(u)) \tag{4-53b}$$

where R_1 and R_2 are rational functions. Otherwise expressed with $\wp(u) = z$ and $\wp'(u) = y$, these equations become reduced to the *polynomial* equations

$$P_2(f(u), z, y) = 0 \tag{4-54a}$$
$$P_1(f(u+\lambda), z, y) = 0 \tag{4-54b}$$
$$y^2 - 4z^3 + g_2z + g_3 = 0 \tag{4-54c}$$

By elimination theory, we can rid ourselves of z and y and reduce (4-54a) to (4-54c) to a *polynomial* equation

$$P_0(f(u+\lambda), f(u)) = 0 \tag{4-55}$$

Now the coefficients in (4-55) will depend on λ. At the same time,

$$P_0(f(u + \lambda), f(\lambda)) = 0 \tag{4-56}$$

with coefficients depending on u alone. If we treat u and λ symmetrically in (4-53*b*), we obtain a symmetric addition theorem (4-48) similarly.

THIRD STAGE: TOPOLOGY

Actually, the full force of Theorem 4-1 is no more significant than (4-55). In fact, a more trivial interpretation of it, extracting a topological observation, is all we wish to remember for later:

> *The period parallelogram has a continuous system of analytic transformations into itself.*

Here we refer to the obvious transformation $u \to u + \lambda$ where λ is a continuously varying complex number. Such transformations will be defined in terms of equivalent Riemann manifolds (Sec. 5-2); for now, let us observe intuitively what is meant by the transformation $u \to u + \lambda$.

Referring to Fig. 4-6*a*, we see that "motion of the parallelogram by displacement through λ" means that the vertically shaded portion and the obliquely shaded portions are "cut off and pasted back" along the vectors $2\omega_1$, $2\omega_2$ as shown. Referring to Fig. 4-6*b*, we see that if λ lies on the vector 0 to $2\omega_2$, then the motion is like the *rigid* (*euclidean*) rotation of the torus about the "dotted" axis of revolution through angle $2\pi\lambda/|2\omega_2|$, preserving azimuthal circles. If λ lies on the vector 0 to $2\omega_1$, the motion is like the (nonphysically realizable) rotation where the longitudinal circles (like 0 to $2\omega_1$) are preserved.

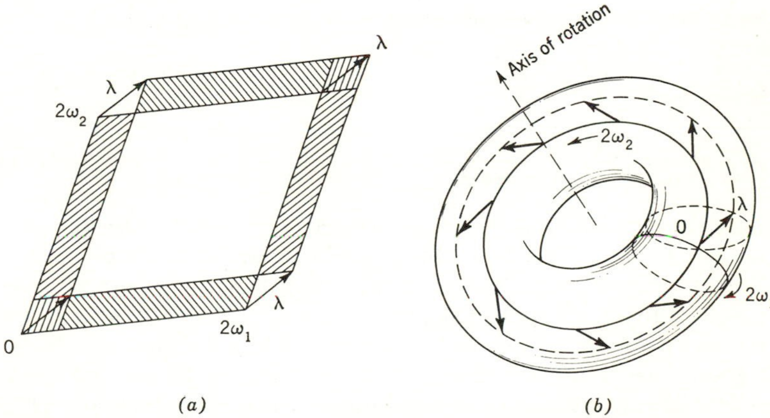

Fig. 4-6 Motion of the period parallelogram by λ (*a*) period parallelogram and (*b*) torus.

We shall find in Sec. 14-1 that according to a theorem of Schwarz (1878), the existence of a continuous system of such analytic transformations *alone* characterizes the period parallelogram (or much simpler manifolds such as sphere, annulus, disk). Hence, this "trivialization" of the addition theorem is not really so trivial!

4-8 Representation Theorems

We now lead up to Theorem 4-2 by two results analogous with Theorems 4-2 and 4-3 of Sec. 2-4.

THEOREM 4-3 *An elliptic function $f(u)$ can be prescribed to have p (>1) poles $\beta_1, \ldots, \beta_p$ and principal parts at these poles, provided only that the total residue is zero. In fact, $f(u)$ can be expressed rationally in terms of $\wp(u)$, $\wp'(u)$ and is uniquely determined up to a constant (additive) term.*

COROLLARY 1 *The elliptic integrals are precisely the same as the integrals of elliptic functions under a suitable change in independent variable by $z = \wp(u)$.*

THEOREM 4-4 *An elliptic function $f(u)$ can be prescribed to have p (>1) poles $\beta_1, \ldots, \beta_p$ and p zeros $\alpha_1, \ldots, \alpha_p$ listed in order of multiplicity, provided only that*

$$\Sigma\alpha_i \equiv \Sigma\beta_i \qquad (\text{mod } \mathbf{\Omega})$$

In fact, $f(u)$ can be expressed rationally in terms of $\wp(u)$, $\wp'(u)$ and is uniquely determined up to a constant (multiplicative) factor.

First let us prove Theorem 4-3. We again introduce the symbol $\omega_3 = -\omega_1 - \omega_2$. Then we restrict β so that $\beta \neq 0, \omega_1, \omega_2, \omega_3$. We have seen that $\wp'(\beta) \neq 0$. Also, generally, $\wp(\beta) = \wp(-\beta)$ and

$$\wp'(\beta) = -\wp'(-\beta)$$

hence, for $t \geq 1$,

$$\text{(4-57}a\text{)} \qquad \gamma \left\{ \frac{\frac{1}{2}[\wp'(u) - \wp'(\beta)]}{\wp(u) - \wp(\beta)} \right\}^t = \frac{\gamma}{(u-\beta)^t} + \frac{\gamma'}{(u-\beta)^{t-1}} + \text{lower terms}$$

Now the indicated function is analytic at $u = -\beta$ by cancellation with the numerator. Therefore, this expression can be made to match the *leading* term without introducing any new singularity except possibly at $u = 0$. Hence, the principal part for any β ($\neq 0, \omega_1, \omega_2, \omega_3$) can be accounted for by an expression of type (4-57a) by just matching the highest order terms each time. In this process, by Lemma 4-1 the total residue always remains zero as we add or subtract the functions (4-57a).

Thus we are left to account for principal parts only for the terms $0, \omega_1, \omega_2, \omega_3$ (again with total residue 0). Analogously with (4-57a), we introduce $\lambda_1(u)$, $\lambda_2(u)$, $\lambda_3(u)$ defined† typically by

(4-57*b*)
$$\lambda_3 = \frac{[\wp(u) - e_1][\wp(u) - e_2]}{\wp'(u)}$$
$$= \tfrac{1}{2}\sqrt{\frac{[\wp(u) - e_1][\wp(u) - e_2]}{\wp(u) - e_3}} \approx \frac{\text{const}}{u(u - \omega_3)}$$

since $\wp(u) - e_i = \wp''(\omega_i)(u - \omega_i)^2/2 + \cdots$ (see Exercise 3). Hence, by subtracting polynomials in λ_3 *we can remove the principal terms for the singularities at* ω_3 (likewise for ω_1 and ω_2). We are left, therefore, with principal terms only for $u = 0$; and since the total residue is zero, there is no term in $1/u$. Thus we must only account for terms like

$$\frac{\gamma_t}{u^t} + \frac{\gamma_{t-1}}{u^{t-1}} + \cdots + \frac{\gamma_2}{u^2} \qquad \gamma_1 = 0$$

Clearly,

(4-57*c*)
$$(-\tfrac{1}{2}\wp'(u))^a(\wp(u))^b = \frac{1}{u^{3a+2b}} + \text{lower poles}$$

Hence a polynomial in $\wp$, $\wp'$ can account for any exponent t of type $(t=)$ $3a + 2b (a \geq 0, b \geq 0)$; thus, by taking care of the highest pole, we take care of all poles of order $t \geq 2$ by induction on t.

Thus we see that the prescribed principal parts can be accounted for by (4-57*a*) to (4-57*c*), inductively (by taking care of the highest powers one at a time). Q.E.D.

To see Corollary 1 (Theorem 4-3), we note that we are merely showing the equivalence of $\int f(u)\,du$ and

$$\int R(\wp,\wp')\,du = \int R(\wp,\wp')\,\frac{d\wp}{\wp'} = \int R(z,y)\,\frac{dz}{y}$$

(where y^2 is the usual cubic $4z^3 - g_2z - g_3$).

Next let us prove Theorem 4-4 by constructing a doubly periodic function to meet the desired specifications. We shall then appeal to Theorem 4-3 for proof of its expressibility in $\wp$, $\wp'$. We introduce a new function

(4-58*a*)
$$\zeta(u) = \frac{1}{u} + \sum_{m,n}{}' \left(\frac{1}{u - \Omega_{m,n}} + \frac{1}{\Omega_{m,n}} + \frac{u}{\Omega_{m,n}^2}\right)$$

formed so that

(4-58*b*)
$$-\zeta'(u) = \wp(u)$$

Taking the logarithmic derivative, we see that the function

(4-59*a*)
$$\sigma(u) = u \prod_{m,n}{}' \left[\left(1 - \frac{u}{\Omega_{m,n}}\right) \exp\left(\frac{u}{\Omega_{m,n}} + \frac{u^2}{2\Omega_{m,n}^2}\right)\right]$$

† In classical elliptic function theory we use the trio of cn u, sn u, and dn u to replace sine and cosine in trigonometry. They are essentially these functions. The addition formula is more easily written in terms of these functions than in terms of $\wp(u)$.

satisfies

$$\frac{\sigma'(u)}{\sigma(u)} = \zeta(u) \tag{4-59b}$$

Now $\zeta(u)$ has but a single pole and $\sigma(u)$ has but a single zero (in a period parallelogram); hence neither can be elliptic, by Lemma 4-2.

We find, from (4-58b), the expression $\zeta(u + 2\omega_1) - \zeta(u)$ has a zero derivative, hence

$$\zeta(u + 2\omega_1) = \zeta(u) + 2\eta_1 \qquad \zeta(u + 2\omega_2) = \zeta(u) + 2\eta_2 \tag{4-58c}$$

for constants η_1, η_2. Furthermore, $\log \sigma(u + 2\omega_1) - \log \sigma(u) - 2\eta_1 u$ by (4-59b) will have a zero derivative, hence will equal a constant, say, $\log \gamma_i$. Thus

$$\sigma(u + 2\omega_i) = \gamma_i \sigma(u) \exp 2\eta_i u \qquad i = 1, 2 \tag{4-58d}$$

Finally, according to the hypotheses of Theorem 3-4, we can express the desired $f(u)$ if we merely take the precaution to adjust the value of, say, α_1 by $\Omega_{m,n}$, a period, writing it as $\alpha_1 + \Omega_{m,n}$ so that $\Sigma\alpha_i = \Sigma\beta_i$ (exactly). Then

$$f(u) = \text{const} \prod_i \frac{\sigma(u - \alpha_i)}{\sigma(u - \beta_i)} \tag{4-59}$$

It is left as Exercise 8 to verify that $f(u)$ actually is doubly periodic.

Q.E.D.

EXERCISES

1 The addition theorem for elliptic functions is often given as follows:

$$\Delta(u,v,w) = \begin{vmatrix} 1 & 1 & 1 \\ \wp(u) & \wp(v) & \wp(w) \\ \wp'(u) & \wp'(v) & \wp'(w) \end{vmatrix} = 0 \tag{4-60}$$

if and only if $u + v + w \equiv 0 \pmod{\Omega}$.

For proof, first assume u, v, w are not poles, and $v \not\equiv \pm w \pmod{\Omega}$. Then $\Delta(u,v,w)$ is of order 3 as a function of u (with 0 as a triple pole), and evidently it has two roots $u_1 = v$, $u_2 = w$. The third root u_3 (by Lemma 4-1) must satisfy $u_1 + u_2 + u_3 \equiv 0 \pmod{\Omega}$.

Complete the proof in order to eliminate unnecessary restrictions on u, v, w. (Can they be poles?)

2 A curve of equation $F(z,w) = 0$ is said to be *unicursal* if a set of substitutions $z = \phi(t)$, $w = \psi(t)$, with ϕ and ψ *rational*, satisfy the equation of the curve, and each complex pair (z,w) satisfying the equation corresponds in one-to-one fashion with the complex values of t over the t sphere (∞ included).

(*a*) Show that the equation $w^2 = 4(z - e_1)(z - e_2)(z = e_3)$ is unicursal if $e_1 = e_2$ [for example, try $w = t(z - e_1)$].

(*b*) Show that the equation is *not unicursal if the roots e_i are distinct.* [*Hint:* $du = dz/w = \psi'(t)\, dt/\phi(t)$.] Show that $\int du$ is finite on the curve while $\int \psi'\, dt/\phi$ must "reach some pole" as t varies on the curve (perhaps at $t = \infty$). (*Remark:* This amounts to saying that u covers the parallelogram while t covers the sphere—and these manifolds are inequivalent topologically.)

3 Verify that $\wp''(\omega_i) \neq 0$.

4 Prove that Theorems 4-3 and 4-4 determine a unique function to within additive (or multiplicative) constants.

5 Show that an *even* elliptic function $f(u) = f(-u)$ is expressible rationally in terms of $\wp(u)$ alone and that an *odd* elliptic function $f(u) = -f(-u)$ is expressible as $\wp'(u)$ multiplied by a rational function of $\wp(u)$ alone.

6 With reference to (4-58a) and (4-58c), show that $\zeta(u) = -\zeta(-u)$, and using $u = -\omega_i$, show that $\eta_i = \zeta(\omega_i)$. Likewise, using $u = -\omega_i$, show $\gamma_i = -\exp 2\eta_i\omega_i$.

7 Integrate $\int_{\mathcal{C}} \zeta(u)\, du$ over $\mathcal{C}$ the perimeter of the period parallelogram, and obtain the Legendre period relation:

(4-61)
$$\int_0^{2\omega_1} d\zeta \int_0^{2\omega_2} du - \int_0^{2\omega_1} d\zeta \int_0^{2\omega_1} du = 2\pi i$$

(or, in easier symbols, $\eta_1\omega_2 - \eta_2\omega_1 = \pi i/2$).

8 Complete the proof that the expression (4-59) is doubly periodic.

CHAPTER 5
MANIFOLDS OVER THE z SPHERE

Historically, Riemann surfaces were introduced as devices (like the cyclic neighborhood in Sec. 3-3) which render certain mappings as one-to-one mappings (when they would not have this property over ordinary regions). Ultimately, through a slow evolution (1850–1900) it was realized that such devices were special cases of a (theoretically) simpler device known as the "Riemann manifold" which includes the Riemann sphere, the period parallelogram, plane (ordinary) and slit regions, n-sheeted surfaces, etc. We shall begin with the abstract concept of a Riemann manifold (over the z sphere) and relate it to the intuitive objects just itemized.

FORMAL DEFINITIONS

5-1 Neighborhood Structure

In principle a geometric object (curve, surface, etc.) *together with a one-to-one parametrization* constitutes a manifold. Specifically, we shall speak only of surfaces and analytic parametrizations.

We consider as a geometric object a set of points $\mathfrak{M} = \{P, Q, \ldots\}$ *lying over (part or all of) the z sphere.*† Thus to each P corresponds its so-called "projection," namely, a value of $z = a_P$ (possibly ∞) lying on the z sphere. Actually, different points P and Q may have the same a_P ($= a_Q$). This is intuitively explained by saying P, Q are on "different sheets."

† Many authors omit mention of the z sphere in defining a Riemann manifold because the important things such as parametrizations and functions are definable in the parameter planes only. *Actually the z sphere is regarded as the place where the points of the Riemann manifold are visualized.* Otherwise, we must think of a Riemann manifold as an equivalence class of parametrizations. This is not at all unreasonable, but it represents a higher form of abstraction than required for our introductory work.

We further require that the points have a *neighborhood structure.* This means that each point P projects onto the center a_P of at least one cyclic disk neighborhood $\mathcal{Z}_P$ over the z sphere, which consists only of points z_P which are projections from $\mathfrak{M}$. This implies that for $z_P \in \mathcal{Z}_P$ we can write

$$\left.\begin{aligned} z_P - a_P &= \phi_P(\zeta_P) && a_P \text{ finite} \\ \frac{1}{z_P} &= \phi_P(\zeta_P) && a_P = \infty \end{aligned}\right\} \quad \zeta_P \in \mathcal{R}_P \tag{5-1a}$$

where the analytic function

$$\phi_P(\zeta_P) = a_P^{(0)}\zeta_P^{e_P} + a_P^{(1)}\zeta_P^{e_P+1} + \cdots \qquad a_P^{(0)} \neq 0 \tag{5-1b}$$

is a power series beginning with the indicated term at the origin $\zeta_P = 0$. Moreover, the equation (5-1a) must map $\mathcal{Z}_P$ in one-to-one fashion onto a disk $\mathcal{R}_P$ about $\zeta_P = 0$. Equations (5-1a) and (5-1b) constitute the local parametrization, and ζ_P is called a *local uniformizing parameter*† (in consistent fashion with the definitions in Sec. 3-3). This point P is called a *branch point of order* $e_P - 1$.

Now to have our concept of neighborhood structure, we are required to be able to say how a cyclic disk neighborhood $\mathcal{Z}_P$ at P *overlaps* a cyclic disk neighborhood $\mathcal{Z}_Q$ at Q (for example, when a point $z_R \in \mathcal{Z}_P \cap \mathcal{Z}_Q$, we say these neighborhoods overlap). Suppose $\mathcal{Z}_P$, $\mathcal{Z}_Q$ overlap and let the corresponding parametrization at Q, say, be

$$\left.\begin{aligned} z_Q - a_Q &= \phi_Q(\zeta_Q) && a_Q \text{ finite} \\ \frac{1}{z_Q} &= \phi_Q(\zeta_Q) && a_Q = \infty \end{aligned}\right\} \quad \zeta_Q \in \mathcal{R}_Q \tag{5-2a}$$

$$\phi_Q(\zeta_Q) = a_Q^{(0)}\zeta_Q^{e_Q} + a_Q^{(1)}\zeta_Q^{e_Q+1} + \cdots \qquad a_Q^{(0)} \neq 0 \tag{5-2b}$$

If we consider $z_R \in \mathcal{Z}_P \cap \mathcal{Z}_Q$, the *inverse* of (5-1$a$) and (5-1$b$) and (5-2$a$) and (5-2$b$) will determine subsets of $\mathcal{R}_P$, $\mathcal{R}_Q$ (say $\mathcal{R}_{P,Q}$, $\mathcal{R}_{Q,P}$) which correspond to all such points R. The overlapping of these neighborhoods $\mathcal{R}_P$, $\mathcal{R}_Q$ will mean that $\mathcal{R}_{P,Q}$ ($\subseteq \mathcal{R}_P$) and $\mathcal{R}_{Q,P}$ ($\subseteq \mathcal{R}_Q$) are mapped into one another by a one-to-one mapping of ζ_P onto ζ_Q over such regions. Symbolically, we write (5-1a) and (5-1b) or (5-2a) and (5-2b) as pointwise or regional mappings

$$\begin{aligned} z_P &= \Phi_P(\zeta_P) \qquad \text{or} \qquad \mathcal{Z}_P = \Phi_P(\mathcal{R}_P) \\ z_Q &= \Phi_Q(\zeta_Q) \qquad \text{or} \qquad \mathcal{Z}_Q = \Phi_Q(\mathcal{R}_Q) \end{aligned} \tag{5-3}$$

† The term *global uniformizing parameter* by contrast means a parameter which permits a one-to-one map over a whole surface and not over just a neighborhood, such as u of elliptic function theory (over the whole period parallelogram) or t of Sec. 4-8, Exercise 2 (over the whole Riemann sphere). A global parameter for a given manifold is generally harder to determine.

Then it is the nature of the overlapping that we equate $z_P = z_Q$ $(= z_R)$; thus symbolically

$$\begin{aligned} \mathcal{Z}_P \cap \mathcal{Z}_Q &= \Phi_P(\mathcal{R}_{P,Q}) = \Phi_Q(\mathcal{R}_{Q,P}) \\ \Phi_P(\zeta_P) &= \Phi_Q(\zeta_Q) \qquad \zeta_P \in \mathcal{R}_{P,Q},\ \zeta_Q \in \mathcal{R}_{Q,P} \end{aligned} \tag{5-4}$$

Hence, denoting inverses by negative powers

$$\Phi_Q^{-1}\Phi_P(\mathcal{R}_{P,Q}) = \mathcal{R}_{Q,P} \qquad \text{or} \qquad \Phi_P^{-1}\Phi_Q(\mathcal{R}_{Q,P}) = \mathcal{R}_{P,Q} \tag{5-5}$$

each map being one-to-one where defined.

Actually, the relationship between different local parameters is implied [as in (5-5)] only where neighborhoods overlap. We call parametrizations at P, Q *coherent* if the relation (5-5) is satisfied by the inverse images (in the ζ_P and ζ_Q planes) of the overlapping part $\mathcal{Z}_P \cap \mathcal{Z}_Q$.

To complete the definition, we consider a set $\mathfrak{M}$ to be *connected* when every two points are part of a *finite* chain of overlapping cyclic disk neighborhoods; e.g., if P, Q are assigned, we are able to find a finite

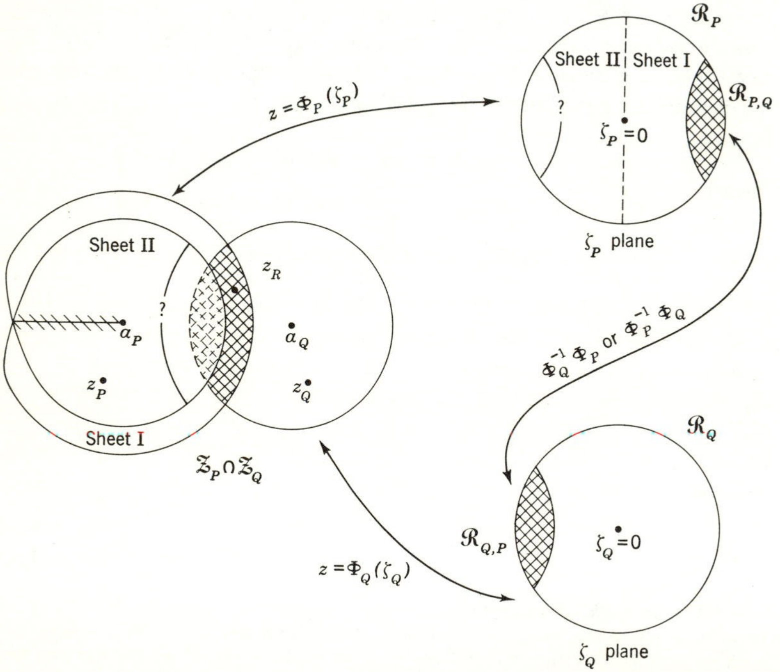

Fig. 5-1 Illustration of the overlapping of neighborhoods. Here $\mathcal{R}_P$ (and $\mathcal{R}_Q$) are the ζ_P (and ζ_Q) parameter disks.

chain of such neighborhoods

$$(\mathcal{Z}_P =)\ \mathcal{Z}_{P_1}, \mathcal{Z}_{P_2}, \ldots, \mathcal{Z}_{P_t}\ (= \mathcal{Z}_Q)$$

such that each $\mathcal{Z}_{P_i}$ overlaps $\mathcal{Z}_{P_{i+1}}$ $(1 \leq i < t)$.

We finally say that a system of *points, neighborhoods, and parametrizations* is a *Riemann manifold* if the points are connected and the parametrizations are coherent. Thus the object $\mathfrak{M}$ (described as a collection of points) is now a system

$$[P, \mathcal{Z}_P, \Phi_P, \mathcal{R}_P] = \mathfrak{M} \tag{5-6}$$

Certainly such a system is awkward to consider, and we shall do well to define workable models which are "equivalent manifolds" very soon.

Very often the "manifold without parametrization" is called the "surface," but there is no clear distinction between Riemann manifold and Riemann surface as enforced by convention. Since only analytic functions Φ_P are used, the term "analytic manifold" is also used.

ILLUSTRATION

Let us illustrate the workings of this concept of parametrization by a simple example. Figure 5-1 shows two sheets near P branching at $z = a_P$, and the neighborhood of Q (at $z = a_Q$) overlaps only the lower Sheet I at $z = a_Q$. The parametrizations might be

$$z_P - a_P = \zeta_P^2 \tag{5-7a}$$

$$z_Q - a_Q = \zeta_Q \tag{5-7b}$$

which determine a region $\mathcal{R}_P$ (and a region $\mathcal{R}_Q$) of the ζ_P (and ζ_Q) parameter planes.

Now the region $\mathcal{R}_{P,Q}$ or $\mathcal{R}_{Q,P}$ is *not necessarily the result of substituting* $z_P = z_Q$ in (5-7a) and (5-7b) (this would include both the crosshatched region and the "?-region"). Of course, the overlapping occurs only on one sheet (not both). We do mean, however, that $\mathcal{R}_{P,Q}$, $\mathcal{R}_{Q,P}$ are mapped onto $\mathcal{Z}_P \cap \mathcal{Z}_Q$ by (5-7a) and (5-7b), respectively, in one-to-one fashion. The induced relations (5-5) take the form

$$a_P + \zeta_P^2 = a_Q + \zeta_Q \tag{5-8}$$

which is one-to-one under the additional restriction that we ignore the ?-region. Thus $\mathcal{R}_{P,Q}$, $\mathcal{R}_{Q,P}$ have a one-to-one relation in (5-8).

EXERCISES

1 Show that a region $\mathcal{R}$ of the z sphere is (trivially) a Riemann manifold. To make things definite call $z_P - a_P = \zeta_P$, and let $\mathcal{R}_P$ be the largest circular disk lying in $\mathcal{R}$. (Take care of ∞ if $\infty \in \mathcal{R}$.)

2 Consider the Riemann surface for $q^2 = 1 - z^2$ as a Riemann manifold. Use cyclic regions about $z = +1$ and $z = -1$ which do not intersect; and use the natural parameters $z - 1 = \zeta_1^2$ and $z + 1 = \zeta_2^2$ for the branch points. Note that two points ∞_1, ∞_2 project over $z = \infty$ (one on each sheet), and each has a different cyclic neighborhood (of degree 1, actually) Z_{∞_1}, Z_{∞_2}.

3 Consider the period parallelogram (see Fig. 4-3), and show it is a Riemann manifold if we draw ordinary neighborhoods in the period u plane (Fig. 4-3a) but identify equivalent points in the parallelogram (Fig. 4-3b).

5-2 Functions and Differentials

We are dealing with an awkward definition of a Riemann manifold in the form of the aggregate

$$\mathfrak{M} = [P, Z_P, \Phi_P, \mathfrak{R}_P] \tag{5-9}$$

Clearly, we should like to say an object *is* a Riemann manifold without having to tie the object forever to some system of parameters. For example, we ask, can the same set of points $\{P, Q, \ldots\}$ be associated with another set of items $\{Z_P\}$, $\{\Phi_P\}$, $\{\mathfrak{R}_P\}$ somehow and still produce the same manifold?

We are asking in effect how we can make a manifold $\mathfrak{M}$ correspond *trivially* to itself. It is no harder to ask how we can make a manifold $\mathfrak{M}$ correspond *nontrivially* to (itself or) another manifold $\mathfrak{M}^0$. We therefore consider another manifold

$$\mathfrak{M}^0 = [P^0, Z_{P^0}, \Phi_{P^0}, \mathfrak{R}_{P^0}] \tag{5-10}$$

(over, possibly, another sphere) and we define

$$\mathfrak{M} \approx \mathfrak{M}^0$$

(equivalence) as follows: There exists a one-to-one correspondence of points P, P^0. The parameters ζ_P, ζ_{P^0} for each such pair correspond biuniquely by analytic transformations, g and g^0, in some new neighborhoods $\mathfrak{R}_P^* \subseteq \mathfrak{R}_P$ (and $\mathfrak{R}_{P^0}^* \subseteq \mathfrak{R}_{P^0}$) containing $\zeta_P = 0$ (and $\zeta_{P^0} = 0$); that is,

$$\begin{aligned} \zeta_P &= g(\zeta_{P^0}) \qquad \zeta_{P^0} \in \mathfrak{R}_{P^0}^* \subseteq \mathfrak{R}_{P^0},\ g'(0) \neq 0 \\ \zeta_{P^0} &= g^0(\zeta_P) \qquad \zeta_P \in \mathfrak{R}_P^* \subseteq \mathfrak{R}_P,\ g^{0\prime}(0) \neq 0 \end{aligned} \tag{5-11}$$

and the relations (5-11) are consistent with the mappings:

$$Z_P = \Phi_P(\mathfrak{R}_P) \qquad Z_{P^0} = \Phi_{P^0}(\mathfrak{R}_{P^0}) \tag{5-12}$$

This last condition is the following when $\mathfrak{M}$ and $\mathfrak{M}^0$ are different parametrizations of the same surface:

$$\begin{aligned} \Phi_P(g(\zeta_{P^0})) &= \Phi_{P^0}(\zeta_{P^0}) \qquad \zeta_{P^0} \in \mathfrak{R}_{P^0}^* \\ \Phi_{P^0}(g^0(\zeta_P)) &= \Phi_P(\zeta_P) \qquad \zeta_P \in \mathfrak{R}_P^* \end{aligned} \tag{5-13}$$

Thus $z_P = \Phi_P(\zeta_P)$ and $z_{P^0} = \Phi_{P^0}(\zeta_{P^0})$ correspond under (5-11), inducing a biunique correspondence of cyclic neighborhoods in Z_P, Z_{P^0} about P, P^0.

The relation $\mathfrak{M} \approx \mathfrak{M}^0$ is genuinely an equivalence relation (see Exercise 7). It divides Riemann manifolds into equivalence classes (often called "conformal" equivalence classes).

We can now define a function $w(P)$ *meromorphic* on $\mathfrak{M}$ as a function on the points P which is meromorphic ($= w_P(\zeta_P)$) in ζ_P in any region $\mathfrak{R}_P$ of the parameter plane. By the equivalence relation (5-11) such a function has values (or *is*) independent of the parametrization; indeed, it is defined on all equivalent manifolds at once. Its orders of zeros and poles is clearly an "intrinsic" property of w on $\mathfrak{M}$ (independent of the parametrization). If no poles occur, of course, $w(P)$ is analytic.

The above definition of the meromorphic function $w(P)$ implies quite a bit, mechanically speaking. The values of $w(P) = w_P(\zeta_P)$ for ζ_P in $\mathfrak{R}_P$ and $w(Q) = w_Q(\zeta_Q)$ for ζ_Q in $\mathfrak{R}_Q$ must agree *or be coherent* where $\mathfrak{R}_P$ and $\mathfrak{R}_Q$ overlap. Thus, using the functions Φ_P, Φ_Q [see (5-5), of Sec. 5-1 above], we find the power series w_P, w_Q to be related by

(5-14) $$w_Q(\Phi_Q^{-1}\Phi_P(\zeta_P)) = w_P(\zeta_P) \qquad \text{where } \zeta_P \in \mathfrak{R}_P \cap \mathfrak{R}_Q$$

Here we recall $\Phi_P(\zeta_P) = \Phi_Q(\zeta_Q)$ where $\mathfrak{R}_P$ and $\mathfrak{R}_Q$ overlap.

Our next result shows that our concept of a Riemann manifold is essentially nonimprovable in the realm of complex analysis.

THEOREM 5-1 *The meromorphic function w on a Riemann manifold $\mathfrak{M}$ gives rise to an equivalent manifold $\mathfrak{M}^0$ over the w sphere. Symbolically,*

$$\mathfrak{M} \approx \mathfrak{M}^0 \qquad \textit{when}\dagger\ \mathfrak{M}^0 = w(\mathfrak{M})$$

The proof is almost entirely a matter of definition. Let w be defined on $\mathfrak{M}$. Then we have a biunique mapping of $\mathfrak{M}$ onto its image under $w(\mathfrak{M})$, once we choose each $\mathfrak{R}_P$ on $\mathfrak{M}$ sufficiently small that $w(\mathfrak{R}_P)$ has a one-to-one image onto a *cyclic* w neighborhood (if necessary). Therefore a biunique mapping of manifolds is correctly called "conformal" when referred to parameter planes. We must now extend our thinking to differentials in order to properly understand functions.

We define analytic or meromorphic *differentials* on $\mathfrak{M}$ in a natural fashion by considering $w(\zeta)\,d\zeta$, where $w(\zeta)$ is an analytic or meromorphic function on $\mathfrak{M}$, and $w(\zeta)\,d\zeta$ is coherent with regard to correspondences of points common to $\mathfrak{R}_P$ and $\mathfrak{R}_Q$. Thus as an analog to (5-14), we have $w(\zeta_Q)\,d\zeta_Q = w(\zeta_P)\,d\zeta_P$ which expands to

(5-15a) $$w(\Phi_Q^{-1}\Phi_P(\zeta_P))\,d\zeta_Q = w(\zeta_P)\,d\zeta_P$$

We must go further and note that $\Phi_Q(\zeta_Q) = \Phi_P(\zeta_P)$ where $\mathfrak{R}_P$ and $\mathfrak{R}_Q$ overlap or

(5-15b) $$d\zeta_Q\,\Phi_Q'(\zeta_Q) = d\zeta_P\,\Phi_P'(\zeta_P)$$

† Here $\mathfrak{M}^0$ lies over the w sphere while $\mathfrak{M}$ lies over the z sphere, but clearly the relabeling of a generic sphere is no difficulty.

We could divide (5-15a) by (5-15b) since "differentials do not really exist" and then substitute $\zeta_Q = \Phi_Q^{-1}\Phi_P(\zeta_P)$, but we would get the ponderous relation

$$\text{(5-15c)} \qquad \frac{w(\Phi_Q^{-1}\Phi_P(\zeta_P))}{\Phi_Q'(\Phi_Q^{-1}\Phi_P(\zeta_P))} = \frac{w(\zeta_P)}{\Phi_P'(\zeta_P)} \qquad \zeta_P \in \mathcal{R}_P \cap \mathcal{R}_Q$$

Certainly a relation as difficult as (5-15c) is not likely to be applied in practice! Usually we write the differential as $w(z)\,dz$, and we depend on a sign of a radical or some other device to tell us on which sheet of a Riemann manifold the variable z happens to be (if there is more than one sheet).

If we define an equivalent manifold $\mathfrak{M}^0$ by relation (5-11), i.e., by a set of functions $\zeta_P = g(\zeta_P{}^0)$, then formally $w^0(\zeta_P{}^0)\,d\zeta_P{}^0$ is the *differential equivalent to* $w(\zeta_P)\,d\zeta_P$ exactly when they are formally equal, or (canceling $d\zeta_P{}^0$ from both sides) when

$$\text{(5-15d)} \qquad w^0(\zeta_P{}^0) = w(g(\zeta_P{}^0))g'(\zeta_P{}^0)$$

CLASSICAL ILLUSTRATIONS

Take the following special cases of a Riemann manifold:

(*a*) *The z sphere itself.* Here the meromorphic functions are the rational functions of z. The analytic functions are constants by Liouville's theorem. The meromorphic differentials are $R(z)\,dz$ with $R(z)$ rational. There are no analytic differentials but 0 (see Exercise 4).

(*b*) *The period parallelogram* (in the u plane). Here the meromorphic functions are the elliptic functions of u. The analytic functions (as before) are constants. The meromorphic differentials are $f(u)\,du$ where $f(u)$ is an elliptic function. There is, however, a nonzero analytic differential, namely du itself.

(*c*) *The fundamental domain* (in the τ plane) *for the modular function.* Here we reduce the problem to (*a*) by using Exercise 10 of Sec. 4-6. We map the domain onto the J plane, and we compactify the mapping (to the J sphere) by $J(i\infty) = \infty$. Then if $J(\tau_0) = J_0$, the neighborhood of τ_0 is generally parametrized† by $\zeta = J - J_0$. The meromorphic functions on the (compactified) domain are rational functions of J, following the pattern of (*a*).

EXERCISES

1 Consider the manifolds $\mathfrak{M}_w$, $\mathfrak{M}_q$ equivalent to $\mathfrak{M}_z$ (= the z sphere) under the transformations $w = z^2$, $q = (w - 1)^2$. Discuss the choice of disks $\mathcal{R}_P$ for the parametrization of $\mathfrak{M}_w$ and $\mathfrak{M}_q$.

† Of course, the neighborhood of $\tau = i\infty$ is parametrized by $1/J$, the neighborhood of $\tau = i$ is parametrized by $(J - 1)^{1/2}$, and the neighborhood of $\tau = \rho_1$ is parametrized by $J^{1/3}$ (see Fig. 4-5).

2 Show that the property of a differential on $\mathfrak{M}$ as analytic versus meromorphic is preserved under equivalence transformations of $\mathfrak{M}$.

3 Show that the ratio of two meromorphic differentials on $\mathfrak{M}$ is a meromorphic function on $\mathfrak{M}$.

4 In illustration (*a*), (*z* sphere) show there are no nonzero analytic differentials. This readily reduces to showing $1 \cdot dz$ is not analytic. (What happens when it is parametrized as $\zeta = 1/z$ for ∞?)

5 If we consider the *z* plane (instead of the *z* sphere), show that the analytic functions are the entire functions and dz is an analytic differential, in fact, that $f(z)\,dz$ is an analytic differential if $f(z)$ is entire.

6 Consider the usual Riemann surface for $y^2 = 4(z - e_1)(z - e_2)(z - e_3)$ with e_i different roots summing to 0. According to Lemma 4-3, the function $z = \wp(u)$ provides a biunique mapping onto the *u* parallelogram. From this, verify that the meromorphic functions are $R(z,y)$ (rational in *z* and *y*) and the meromorphic differentials are $R(z,y)\,dz$. Show the analytic differentials are (constant) dz/y [the abelian integrals of the first kind (see Sec. 4-2)].

7 Show that the relation $\mathfrak{M} \approx \mathfrak{M}^0$ is transitive, symmetric, and reflexive.

TRIANGULATED MANIFOLDS

5-3 Triangulation Structure

In order to think of the systems of neighborhoods, parametrizations, etc., in a practical way, we need a model. The important model, of course, is the *n*-sheeted Riemann surface, but it must be explained in terms of earlier artifacts. A simple explanation is to cut up the *n* sheets into a finite number of curvilinear euclidean triangles (some of which may be infinitely large) and to explain that each triangle is a separate entity, but they "happen to fit together so as to become interesting." This approach was first formalized by Weyl (1913). We shall start with triangles, and in the succeeding sections of the chapter, we shall show how to build up the *n*-sheeted Riemann surface from them.

Begin by considering a finite or infinite sequence of (open) curvilinear triangles which may be finite or infinite

$$\Delta_1, \Delta_2, \ldots \tag{5-16}$$

with the property that Δ_i lies on the u_i sphere. Each triangle has boundary points P_{i1}, P_{i2}, P_{i3} and (open) boundary segments $\mathfrak{L}_{i1}, \mathfrak{L}_{i2}, \mathfrak{L}_{i3}$ where in chain notation

$$\partial\Delta_i = \mathfrak{L}_{i1} + \mathfrak{L}_{i2} + \mathfrak{L}_{i3} \tag{5-17}$$

and $\mathfrak{L}_{ij}$ is the directed arc from $P_{i(j+1)}$ to $P_{i(j+2)}$ where $j = 1, 2, 3$ (and the notation is cyclic "modulo 3," for example, $P_{i4} = P_{i1}$, $P_{i5} = P_{i2}$). Here too, P_{ij} and $\mathfrak{L}_{ij}$ are not necessarily in the finite plane.

The *triangulated* (*Riemann*) *manifold* $\mathfrak{M}$ is then all points in the interior of the Δ_i together with some or all of the (open) boundary arcs

$\mathfrak{L}_{ij}$ and some or all of the end points P_{ij}, provided certain conditions for inclusion of the boundaries are satisfied. We summarize these conditions and then express them in detail:

(*a*) An arc $\mathfrak{L}_{ij}$ of Δ_i is included in $\mathfrak{M}$ precisely when it is paired with exactly one other arc $\mathfrak{L}_{IJ}$ of Δ_I $(i \neq I)$, so that Δ_i and Δ_I fit together on $\mathfrak{L}_{ij}$ and $\mathfrak{L}_{IJ}$. The points of $\mathfrak{L}_{ij}$ and $\mathfrak{L}_{IJ}$ which fit together are identified to produce a single arc in $\mathfrak{M}$. Two triangles are to fit together at only one pair of sides (if at all).

(*b*) A point P_{ij} of Δ_i is included in $\mathfrak{M}$ precisely when it is identified with a finite sequence $P_{i'j'}$ of $\Delta_{i'}$, $P_{i''j''}$ of $\Delta_{i''}$, etc., such that n triangles $\Delta_i, \Delta_{i'}, \ldots, \Delta_{i^{(n-1)}}$ all fit together so as to identify the common vertex P formed by identifying all n points $P_{ij}, P_{i'j'}$, etc., with one another. Then P belongs to $\mathfrak{M}$. (See Fig. 5-2.)

(*c*) Enough arcs are paired together so that the set of triangles is connected. To be precise, any two triangles can be connected by a chain of intermediary triangles, each one of which fits together with its neighbor in the chain along a common side.

It is clear that for conditions (*a*) and (*b*) we need only define the fitting together of Δ_i and Δ_I on $\mathfrak{L}_{ij}$ and $\mathfrak{L}_{IJ}$. Condition (*a*) means that two functions $f_i(u_i)$, $f_I(u_I)$ exist which are biunique and meromorphic† and which are written

$$z = \begin{cases} f_i(u_i) & u_i \in \Delta_i \cup \mathfrak{L}_{ij} \\ f_I(u_I) & u_I \in \Delta_I \cup \mathfrak{L}_{IJ} \end{cases} \tag{5-18}$$

† If we assume Δ_i and Δ_I are finite, we can take all functions to be holomorphic.

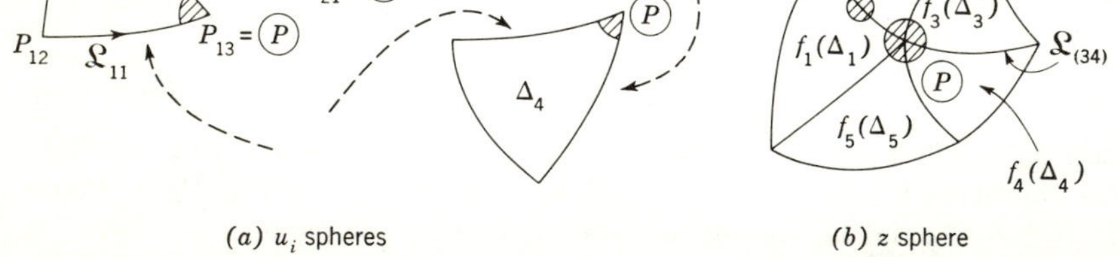

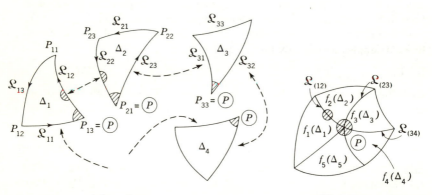

Fig. 5-2 Illustration of fitting of triangles at a vertex. The cyclic region is taken to have order 1 for convenience. The neighborhoods of points on sides and vertices are indicated.

These functions must have the property that the images $f_i(\Delta_i)$, $f_I(\Delta_I)$ lie on opposite sides of the common image $\mathfrak{L}_{(iI)}$ (adjoining Δ_i and Δ_I)

$$\mathfrak{L}_{(iI)} = f_i(\mathfrak{L}_{ij}) = f_I(\mathfrak{L}_{IJ}) \tag{5-19}$$

so that the orientations derived from these functions cancel. The variable z defined by (5-18) has to be biunique in the further sense that

$$f_i(u_i) \neq f_I(u_I) \tag{5-20}$$

except for the values in (5-19), but condition (5-20) must be interpreted more broadly as being valid in a cyclic neighborhood of order e near one of the end points, say $z = a$. This means that from (5-18),

$$\zeta = (z - a)^{1/e} \qquad (\text{or } z^{-1/e} \text{ if } a = \infty) \tag{5-21}$$

provides a biunique map of $\Delta_i \cup \Delta_I \cup \mathfrak{L}_{ij}$.

It is admittedly awkward but still necessary to allow the triangles to fit together on *cyclic* neighborhoods, or else we could not account for branch points of Riemann manifolds!

In this way to clarify condition (*b*), we can also fit together n triangles with a common vertex P so that P shall also belong to the manifold. For ease of notation let us suppose, as in Fig. 5-2, that we are fitting together

$$\begin{array}{ll} \Delta_1 \text{ and } \Delta_2 & \text{at } \mathfrak{L}_{12} \equiv \mathfrak{L}_{22} \\ \Delta_2 \text{ and } \Delta_3 & \text{at } \mathfrak{L}_{23} \equiv \mathfrak{L}_{31}, \text{ etc.} \end{array}$$

Then we have a sequence of meromorphic functions formed by successive pairs of type (5-18)

$$z = \begin{cases} f_1(u_1) & u_1 \in \Delta_1 \cup \mathfrak{L}_{12} \cup P_{13} \\ f_2(u_2) & u_2 \in \Delta_2 \cup \mathfrak{L}_{22} \cup P_{21} \end{cases}$$

$$z = \begin{cases} f_2(u_2) & u_2 \in \Delta_2 \cup \mathfrak{L}_{23} \cup P_{21} \\ f_3(u_3) & u_3 \in \Delta_3 \cup \mathfrak{L}_{31} \cup P_{33}, \text{ etc.} \end{cases} \tag{5-22}$$

Not only do adjacent sides (5-19) agree, but also so do the vertices

$$f_1(P_{13}) = f_2(P_{21}) = f_3(P_{33}) = \cdots = a_P \tag{5-23}$$

We require that for a_P the projection of P, and for some integer e,

$$\zeta = (z - a_P)^{1/e} \qquad (\text{or } z^{-1/e} \text{ if } a_P = \infty) \tag{5-24}$$

be definable so that when applied to (5-22), we have images of the domain shown $(\Delta_1 \cup \mathfrak{L}_{12} \cup P_{13} \cup \mathfrak{L}_{22} \cup \Delta_2 \cup P_{21} \cup \cdots)$ which provide a biunique covering of the origin in the ζ plane [except for the identification of sides (5-19) and vertices (5-23)].

Note that under this definition of fitting together, it is possible that for triangle Δ_1 the vertex P_{13} can belong to $\mathfrak{M}$ but not P_{11} (and consequently not P_{23} as shown in Fig. 5-2). It is also necessary that several different functions be used to fit together the same pair of sides. For example, suppose $P_{11} \equiv P_{23}$ did belong to $\mathfrak{M}$, then another system like (5-22) would also apply to $\mathcal{L}_{12} \equiv \mathcal{L}_{22}$ (in order to fit together the triangles surrounding $P_{11} \equiv P_{22}$). We must therefore provide that all systems are coherent, e.g., the identifications of points (5-19) of $\mathcal{L}_{ij}$ and $\mathcal{L}_{IJ}$ be well defined regardless of how many different times we had to fit these sides together.

The triangulation structure now determines a Riemann manifold. The *points* are interior points of Δ_i and *only those boundary points* (on sides or vertices) *for which identifications have been arranged* (as described above). The neighborhoods are the natural ones for points interior to Δ_i but are described on the z spheres (or ζ planes) for the boundary points of $\mathcal{L}_{ij}$ or P_{ij} (see Fig. 5-2). The parametrizations come from z or ζ. Thus, we have a Riemann manifold $\mathfrak{M}$ over the z sphere.

Various examples of triangulated manifolds will be left as exercises. There may be a finite or an infinite set of triangles, naturally.

RADO'S THEOREM (1925) *Every Riemann manifold can be represented as a triangulated manifold.*

On the basis of this result, we are losing nothing by restricting ourselves to triangulated manifolds. Actually, we shall consider large classes of problems where the triangulated structure is apparent from the geometrical or physical problem. Therefore, the proof of this theorem can be left to advanced texts.

In the illustrations given here we show for simplicity that the triangles Δ_i adjoin so as to cancel the orientations of $\partial\Delta_i$, but this will be shown to be a general property called "orientability" in Theorem 7-1.

We shall even make a more severe restriction, to confine our attention initially. By implication, a point of $\partial\Delta_i$ which is not part of an identification scheme does not belong to the manifold $\mathfrak{M}$. Such points are called "boundary points" of $\mathfrak{M}$ (they are a subset of all the boundary points of the Δ_i). There can also be other ideal boundary points (such as the point at ∞). For example, if there are *infinitely* many Δ_k (arranged in sequence), then for a choice of $Q_k \in \Delta_k$, the sequence $Q_1, Q_2, \ldots$ will have no limit point as such but will determine an ideal limit (defined in an ad hoc manner). We banish both types of occurrence by ultimately restricting ourselves (see Sec. 6-2) to *compact* manifolds or finitely triangulated manifolds without boundaries.

Very often we speak of "polygonal" manifolds where convenient, and we even speak of different sides of the same polygon being identified (omitting triangulations) when the meaning is obvious, for example, as in the case of the period parallelogram. Usually two triangles will be given in u spheres in a manner exactly consistent with their position in the z sphere, so that the functions (5-18) and (5-22) are generally identity

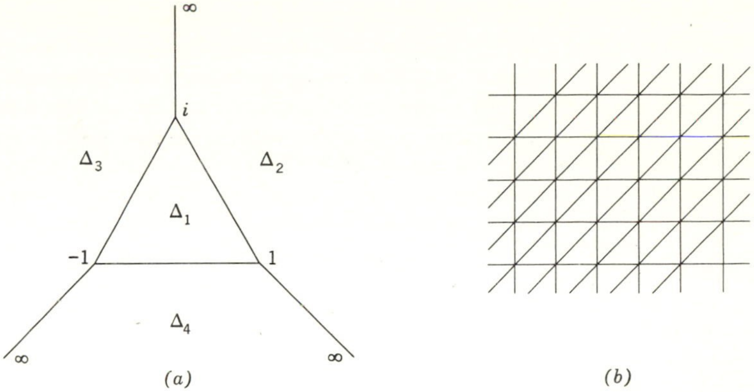

Fig. 5-3 Triangulations of the z sphere and z plane.

functions. We can make the further restriction for future convenience that all angles of the Δ_i are between 0 and π (so no "cusps" occur).

EXERCISES

1 Consider a finite triangulation of the z sphere. Explain the mappings (5-22) which must be used to fit the triangles together in Fig. 5-3*a*.

2 Explain Fig. 5-3*a* and *b* as a triangulation of the z plane. Note how ∞ is an actual boundary point in Fig. 5-3*a* and an ideal boundary point in Fig. 5-3*b*.

3 Returning to Fig. 5-3*a*, explain $\mathfrak{R}$ the exterior of the slit $-1 \leq z \leq +1$ as a triangulated Riemann manifold. (Here ∞ and i belong to $\mathfrak{R}$ but not $+1$ or -1.)

4 Explain the period parallelogram (Fig. 5-4*a*) as a triangulated manifold. Using period translations, justify the set of triangles surounding A (Fig. 5-4*b*) and likewise for B or C.

5 Explain a disk (say $|z| < 1$) as a triangulated manifold (cut it up into three 120° sectors like a pie).

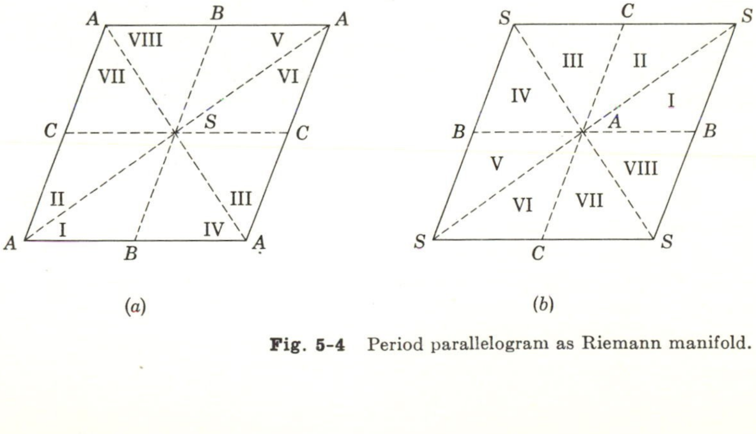

Fig. 5-4 Period parallelogram as Riemann manifold.

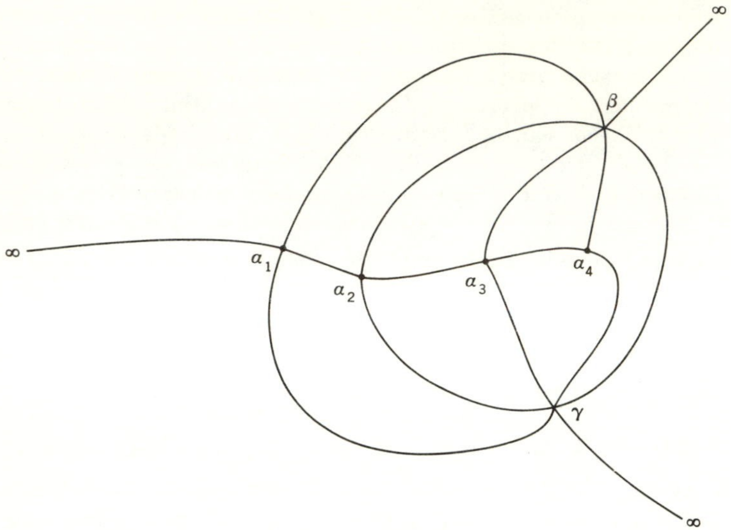

Fig. 5-5 Triangulation for each sheet of $w^2 = (z - a_1) \cdots (z - a_4)$.

6 Explain the punctured disk $(0 < |z| < 1)$ as a triangulated manifold.

7 Show how to triangulate an n-tuply connected region, systematically without regard to the value of n. (How many triangles are needed?)

8 Show how to triangulate the Riemann surface for $w^2 = (z - a_1) \cdots (z - a_4)$. (See Fig. 5-5.) Label the triangles and indicate how they will fit together about the points a_1, a_2, a_3, a_4 and $\infty_{\text{I}}, \infty_{\text{II}}, \beta_{\text{I}}, \beta_{\text{II}}, \gamma_{\text{I}}, \gamma_{\text{II}}$ (on Sheets I and II).

5-4 Algebraic Riemann Manifolds

Consider an algebraic function† defined as the set of convergent power series $w(\zeta)$ [in $\zeta = (z - a)^{1/e}$ or $1/z^e$] satisfying

$$\Phi(z,w) = w^n + \phi_1(z)w^{n-1} + \cdots + \phi_n(z) \tag{5-25}$$

where $\phi_t(z)$ $(1 \leq t \leq n)$ are rational functions of z (in complex coefficients) of maximum degree (in numerator or denominator of ϕ_t) equal to m. We shall sketch the algebraic background required for the direct construction of the Riemann surface for $w(z)$ as a compact manifold $\mathfrak{M}$ over the z sphere.

We assume Φ to be irreducible. We shall prove that the number of sheets of $\mathfrak{M}$ is n except at branch points (and there the number of sheets is still n if we use a reasonable counting procedure). By custom, algebraic Riemann manifolds are also called "Riemann surfaces" since they

† This section is unnecessary for continuity with the rest of the text. It represents, however, the focal point of the pure mathematician's interest.

were Riemann's original illustrations (from which the whole concept of "manifold" arose).

We require one result which is predominantly algebraic.

LEMMA 5-1 *Consider, in addition to* $\Phi(z,w)$, *a second polynomial (with rational functions as coefficients) but (unlike* Φ*) not necessarily irreducible,*

$$\Psi(z,w) = w^q + \psi_1(z)w^{q-1} + \cdots + \psi_q(z) \tag{5-26}$$

Assume the degree $q < n$. *Then there exist two polynomials in* w *whose coefficients are rational functions of* z *(possibly one is* 0*)*

$$\begin{aligned} A(z,w) &= \alpha(z)w^r + \cdots + \alpha_r(z) \\ B(z,w) &= \beta(z)w^s + \cdots + \beta_s(z) \end{aligned} \tag{5-27}$$

such that the following identity holds in variables z *and* w:

$$A(z,w)\Phi + B(z,w)\Psi = 1 \tag{5-28}$$

PROOF Consider all polynomials in w of type $F(w) = A\Phi + B\Psi$ where A and B are of type (5-27) (including $A = 0$ and $B = 0$). There is evidently a polynomial $F(w)$ (*other than zero*) of lowest degree d (in w).

Clearly, $d \leq q < n$ (try $A = 0$, $B = 1$). We must show $d = 0$.

For if $d > 0$, we can find some polynomials A_0, B_0 in w (with rational coefficients in z), such that

$$F_0(w) = A_0\Phi + B_0\Psi = 1 \cdot w^d + \text{lower terms in } w \tag{5-29}$$

[We can choose the leading coefficient in $F_0(w)$ as 1 by dividing out the leading coefficient of the term of minimal degree in F_0 and inserting this coefficient into A and B.]

We next perform (ordinary) division of Φ by F_0, which leads to a quotient $Q(z,w)$ and a *nonzero* remainder $R(z,w)$ of degree d' in w, where $0 \leq d' < d$ (or else F_0 divides Φ in contradiction to irreducibility of Φ). Hence we have the so-called "euclidean algorithm,"

$$\Phi = QF_0 + R \tag{5-30a}$$

and substituting (5-29) for F_0 we obtain from (5-30*a*)

$$R(z,w) = (1 - QA_0)\Phi + (-QB_0)\Psi \tag{5-30b}$$

a polynomial in w of lower degree (d'). To avoid this contradiction, we conclude $d = 0$, or F_0 must be a constant (in w). Hence $F_0 = 1$ in (5-29).

Q.E.D.

LEMMA 5-2 *Let* $\Phi(z,w)$ *be a polynomial in* z *and* w *satisfying* $\Phi(\alpha,\beta) = 0$ *and let* $\partial\Phi/\partial w \neq 0$ *at* $z = \alpha$, $w = \beta$. *Then an analytic function* $w(z)$ *is determined near* $z = \alpha$ $(w = \beta)$ *satisfying* $\Phi(z,w) = 0$.

This result has already been proved by Theorem 3-3.

LEMMA 5-3 *With the exception of, at most, a finite number of points* $\alpha = b_1, b_2, \ldots, b_r \ (= \infty)$ *on the z sphere, the equation* (5-25) *is solvable for n different analytic power series* $w(z)$ *near each point* α *of the z sphere in the variable* $(z - \alpha)$.

PROOF In Lemma 5-1, let $\Psi = \partial\Phi/\partial w$. Then from (5-28) it will follow that Φ and Ψ can vanish simultaneously ($z = \alpha$, $w = \beta$) only where $A(\alpha,\beta)$ and $B(\alpha,\beta)$ are infinite, i.e., only at points $z = b_1, b_2, \ldots, b_{r-1}$ where the denominator of $\alpha(z)$, of $\beta(z)$, or of a coefficient of $\Phi(z)$ is 0 and possibly at $z = b_r \ (= \infty)$. Hence, excluding a *finite* number of points, there are n different values of w for each $z = \alpha$. Thus, there are (by Lemma 5-2) n different power series $w(z)$. Q.E.D.

Now, not every one of the points $z = b_1, b_2, \ldots, b_r$ will have to be a branch point, but having isolated such points we join them by a simple curve $\mathcal{C}$ (see Fig. 5-6). We let $\mathcal{R}^*$ denote what is left of the z sphere with this curve $\mathcal{C}$ removed (as a slit). Since $\mathcal{R}^*$ is now simply-connected,† we can invoke the monodromy theorem (of Sec. 3-1) to define n power series (or root functions) by analytic continuation uniquely in $\mathcal{R}^*$.

$$w_1(z), w_2(z), \ldots, w_n(z) \tag{5-31}$$

By analytic continuation *across* the slit, the relation $\Phi(z,w_k(z)) = 0$ is always preserved. Thus, we find that each $w_k(z)$ can become only another $w_l(z)$ (possibly $k = l$). In Fig. 5-6, for instance, if each $w_k(z)$ is extended across arc $b_t b_{t+1}$ by the path $\mathcal{C}_t$ ($t = 1, 2, \ldots, r - 1$), we

† We can think of the slit as having an unmatched upper and lower side in the sense of a Riemann manifold (see Exercise 3 of Sec. 5-3), or else we can think of the slit being isolated by an arbitrary small covering so that $\mathcal{R}^*$ is bounded by a simple curve $\mathcal{C}^*$ (see Fig. 5-6).

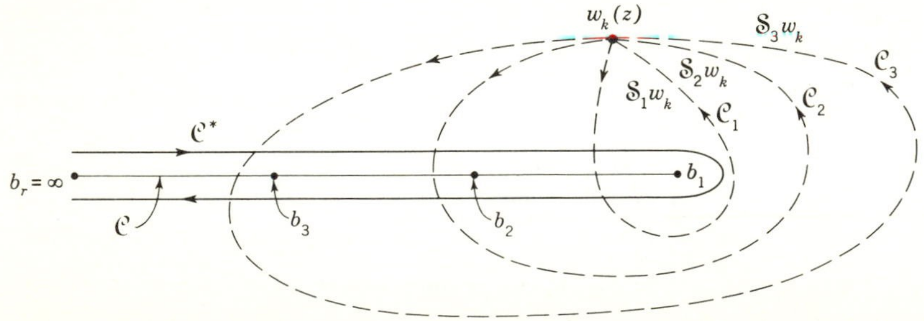

Fig. 5-6 Paths of analytic continuation of $w_k(z)$.

have a rearrangement

$$\begin{array}{ccc} w_1 & \text{becomes} & w_{a_1} \\ w_2 & \text{becomes} & w_{a_2} \\ \cdots & \cdots & \cdots \\ w_n & \text{becomes} & w_{a_n} \end{array}$$

where $a_1, \ldots, a_n$ is a rearrangement of integers $1, 2, \ldots, n$. The rearrangement is without repetition, or $a_i \neq a_j$ when $i \neq j$, because otherwise if w_i and w_j extended themselves to the same value w_k on, say, $\mathcal{C}_1$, we should find that the unique (reverse) continuation on $-\mathcal{C}_1$ could not make w_i and w_j unequal again. Because no repetition occurs, all roots of (5-25) are available by analytic continuation of any set of n root functions to any ordinary point (not $b_1, \ldots, b_r$). We write

$$\mathcal{S}_t = \begin{pmatrix} 1 & 2 & \cdots & n \\ a_1 & a_2 & \cdots & a_n \end{pmatrix} \tag{5-32}$$

as the *monodromy substitution* along $\mathcal{C}_t$ (see Fig. 5-6). As can be seen, these substitutions generate a group. This group is called the *monodromy group*. It consists of all permutations of $w_i(z)$ under analytic continuation over any closed path† in the complement of the set $\{b_t\}$.

LEMMA 5-4 *The equation in w,*

$$\Phi(w) = \phi_0 w^n + \phi_1 w^{n-1} + \cdots + \phi_n = 0$$

with constant complex coefficients ($\phi_0 \neq 0$) *has all roots bounded by:*

$$w_t \leq \frac{|\phi_0| + \cdots + |\phi_n|}{|\phi_0|} = K \tag{5-33}$$

PROOF Note that if $|w| > K$, then $|w| > 1$, and

$$|w| < |w^2| < \cdots < |w^{n-1}|$$

meanwhile,

$$|\phi_0 w^n| > K|\phi_0| \cdot |w|^{n-1} = (|\phi_1| + \cdots + |\phi_n|)|w|^{n-1} \geq |\phi_1 w^{n-1}| + |\phi_2 w^{n-2}| + \cdots + |\phi_n| \geq |\phi_1 w^{n-1} + \phi_2 w^{n-2} + \cdots + \phi_n|$$

and

$$|\Phi(w)| \geq |\phi_0 w^n| - |\phi_0 w^{n-1} + \phi_2 w^{n-2} + \cdots + \phi_n| > 0$$

Q.E.D.

LEMMA 5-5 *The solutions $w(z)$ of the algebraic equation* (5-25) *in* $\mathfrak{R}^*$ *are all expressible in the neighborhood of any point of the z sphere as (meromorphic) power series in* $(z - a)^{1/e}$ *or* $1/z^{1/e}$ *by analytic continuation; such*

† This last statement can be seen intuitively from the fact that any closed path not containing any b_t can be deformed into a succession of loops $\mathcal{C}_t$. (Imagine the arbitrary path as a string in the z plane and the points b_t as "impassable" pegs.)

power series provide the complete analytic continuation of all n root functions $w_i(z)$ over any point of the z sphere and over cyclic neighborhoods over special points $b_1, \ldots, b_r$.

PROOF We need only verify behavior at the possible branch points $b_1, \ldots, b_r$.

Near any exceptional point b_t the sheets representing the branches $w_1(z), w_2(z), \ldots, w_n(z)$ are uniquely defined but still may be joined in a generally complicated fashion. To understand this more clearly, take an exceptional point b_t projected in the z plane and surrounded by the disk $\mathfrak{D}$

$$|z - b_t| \leq R \tag{5-34}$$

in which no other branch point occurs. Then by analytic continuation the branches $w_1(z), \ldots, w_n(z)$ transform into new branches $w_{a_1}(z), \ldots, w_{a_n}(z)$ on analytic continuation around the circle $|z - b_t| = R$. The transformation can be broken up into so-called "cycles" by relabeling branches.

Let $w_1(z) \to w_2(z)$; for example, call $w_2(z)$ the branch which $w_1(z)$ happens to transform into by continuation about b_t. Similarly, let $w_2(z) \to w_3(z)$, etc. We find that ultimately (say, at e steps) $w_e(z) \to w_1(z)$. (Here e might be $<n$.) Next, w_1 is considered as a function of $\zeta = (z - b_t)^{1/e}$. We set

$$z - b_t = \zeta^e \tag{5-35}$$

and we see that when ζ encircles the origin once, z encircles b_t a total of e times and $w_1(z)$ returns to its original value [likewise for $w_2(z), \ldots, w_e(z)$].

Here $(w_1, w_2, \ldots, w_e)$ is said to constitute a "cycle" at z. There are possibly several cycles since $e \leq n$. Each cycle corresponds to a set of root functions; hence each cycle corresponds to a single point of the Riemann manifold.

Now $w_1, w_2, \ldots, w_e$ are single-valued in ζ, around $\zeta = 0$. If we expand w_i into a Laurent series, moreover, it must terminate in the negative powers (or have a pole), since the roots w_i are bounded by "polar" behavior from Lemma 5-4. Thus, if (at worst) b_t is a pole of some $\phi_i(z)$ of degree g, then $|w_t| <$ constant $|z - b_t|^g$, displaying a pole (and not an essential singularity).

Similar considerations hold for $z = \infty$ by use of the new variable $1/z$ (and $\zeta = 1/z^e$ if needed). Q.E.D.

We now take our n replicas of the slit z sphere $\mathfrak{R}^*$ and triangulate them [in some trivial way (see Exercise 8 of Sec. 5-3)]. We join the triangles according to the pattern of analytic continuation. This provides, in particular, a joining of different triangles across $\mathfrak{R}^*$ via the monodromy substitutions (5-32); but by Lemma 5-5, this process can be imagined to take place in cyclic neighborhoods.

LEMMA 5-6 *The triangulation system of the n replicas of* $\mathcal{R}^*$ *determined by analytic continuation is connected.*

PROOF Otherwise, this would mean that of the n sheets only N would be *connected* in the sense that the N branches ($N < n$)

$$w_1, w_2, \ldots, w_N$$

are attainable by analytic continuation from $w_1(z)$ across the cut in $\mathcal{R}^*$. We show this cannot happen. For otherwise the elementary *symmetric functions* (for $z \neq b_t$) of $w_1, \ldots, w_N$, namely,

$$\begin{aligned} S_1 &= w_1 + \cdots + w_N \\ S_2 &= w_1w_2 + \cdots + w_{N-1}w_N \\ &\cdots\cdots\cdots\cdots \\ S_N &= w_1w_2 \cdots w_N \end{aligned} \tag{5-36}$$

would all be single-valued in z under analytic continuation (as the w_i are then only *rearranged*). The S_t would otherwise behave as poles, hence they are rational in z. Thus the function

$$\begin{aligned} \Lambda(z,w) &= (w - w_1) \cdots (w - w_N) \\ &= w^N - S_1w^{N-1} + S_2w^{N-2} - \cdots \pm S_N \end{aligned}$$

would be a polynomial in w with *rational* functions of z as coefficients, and $\Phi(z,w) = (w - w_1) \cdots (w - w_n)$ would cease to be *irreducible*.

Q.E.D.

We now can interpret the triangulation of the replicas $\mathcal{R}^*$ (joined by analytic continuation) as a triangulated Riemann manifold $\mathfrak{M}$. The important thing is now that each point P of it corresponds to some unique power series (with possible pole) in $(z - a_P)^{1/e}$ (or $1/z^{1/e}$) if P lies over a_P (or ∞).

THEOREM 5-2 *Any solution $w(z)$ of an irreducible algebraic equation of degree n in w is meromorphic on a compact Riemann surface* $\mathfrak{M}$ *over the z sphere with n sheets in the sense that for all points P on* $\mathfrak{M}$ *lying over a given* a*, the orders e of the uniformizing parameter satisfy* $\Sigma e = n$. *A manifold of fewer than n sheets over the z sphere cannot have $w(z)$ as a meromorphic function.*

Otherwise expressed, any algebraic function can be uniquely extended analytically over a Riemann surface of exactly n sheets on the z sphere, if n is the degree of the irreducible defining equation. Thus, a meromorphic function on $\mathfrak{M}$ can be written as $F(z,w)$; that is, it is determined by the z and the *number* of the sheet on which w lies.

Very often, a point on a Riemann surface is expressed as (z,w) to indicate that for each z the function w generally tells us which sheet we are talking about. [Thus the *power series* $w(z)$ and not the *value* of w is intended by the symbol (z,w).]

EXERCISES

1 Show that all function elements $w(z)$ or $w(\zeta)$ arising from an algebraic function can be contained in a finite collection of power series (including, of course, series about $z = \infty$).

2 Consider $w^3 + z^3 - 3azw = \Phi(z,w) = 0$ (the folium of Descartes). Show the common roots of $\Phi = \partial\Phi/\partial w = 0$ (by elimination of w) are only $z = 0$, $z^3 = 4a^3$, $z = \infty$. Show that these power series are valid:

$$\text{at } z = \infty \begin{cases} w_1 = -z + c_1 + \dfrac{c_2}{z} + \cdots \\ w_2 = -z\rho_1 + c_1' + \dfrac{c_2'}{z} + \cdots \\ w_3 = -z\rho_1^2 + c_1'' + \dfrac{c_2''}{z} + \cdots \end{cases} \qquad \text{where} \qquad \rho_1 = \frac{-1 + \sqrt{-3}}{2}$$

$$\text{at } z = 0 \begin{cases} w_1 = \dfrac{z^2}{3a} + d_1 z^3 + d_2 z^4 + \cdots \\ w_2\ (= w_3) = \zeta\sqrt{3a} + e_1\zeta^2 + \cdots \end{cases} \qquad \zeta^2 = z$$

$$\text{at } z = \sqrt[3]{4}\,a\rho \begin{cases} w_1 = -\sqrt[3]{16}\,\rho^2 a + f_1(z - \sqrt[3]{4}\,a\rho +) \cdots \\ w_2\ (= w_3) = \sqrt[3]{2}\,\rho^2 a + g_1\zeta + \cdots \end{cases} \qquad \begin{array}{l} \zeta^2 = z - \sqrt[3]{4}\,a\rho, \\ \text{for } \rho = 1, \rho_1, \rho_1^2 \end{array}$$

(here the numbering w_1, w_2, w_3 need not be consistent with the root functions as defined earlier). [*Hint:* For $z = 0$, set $w = z^2k(z)$ and verify that $k(z) = 1/3a + z^3k^3(z)/3a$; set $w = \zeta k(\zeta)$ and verify that $k(\zeta) = \zeta^3/3a + k^3(\zeta)/3a$, etc.]

3 (*Puiseaux series*) Show that every power series $w = \zeta^t(A_0 + A_1\zeta + \cdots)$ solution to $\Phi(z,w) = 0$, where $\zeta = z^{1/e}$, has an order t/e determined as follows: Set

$$\Phi(z,w) = \Sigma a_{ij} z^i w^j \tag{5-37}$$

and substitute $w = z^{t/e}$. Then for at least two (nonzero) coefficients a_{ij}, a_{IJ} *appearing in the sum* (5-37) cancellation occurs, or,

$$i + \left(\frac{t}{e}\right) j = I + \left(\frac{t}{e}\right) J = E \tag{5-38}$$

and for all other (nonzero) coefficients $a_{i'j'}$, $i' + (t/e)j' \geq E$. Apply this method to Exercise 2.

(*Remark:* Thus we see it might take more effort to prove the *existence* of the power series from Puiseaux's method alone than by Lemma 5-2.) We refer the reader to treatises on algebraic functions for more details.

4 Show the folium equation (of Exercise 2) is reduced (in w). Hence, all branches are connected by analytic continuation, and some branch is regular over each point of the z sphere, although the continuation is not valid over an *arbitrary* path (without meeting some branch point). (See remark after monodromy theorem, Sec. 3-1.) [*Hint:* If the equation $\Phi = 0$ is factorable, one factor $[w - \phi(z)]$ is linear. Then the substitution $w = \phi(z)$ (rational) will satisfy $\Phi(z,w) = 0$. Next, parametrize the folium by writing $w = tz$, thus $z = 3at/(t^3 + 1)$, $w = 3at^2/(t^3 + 1)$. Moreover

$w = \phi(z)$ must be satisfied identically in t by this parametrization. Obtain a contradiction from the fact that each z determines *three* t, hence each z could not determine one w as from $w = \phi(z)$.]

5 Consider the conjugates $w_1 = +\sqrt{z} + \sqrt{z-1}$, $w_2 = -\sqrt{z} + \sqrt{z-1}$, $w_3 = +\sqrt{z} - \sqrt{z-1}$, $w_4 = -\sqrt{z} - \sqrt{z-1}$ defined in $\mathfrak{R}^*$, the exterior of the slit from $b_1 = 1$ to $b_2 = 0$ to $b_3 = -\infty$. (Refer to Fig. 5-6.) Show that, in the notation of (5-32),

$$\mathfrak{S}_1 = \begin{pmatrix} 1 & 2 & 3 & 4 \\ 3 & 4 & 1 & 2 \end{pmatrix} \qquad \mathfrak{S}_2 = \begin{pmatrix} 1 & 2 & 3 & 4 \\ 4 & 3 & 2 & 1 \end{pmatrix}$$

Find the algebraic equation (of degree 4) for these root functions.

6 Do likewise for $w = \sqrt{(\sqrt{z} + 1)}$.

7 (*a*) Show that $\mathfrak{S}_t = \mathfrak{S}_{t+1}$ if and only if b_{t+1} is not a branch point. (*b*) Show that the substitution $\mathfrak{S}_t\mathfrak{S}_{t+1}^{-1} = \mathfrak{J}$ has an order which is divisible by each of the orders e of the cyclic neighborhoods of points lying over b_{t+1}. (The order of a substitution $\mathfrak{J}$ is the least positive integer m such that $\mathfrak{J}^m$ is the identical substitution.)

8 The Riemann surface for the folium (Exercise 2) can be drawn as follows: Sheet I has a neighborhood of order 2 only at 0 and $\sqrt[3]{4}\,a$, and it joins Sheet II (on a slit) at these points. Sheet III likewise has a neighborhood of order 2 only at $\rho_1\sqrt[3]{4}\,a$ and $\rho_1^2\sqrt[3]{4}\,a$, and it also joins Sheet II (on a slit) at these points. Thus Sheet I *or* Sheet III has a regular root function for each point without exception.

To establish this picture, set up $\mathfrak{C}$ (as in Fig. 5-6) as

$$\mathfrak{C} = (b_5,b_4,b_3,b_2,b_1) = (-\infty, \rho_1\sqrt[3]{4}\,a, \rho_1^2\sqrt[3]{4}\,a, 0, \sqrt[3]{4}\,a)$$

By Exercise 2, $e = 1$ or 2 (only), hence by Exercise 7, $\mathfrak{S}_t\mathfrak{S}_{t+1}^{-1}$ is always exactly of order 2. This leads (with suitable labeling) to

$$\mathfrak{S}_1 = \begin{pmatrix} 1 & 2 & 3 \\ 2 & 1 & 3 \end{pmatrix} \quad \mathfrak{S}_2 = \begin{pmatrix} 1 & 2 & 3 \\ 1 & 2 & 3 \end{pmatrix} \quad \mathfrak{S}_3 = \begin{pmatrix} 1 & 2 & 3 \\ 1 & 3 & 2 \end{pmatrix} \quad \mathfrak{S}_4 = \begin{pmatrix} 1 & 2 & 3 \\ 1 & 2 & 3 \end{pmatrix}$$

CHAPTER 6
ABSTRACT MANIFOLDS

The purpose of this short chapter is to reorient the subject matter. The last theorem (more precisely a construction) reveals the fact that for any algebraic function, a Riemann surface exists which makes this function single-valued under analytic continuation. We have previously seen, however, that under analytic continuation an abstract construction of Weierstrass can be performed (see Sec. 3-1) for such a manifold, so one may wonder what is accomplished by this elegant artifact of Riemann.

In a succession of major conceptualizations, it became clear from 1850 to 1900 that the abstract geometric manifold was the real starting point of investigation and not the answer. From the geometric manifold, there arises an algebraic field, and with it, conformal equivalence of Riemann surfaces acquires the algebraic meaning of equivalent fields.

6-1 Function Field on $\mathfrak{M}$

In the last chapter we started with $w(z)$ the root function(s) of an irreducible algebraic equation of degree n, namely

$$\Phi(z,w) = w^n + \phi_1(z)w^{n-1} + \cdots + \phi_n(z) = 0 \tag{6-1}$$

We have seen how to construct an algebraic manifold $\mathfrak{M}$ for $w(z)$. We ask, next: To what extent does $\mathfrak{M}$ correspond uniquely to $w(z)$? For instance,

$$w_1(z) = w(z) + R(z) \qquad R(z) \text{ rational}$$

will evidently also determine the same $\mathfrak{M}$ (indeed the same monodromy substitutions). On the other hand,

$$w_2(z) = w^2(z)$$

will not determine the same $\mathfrak{M}$ if $w(z)$ is a square root. Starting on another tack, we can regard the equation $\Phi(z,w) = 0$ as determining $z(w)$, hence we obtain a Riemann surface for $z(w)$ over the w sphere from the same equation. Where shall we find a unique correspondence? *The answer* *is that there is a unique correspondence between the algebraic function field generated by $\Phi(z,w)$ and the class of Riemann manifolds equivalent to $\mathfrak{M}$ under mappings meromorphic on $\mathfrak{M}$.* Clearly such a correspondence belies any artificial separation of algebra from analysis; it is therefore easy to appreciate its central importance.

We define the *function field* $\mathfrak{F}(\Phi)$ of the irreducible equation (6-1) as the set of rational functions $R(z,w)$. The set is viewed abstractly. For example, another equation $\Psi(z,w)$ [say, the one defining

$$w_1(z) = w(z) + R(z)]$$

could determine the same field; that is, $\mathfrak{F}(\Phi) = \mathfrak{F}(\Psi)$. We define† $\mathfrak{M}(\Phi)$ as the (algebraic) Riemann surface of $w(z)$, and again we could have $\mathfrak{M}(\Phi) = \mathfrak{M}(\Psi)$.

A more conventional notation for $\mathfrak{F}(\Phi)$ is actually

$$\mathfrak{F}(\Phi) = C(z,w) \qquad \text{where } \Phi(z,w) = 0$$

Here C represents the complex numbers, and z and w are adjoined to C together with all quantities obtainable by rational operations on z and w [related by $\Phi(z,w) = 0$].

The individual rational functions are conventionally denoted by $R(z,w)$ so that we can also describe the field as a set

$$\mathfrak{F}(\Phi) = \{R(z,w)\} \qquad \text{where } \Phi(z,w) = 0$$

THEOREM 6-1 *The function field $\mathfrak{F}(\Phi)$ is the same as the set of functions meromorphic on $\mathfrak{M}(\Phi)$.*

PROOF It is trivial that $R(z,w)$ is meromorphic [as is $w(z)$] on $\mathfrak{M}$. For the nontrivial part, let $F = F(z,w)$ be meromorphic on $\mathfrak{M}$. Then, using the slit $\mathfrak{C}$ in Fig. 5-6, we have $\mathfrak{M}$ rendered into n sheets and the functions $F_1, \ldots, F_n$ defined uniquely on each sheet and labeled so that $F_i = F(z,w_i)$. Then we write down n equations in $\mathfrak{R}^*$ (complement of $\mathfrak{C}$)

$$\begin{aligned} F_1 &= A_0 + A_1 w_1 + \cdots + A_{n-1} w_1^{n-1} \\ &\cdots\cdots\cdots\cdots\cdots \\ F_n &= A_0 + A_1 w_n + \cdots + A_{n-1} w_n^{n-1} \end{aligned} \tag{6-2}$$

in the n unknowns $A_0, A_1, \ldots, A_{n-1}$. We ask at present only that the $A_i(z)$ be defined on $\mathfrak{R}^*$.

According to a well-known theorem of Vandermonde, the determinant of w_i^j $(i = 1, 2, \ldots, n; j = 0, \ldots, n-1)$ has value $\pm \prod_{j>J} (w_j - w_J)$ over $n(n-1)/2$ distinct pairs $(j \neq J)$. Thus, as long as $w_i \neq w_j$ (in $\mathfrak{R}^*$),

† Actually, the construction of Sec. 5-4 is not unique (e.g., in drawing $\mathfrak{C}$, in labeling and triangulating the sheets, we can introduce arbitrariness). Nevertheless, we shall ultimately speak of equivalence classes so that the nonuniqueness of $\mathfrak{M}(\Phi)$ can be ignored for convenience of notation.

we can solve for the functions $A_i(z)$ in $\mathcal{R}^*$. The various relationships (6-2) are valid when extended across $\mathcal{C}$ (only the equations are permuted by the monodromy substitutions). Hence $A_i(z)$ are defined everywhere but at $\{b_t\}$, and these functions are single-valued. Also, the behavior of w_i and F_i is given by power series at zeros, poles, etc., so that the $A_i(z)$ behave (at worst) like poles. Therefore, the $A_i(z)$ are rational and

$$F(z,w) = A_0 + A_1 w + \cdots + A_{n-1} w^{n-1} \tag{6-3}$$

Q.E.D.

THEOREM 6-2 *The function $F(z,w)$, meromorphic on $\mathfrak{M}$, satisfies an algebraic equation*

$$G(z,F) = F^n + f_1(z)F^{n-1} + \cdots + f_n(z) = 0 \tag{6-4}$$

where $f_i(z)$ are rational functions of z. This algebraic equation is either irreducible, or else $G(z,F) = H(z,F)^m$ where

$$H(z,F) = F^k + h_1(z)F^{k-1} + \cdots + h_k(z) \tag{6-5}$$

and $h_i(z)$ are again rational and (naturally) $n = mk$.

PROOF All we need do is to define $f_i(z)$ as the elementary symmetric functions of $F_i = F(z,w_i(z))$. For instance,

$$-f_1 = F_1 + \cdots + F_n \qquad +f_2 = F_1F_2 + F_2F_3 + \cdots + F_1F_n$$

By the usual recourse to power series behavior, the single-valued functions $f_i(z)$ are seen to be rational. This establishes (6-4).

To determine the irreducibility of (6-4) consider the set of function elements formed by $F_i = F(z,w_i)$ over $\mathfrak{M}$. The set is attained by analytic continuation regardless of whether we start with F_1, F_2, etc., since w_1, w_2, etc., lie on connected sheets. Let us suppose that in $\mathcal{R}^*$ (on the z plane) we find that k $(\leq n)$ values, say,

$$F(z,w_1), \ldots, F(z,w_k) \tag{6-6}$$

are sufficient (and nonsuperfluous) in producing all the possible function elements by analytic continuation. This means that any two functions in (6-6) differ somewhere in $\mathcal{R}^*$, and any other $F(z,w_K)$ with $K > k$ (when $k < n$) would be a repetition of an item in (6-6). Then starting with $F(z,w_{k+1})$, we must obtain exactly k others by analytic continuation

$$F(z,w_{k+1}), \ldots, F(z,w_{2k}) \tag{6-7}$$

These are repetitions of (6-6) since w_{k+1} has the same range as w_1 (as just noted). If (6-6) and (6-7) do not exhaust all n of the F_i, we continue until we have m $(=n/k)$ rows like (6-6) and (6-7). This proves (6-5).

Q.E.D.

COROLLARY 1 (*Liouville's theorem*) *If $F(z,w)$ has no poles on $\mathfrak{M}$, it is constant.*

For proof, refer to the symmetric functions in (6-6).

COROLLARY 2 *We can define a function $F(z,w)$ on $\mathfrak{M}$ uniquely by one of the following:*

(*a*) *giving the principal terms of its Laurent expansion in powers of the local uniformizing parameter ζ at each pole on $\mathfrak{M}$*

(*b*) *giving the order of the zero or pole in the local uniformizing parameter ζ wherever they occur on the Riemann surface $\mathfrak{M}$*

This result follows from Corollary 1.

In method (*b*) we can use *divisor* notation. Let P denote *symbolically* a point of the Riemann surface. Then if $F(z,w) = \zeta^s \cdot \phi(z,w)$ where $\phi(z,w) \neq 0$ or ∞ near P, we say that F has the divisor P^s if P is finite or P^{-s} if P is infinite. In practice, method (*b*) is more convenient than (*a*), but we defer details to Chap. 14 for an interesting reason, which we can paraphrase for the present as follows:

The above *uniqueness* statements are easy. Yet they are deceptive because the *existence* theorems are generally drastically restricted (far more than for elliptic functions)! It is *not true* that every assignment of principal parts is realized by a function in $\mathfrak{F}$; in fact, it is not even possible to assign the *location of poles* arbitrarily. Existence theorems associated with Corollary 2 are very difficult and very unsatisfactory by comparison with the Riemann sphere (Sec. 2-4) and the period parallelogram (Sec. 4-8). Method (*b*) will still be easier to handle.

COROLLARY 3 *If the equation* (6-4) *for $F(z,w)$, namely $G(z,F) = 0$, is irreducible, then z and $F(z,w)$ determine the same function field as z and w or (variously)*

(6-8*a*) $$\mathfrak{F}(\Phi) = \mathfrak{F}(G)$$

(6-8*b*) $$C(z,w) = C(z,F(z,w))$$

(6-8*c*) $$\{R(z,w)\} = \{R(z,F(z,w))\}$$

PROOF We need only show that w is rational in z and F [since clearly $F(z,w)$ is necessarily rational in z and w]. But this is a consequence of the fact that the n values of F_i are different (except at isolated points). Thus, we can write $w_i = A_0 + A_1F_i + \cdots + A_{n-1}F_i^{n-1}$ and solve for *rational* $A_j(z)$ as we did in Theorem 6-1. Q.E.D.

If F and z determine $\mathfrak{F}$ [that is, $\mathfrak{F}(\Phi) = \mathfrak{F}(G)$, etc.], then F is called a *primitive element of $\mathfrak{F}$ over z.*

Ultimately we shall see that *for any $F(z,w)$ primitive or not* (*as long as it is not constant*), *we can find another*† *function $Z = F_1(z,w)$ in $\mathfrak{F}(\Phi)$*

† The function Z might be the same as F.

such that F and Z together determine the function field $\mathfrak{F}(\Phi)$. This is actually one of the more advanced results (see Lemma 13-6 of Sec. 13-2). Assuming this result, we see that the function field $\mathfrak{F}$ can be visualized in "invariant" fashion in terms of all pairs (Z,W) for which the polynomial $\Psi(Z,W) = 0$ is an algebraic equation in W with rational coefficients in Z and degree N (not easily related to n, the original degree of Φ).

Thus, any two functions W_1, W_2 in $\mathfrak{F}$ determine a field $C(W_1,W_2)$ which equals $\mathfrak{F}$ or a subfield of $\mathfrak{F}$. In either case, W_1 and W_2 are algebraically related. All we need do is use the fact that Z_1 exists in $\mathfrak{F}$ such that $\mathfrak{F} = C(W_1,Z_1)$. Since $W_2 \in \mathfrak{F}$, then we apply Theorem 6-2, with W_1 playing the role of z and Z_1 playing the role of w. Thus, for some irreducible function, $\Phi_1(W_1,W_2) = 0$.

Finally, consider an arbitrary function Z meromorphic on $\mathfrak{M}$. It maps $\mathfrak{M}$ onto a manifold $\mathfrak{M}'$ over the Z sphere. The manifold $\mathfrak{M}'$ must serve as a Riemann surface for the primitive element $W(Z)$ which together with Z determines $\mathfrak{F}$. This is almost a matter of definition: All functions meromorphic on $\mathfrak{M}'$ (hence on $\mathfrak{M}$) are determined by W and Z (as well as w and z). Indeed, if W is of degree N in Z, the manifold $\mathfrak{M}'$ is N-sheeted over the Z sphere. The pairs (z,w) and (Z,W) are related by rational transformations and inverses since $\{R(z,w)\} = \{R(Z,W)\}$. Thus,

$$z = R_1(Z,W) \quad w = R_2(Z,W) \qquad \text{and} \qquad Z = R_3(z,w) \quad W = R_4(z,w) \tag{6-9}$$

We say that $\mathfrak{M}$ and $\mathfrak{M}'$ are "birationally equivalent" to emphasize the algebraic character of the conformal equivalence relationship.

Thus, the equivalent representations of the function field $\mathfrak{F}$ [originally introduced by $\Phi(z,w) = 0$] are the same as equivalent Riemann surfaces† [originally introduced for $w(z)$].

6-2 Compact Manifolds Are Algebraic

We shall define a compact Riemann manifold $\mathfrak{M}$ formally in the natural way. The use of a local uniformizing parameter ζ_P in $\mathfrak{R}_P$ (for each point) provides a system of neighborhoods about each point, namely, the neighborhoods of $\zeta_P = 0$ in $\mathfrak{R}_P$. Thus, we can speak of limit points of sets and sequences on $\mathfrak{M}$.

A *compact* manifold $\mathfrak{M}$ is one for which every infinite sequence has a limit point. *A triangulated manifold* $\mathfrak{M}$ *is now seen to be compact if and only if every boundary of a triangle lies in* $\mathfrak{M}$ *and if there are only a finite number of triangles in* $\mathfrak{M}$. It is clear that the boundaries of the triangles must be included. The rest of the criterion is seen by postulating an

† In the case of the Riemann sphere the field $\mathfrak{F}$ is determined by the triviality $\Phi \equiv w - z\ (= 0)$, so $\mathfrak{F}$ is the collection of rational functions of z (alone). Compare Theorem 13-2.

infinitude of triangles $\Delta_1, \Delta_2, \ldots$. If we take one point P_i interior to each Δ_i, we obtain a sequence which cannot have a limit point. Indeed, any such limit P_0 would have to be in some Δ_0 (possibly a Δ_i), but there could not be an infinitude of elements of $\{P_i\}$ arbitrarily close to P_0 [since there is *at most* one point of $\{P_i\}$ in Δ_0 and each triangle adjacent to Δ_0].

Our previous illustrations of compact manifolds were the z sphere, algebraic Riemann manifolds (surfaces), and the period parallelogram or torus (which was seen to be equivalent to the Riemann surface for $y^2 =$ cubic in z). A noncompact manifold might, for example, be a proper subregion of one of these manifolds.

RIEMANN'S FUNDAMENTAL THEOREM (1852) *Every irreducible algebraic function determines (under analytic continuation) some compact Riemann manifold, and conversely, every compact Riemann manifold is equivalent to the Riemann surface determined by an appropriate algebraic function.*

The nature of this theorem is that it presupposes the so-called "abstract" Riemann manifold, defined by neighborhood systems or triangulations with no reference to a function which shall be meromorphic on it. The abstract n-sheeted manifold, for instance, consists of an assignment of branch points

$$b_1, b_2, \ldots, b_r \ (= \infty) \tag{6-10}$$

and monodromy substitutions for continuation clockwise across $b_t b_{t+1}$

$$\mathfrak{S}_t = \begin{pmatrix} 1 & 2 & \cdots & n \\ a_1 & a_2 & \cdots & a_n \end{pmatrix} \qquad t = 1, 2, \ldots, r-1 \tag{6-11}$$

According to Riemann's theorem, with the data (6-10) and (6-11) assigned in advance, we can find an algebraic function $w(z)$ of degree n whose n values determine the n-sheeted Riemann surface thus described.

The most remarkable feature of Riemann's theorem is not even the depth of the proof, which will occupy practically the remainder of the text! The most remarkable feature of the theorem is that at one time it represented almost all of mathematics in microcosm. A researcher in almost any field from algebra to hydrodynamics was *compelled* to take interest in it. Indeed, some of the "wordly" instincts of the hydrodynamicist (as we shall see Chap. 9) were as crucial in this theory as the refined asceticism of the algebraist. It took from 1850 to 1900 for mathematicians to be satisfied that they could prove (and also understand) this theorem. In the meantime, some of the major modern fields of mathematics attained their identity mainly as a result of this one problem. This list of fields includes such unlikely companions as calculus of variations, most facets of topology, function spaces, algebraic geometry, special functions, and major subfields of applied mathematics and number theory!

We shall gain some insight into the significance of this bygone euphoric state of affairs by two observations. The direct part of the theorem, the construction of a geometric object from an equation, was primarily algebraic (see Sec. 5-4). Yet the converse part of the theorem, the construction of an equation from the geometric object, will be basically "applied" in attitude.

We shall content ourselves for now with one last observation on the fundamental theorem, namely that the full force of the fundamental theorem requires the idea of the Riemann manifold in the form of triangulation over the u_i planes, *not* over the z sphere. This was first brought out clearly by Klein (1890), and it led to the "attitude of abstraction" in the modern definition of Riemann manifold. Klein showed many transcendental functions derive an existence proof in peremptory fashion from the concept of a triangulated manifold, and such functions need not be known in closed form. We conclude with some famous illustrations (in Sec. 6-3).

6-3 Modular Functions

In Chap. 4 we saw that any field $\mathfrak{F}$ of elliptic functions is precisely the field of some equation

$$\Phi = w^2 - P(z) = 0 \qquad [P(z) = \text{cubic with distinct roots}] \tag{6-12}$$

The equation (6-12), however, does not belong uniquely to $\mathfrak{F}$ [in Sec. 6-1 we saw how we could write $\mathfrak{F} = \mathfrak{F}(\Phi) = \mathfrak{F}(\Psi) = \cdots$ for various polynomials $\Phi, \Psi, \ldots$]. In this section we shall discuss the complex quantity J [defined in (6-16)] which is *the* invariant of $\mathfrak{F}$, as we shall see in Theorem 6-3.

To recapitulate, the field $\mathfrak{F}$ is determined by Weierstrass' normal form (Sec. 4-5) starting with the double-period module $\mathbf{\Omega}$, where

$$\mathbf{\Omega} = \{\Omega_{m,n}\} = \{2\omega_1 m + 2\omega_2 n\} \qquad m,n \text{ integers} \tag{6-13}$$

$$\operatorname{Im} \frac{\omega_2}{\omega_1} > 0 \tag{6-14}$$

Once we are given this field $\mathfrak{F}$, can we recover $\mathbf{\Omega}$? The field, of course, is given abstractly as $\mathfrak{F} = R(z,w) = R(Z,W) = \cdots$ for various pairs of complex variables (z,w), (Z,W), etc., so in theory, we cannot even pick out the z variable. Nevertheless, the abelian integrals (and hence) differentials of the first kind are well defined. We know $du = dz/w$ is such a variable, but then is it unique? Fortunately, all such differentials are proportional† so that any other period system $\mathbf{\Omega}'$ would belong to

† This is a very general result proved in Chap. 13 but also verified in Exercise 5, Sec. 8-5, for the case at hand.

$du' = \rho\, du$ (ρ = const $\neq 0$). Thus

(6-15) $$\mathbf{\Omega}' = \rho\mathbf{\Omega} = \{\rho\Omega_{m,n}\} \qquad \rho\Omega_{m,n} = (2\omega_1\rho)m + (2\omega_2\rho)n$$

We say two modules $\mathbf{\Omega}$, $\mathbf{\Omega}'$ have the *same shape* when (6-15) holds. Thus the *sole invariant of an elliptic function field is the shape of the double-period module.*

Now for a module $\mathbf{\Omega}$, (in Sec. 4-6) we defined g_2 and g_3 in terms of sums over $\mathbf{\Omega}$ and from these we defined $J = J(\mathbf{\Omega})$ as

(6-16) $$J = \frac{g_2{}^3}{g_2{}^3 - 27g_3{}^2}$$

Actually J is not only a function of the module $\mathbf{\Omega}$ itself, it is also a function of the aggregate of modules $\rho\mathbf{\Omega}$ since g_2 and g_3 are homogeneous in ω_1 and ω_2 of degrees -4 and -6, respectively. Thus, $J(\rho\mathbf{\Omega}) = J(\mathbf{\Omega})$, so J is dependent only on the *shape* of the module $\mathbf{\Omega}$.

THEOREM 6-3 *The invariant J (just defined for an elliptic function field $\mathfrak{F}$) corresponds biuniquely to $\mathfrak{F}$; therefore, to every value of J in the finite J plane there corresponds a unique field $\mathfrak{F}$ (and a unique equivalence class of Riemann surfaces).*

The value $J = J(\omega_2/\omega_1)$ is called the *module*† *of $\mathfrak{F}$ (or of the surface $\mathfrak{M}$).* In one direction, Theorem 6-3 is evident, namely, two period parallelograms with different modules J have different shapes, hence they belong to different fields (and, according to Sec. 6-1, to conformally inequivalent Riemann manifolds). Furthermore, we can prescribe J_0 as an *arbitrary* (finite) complex number and then determine a parallelogram $(2\omega_1, 2\omega_2)$ so that $J(\omega_2/\omega_1) = J_0$ (see Sec. 4-6). What remains to be proved is that two period parallelograms with the same value of $J(\omega_2/\omega_1)$ belong to modules $\mathbf{\Omega}$ which have the same shape.

LEMMA 6-1 *Let the module $\mathbf{\Omega}$ be given by* (6-13), *and let the module $\mathbf{\Omega}'$ be given as $\{\Omega'_{m,n}\}$ where $\Omega'_{m,n} = 2\omega'_1 m + 2\omega'_2 n$. Then $\mathbf{\Omega}$ and $\mathbf{\Omega}'$ have the same shape precisely when*

(6-17) $$\frac{\omega_2}{\omega_1} = \frac{a(\omega'_2/\omega'_1) + b}{c(\omega'_2/\omega'_1) + d} \qquad \textit{(integers } a, b, c, d \textit{ satisfy } ad - bc = 1)$$

assuming we label ω'_1 and ω'_2 *in such an order*‡ *that* $\mathrm{Im}\,(\omega'_2/\omega'_1) > 0$.

† The use of the word "module" in the context of $\mathbf{\Omega}$ and J may seem confusing at first but, generically speaking, a module connotes "an aggregate of equivalent things." Thus, we speak of "equivalent fields $\mathfrak{F}$ or Riemann manifolds $\mathfrak{M}$." We also spoke of "equivalent parts of the u plane" as those differing by an element of $\mathbf{\Omega}$; this constitutes the double-period structure $u \equiv u + \Omega_{m,n}$.

‡ The reader should recall that if $\mathrm{Im}\,(\omega'_2/\omega'_1) < 0$, then $\mathrm{Im}\,(\omega'_1/\omega'_2) > 0$. Also, if $\mathrm{Im}\,\tau > 0$, then $\mathrm{Im}\,[(\alpha\tau + \beta)/(\gamma\tau + \delta)] > 0$ for any real numbers α, β, γ, δ, not necessarily integers, as long as $\alpha\delta = \beta\gamma > 0$.

PROOF We have $\mathbf{\Omega} = \lambda\mathbf{\Omega}'$ precisely when the elements $2\omega_1'\lambda$, $2\omega_2'\lambda$ in $\lambda\mathbf{\Omega}'$ can be used as a basis of $\mathbf{\Omega}$. They might not be $2\omega_1$ and $2\omega_2$, but by a well-known theorem of algebra, $2\omega_1'\lambda$ and $2\omega_2'\lambda$ are basis elements of $\mathbf{\Omega}$ precisely when integers a, b, c, d exist for which

$$\begin{aligned} 2\omega_1 &= \lambda(2\omega_1'd + 2\omega_2'c) \\ 2\omega_2 &= \lambda(2\omega_1'b + 2\omega_2'a) \end{aligned} \tag{6-18}$$

where $ad - bc = \pm 1$ [actually only $+1$ occurs by the sign of Im (ω_2/ω_1)].

Q.E.D.

We next return to consideration of $\mathfrak{M}$ the fundamental domain for $J(\tau)$, as drawn in Fig. 4-5 with suitable boundary identifications,

$$\mathfrak{M}: \quad \begin{cases} -\frac{1}{2} \leq \operatorname{Re}\tau < \frac{1}{2} \\ 1 \leq |\tau| \qquad \text{(equality only if } \operatorname{Re}\tau \leq 0) \end{cases} \tag{6-19}$$

We saw that $J(\tau)$ maps $\mathfrak{M}$ onto the whole finite J plane uniquely.

LEMMA 6-2 *If τ_2 is any complex number in the upper half τ plane, then there exist integers a, b, c, d such that a unique τ_1 exists given by*

$$\tau_1 = \frac{a\tau_2 + b}{c\tau_2 + d} \qquad ad - bc = 1 \tag{6-20}$$

with $\tau_1 \in \mathfrak{M}$.

The proof is given as Exercises 1 and 2. In the course of the proof we note that τ_1 is unique (allowing boundary identifications in $\mathfrak{M}$) while the integers a, b, c, d (which produce τ_1 from τ_2) need not be unique for special τ_2, such as i or $(-1 + i\sqrt{3})/2$. Thus Theorem 6-3 reduces to the next result.

LEMMA 6-3 *Let τ_1 and τ_2 lie in the upper half plane. Then a necessary and sufficient condition for* (6-20) *is that $J(\tau_1) = J(\tau_2)$.*

PROOF The necessity follows from Lemma 6-1 if we set $\tau_1 = \omega_2/\omega_1$, $\tau_2 = \omega_2'/\omega_1'$. For the sufficiency, assume $J(\tau_1) = J(\tau_2)$. Then a $\tau_* \in \mathfrak{M}$ satisfies $J(\tau_*) = J(\tau_1) = J(\tau_2)$ and (by Lemma 6-2),

$$\tau_* = \frac{a_1\tau_1 + b_1}{c_1\tau_1 + d_1} \qquad \tau_* = \frac{a_2\tau_2 + b_2}{c_2\tau_2 + d_2}$$

where $a_1, \ldots, d_2$ are eight integers satisfying

$$a_1d_1 - b_1c_1 = a_2d_2 - b_2c_2 = 1$$

We can, of course, eliminate τ_* and obtain a relation of type (6-20)

Q.E.D.

SUBGROUPS AND SUPERDOMAINS

The domain $\mathfrak{M}$ (often called "Klein's domain") is quite significant because it enjoys an existence essentially independent of our special knowledge of $J(\tau)$. The property of Lemma 6-2 is the reason $\mathfrak{M}$ is called a "fundamental domain" for the so-called "modular group" consisting of the "unimodular" transformations

$$\tau \to \frac{a\tau + b}{c\tau + d} \tag{6-21}$$

where a, b, c, d are integers for which $ad - bc = 1$. The group operations (6-21) set up equivalence classes of τ on the upper half τ plane (e.g., in Lemma 6-2, τ_1 and τ_2 are equivalent under the modular group). *Only one element of each class lies in* $\mathfrak{M}$.

In similar fashion the period parallelogram is the fundamental domain of the group of translations

$$u \to u + \Omega_{m,n} \tag{6-22}$$

Klein (and Poincaré) realized that the *geometric entity* consisting of the fundamental domain need not "exist for the analyst" who deals with functions such as $J(\tau)$ and $\wp(u)$ which are invariant under the group. Only the *group* need exist. It becomes, therefore, increasingly difficult to distinguish analysis, geometry, algebra, and (in due time) topology.

We consider illustrations of the following general type: The equation (6-21) or (6-22) constitutes infinite groups $\{S(z)\}$ of transformations on the z sphere, written

$$\mathcal{G}: \quad z \to S(z) \tag{6-23}$$

The group $\mathcal{G}$ determines a fundamental domain (such as Klein's domain or the period parallelogram) designated by $\mathfrak{D}$ on the z sphere. (Here z, of course, denotes the familiar τ or u variables.) Let us now consider a subgroup $\mathcal{G}^* \subseteq \mathcal{G}$ of *finite index t*. This means that every transformation $S(z)$ can be put in the form

$$S(z) = S^*(S_i(z)) \tag{6-24}$$

where $S^*(z) \in \mathcal{G}^*$ and $S_i(z)$ is an element of the (minimal) finite set of "representatives"

$$S_1(z), \ . \ . \ . \ , S_t(z) \tag{6-25a}$$

Actually, $S_1(z)$ is usually the identity $(= z)$. [We must be concerned with the order, since $S(S'(z)) \neq S'(S(z))$ in noncommutative groups such as the modular group.] We then can easily see that the set

$$\mathfrak{D}^* = S_1^{-1}(\mathfrak{D}) \cup S_2^{-1}(\mathfrak{D}) \cup \cdots \cup S_t^{-1}(\mathfrak{D}) \tag{6-25b}$$

constitutes a fundamental domain of $\mathcal{G}^*$ if we can arrange things so as to make $\mathfrak{D}^*$ a connected set (this is never difficult).

Assume, now, that $\mathfrak{D}$ (and $\mathfrak{D}^*$) are compact; this condition again will be present in our particular applications. *Then the meromorphic* *functions on* $\mathfrak{D}$ *and* $\mathfrak{D}^*$ *will be algebraic function fields* according to Riemann's fundamental existence theorem! Furthermore, every function meromorphic on $\mathfrak{D}$ is (trivially) meromorphic on $\mathfrak{D}^*$. Thus the function field on $\mathfrak{D}$ is a subfield of the function field† on $\mathfrak{D}^*$. Conversely, however, a function $g(z)$ meromorphic on $\mathfrak{D}^*$ is not necessarily meromorphic on $\mathfrak{D}$ since it may be multivalued. Indeed, each point of $\mathfrak{D}$ corresponds to t points of $\mathfrak{D}^*$ where $g(z)$ may take t different values.

In other words, the purely algebraic concept of subgroup and subfield is tied in with "superdomain"—as a Riemann manifold.

This idea was a crucial step in the process of abstraction of a Riemann manifold from the n-sheeted Riemann surface to the idea of manifold as a collection of neighborhoods with analytic structure. We shall conclude these vague remarks with a brief statement of two very famous examples.

DIVISION OF PERIODS OF ELLIPTIC FUNCTIONS

Consider $\mathcal{G}$ the translation group [see (6-22)] of the module $\mathbf{\Omega}$ and the fundamental domain consisting of the period parallelogram $\mathfrak{D}$ of the vectors $2\omega_1$, $2\omega_2$. Let $\mathcal{G}^*$ be the translation group

(6-26) $$u \to u + N\Omega_{m,n} \qquad \Omega_{m,n} \in \Omega$$

for some fixed integer N. This is a subgroup of $\mathcal{G}$ of index N^2; indeed we take

(6-27) $$S_{jk}(z) = u + j(2\omega_1) + k(2\omega_2) \qquad 0 \le j,\, k \le N - 1$$

The fundamental domain $\mathfrak{D}^*$ for $\mathcal{G}^*$ consists of N^2 translates of $\mathfrak{D}$ by the operations shown in (6-27). Thus $\mathfrak{D}^*$ consists of a "super parallelogram" $\mathfrak{D}^*(\supset\mathfrak{D})$ determined by vectors $2\omega_1 N$, $2\omega_2 N$ (for example, $\mathfrak{D}^*$ has N^2 times the area of $\mathfrak{D}$).

Now let $f(u)$ be meromorphic on $\mathfrak{D}$ [for example, $f(u) = \wp(u)$]. Then consider $g(u) = f(u/N)$. Certainly, $g(u)$ need not be invariant (meromorphic) on $\mathfrak{D}$, but it is invariant on $\mathfrak{D}^*$

$$\left[g(u + N\Omega_{m,n}) = f\left(\frac{u}{N} + \Omega_{m,n}\right) = g(u)\right]$$

Indeed, the *set* of N^2 functions

(6-28) $$g_{jk}(u) = g\left(\frac{u}{N} + \frac{2\omega_1 j}{N} + \frac{2\omega_2 k}{N}\right) \qquad 0 \le j,k \le N - 1$$

† It could be isomorphic to the original field even if it has fewer functions; e.g., the rational functions in z^2 are isomorphic to the more inclusive set of rational functions in z.

is *rearranged* by transformations of $\mathcal{G}$. These functions have symmetric functions which *are* meromorphic on $\mathfrak{D}$, hence we have N^2 roots of an algebraic equation connecting $g(u)$ and $f(u)$ (in complex coefficients).

This so-called "period division" was studied widely around 1800, and it provided some of the concrete background of Galois' and Abel's abstract approach to the solution of equations by groups. A more exotic form of this problem is *complex multiplication* (or "division") where, essentially, N is treated as complex. The first investigations were the work of Abel (and we always offer just the simplest example).

Consider a period module $\mathbf{\Omega}$ (and group $\mathcal{G}$) in which $\omega_2/\omega_1 = 2^{1/2}i$. Then the period parallelogram $\mathfrak{D}$ for u is a rectangle whose sides have ratio $\sqrt{2}$. Let $f(u)$ be meromorphic on $\mathfrak{D}$. Then consider

$$g(u) = f\left(\frac{u}{2^{1/2}i}\right)$$

Clearly, $g(u)$ is invariant under the group $\mathcal{G}^*$, $u \to u + 2^{1/2}i\Omega_{m,n}$. Now *because* $\omega_2/\omega_1 = 2^{1/2}i$, it follows that the group $\mathcal{G}^*$ is a subgroup of index 2. To see this, note

$$\Omega_{m,n} = 2\omega_1 m + 2\omega_2 n$$
$$\Omega^*_{m,n} = 2^{1/2}i\Omega_{m,n} = \Omega_{-2n,m}$$

Thus every $\Omega \in \mathbf{\Omega}$ is equal to a value of $\Omega^*_{m,n}$ or $\Omega^*_{m,n} + 2\omega_1$, and we can now visualize the fundamental domain $\mathfrak{D}^*$ as consisting of two replicas of $\mathfrak{D}$ standing side by side. The function $g(u)$ is a quadratic function over the elliptic function field of $\wp(u)$ and $\wp'(u)$ (referred always to the module $\mathbf{\Omega}$). Further exploration of this problem actually makes algebraic number theory a full-fledged partner in the emergent analysis-algebra-geometry-topology synthesis.

THE MODULAR SUBGROUP OF INDEX 6

Let us return to the modular group $\mathcal{G}$ with the Klein domain $\mathfrak{M}$ over the τ plane. Consider the subgroup $\mathcal{G}^*$ of $\mathcal{G}$ consisting of the substitutions

(6-29) $$\mathcal{G}^*:\quad \tau \to \frac{a\tau + b}{c\tau + d} = S^*(\tau) \qquad ad - bc = 1$$

where a and d are odd while b and c are even.

We use the number-theoretic "congruence" terminology $a \equiv b \pmod q$ to mean q divides $(b - a)$. Although we make no deep use of number theory here, we find the designations convenient. For example, $\mathcal{G}^*$ is the "principal congruence subgroup of $\mathcal{G}$ modulo 2." Thus (6-29) can be written as $a\tau + b \equiv 1\tau + 0$, $c\tau + d \equiv 0 \cdot \tau + 1$, or symbolically, (6-29) is written as

(6-30) $$S^*(\tau) \equiv \tau \pmod 2$$

We can easily see that $\mathcal{G}^*$ is of index 6. For instance, if we consider a, b, c, d as odd or even and if we recall that $ad - bc = 1$, then any transformation of $\mathcal{G}$ is one of six types:

$$S(\tau) = \frac{a\tau + b}{c\tau + d} \equiv S_1(\tau), S_2(\tau), \ldots, S_6(\tau) \tag{6-31}$$

where $S_1(\tau) = \tau \qquad S_2(\tau) = \tau + 1 \qquad S_3(\tau) = -\dfrac{(\tau + 1)}{\tau}$

$$S_4(\tau) = -\frac{1}{(\tau + 1)} \qquad S_5(\tau) = -\frac{1}{\tau} \qquad S_6(\tau) = \frac{\tau}{(1 + \tau)}$$

(see Exercise 3), and $S(\tau) = S^*(S_k(\tau))$ for some $k(S^* \in \mathcal{G}^*)$. In Fig. 6-1, we show six replicas of $\mathfrak{M}$; that is, S_k denotes $S_k(\mathfrak{M}^+)$ where $\mathfrak{M}^+$ is the half of $\mathfrak{M}$ which J maps onto the upper half J plane (see Exercise 10 in Sec. 4-6). The region $\mathfrak{M}^*$ depicted in Fig. 6-1 is the fundamental domain for $\mathcal{G}^*$.

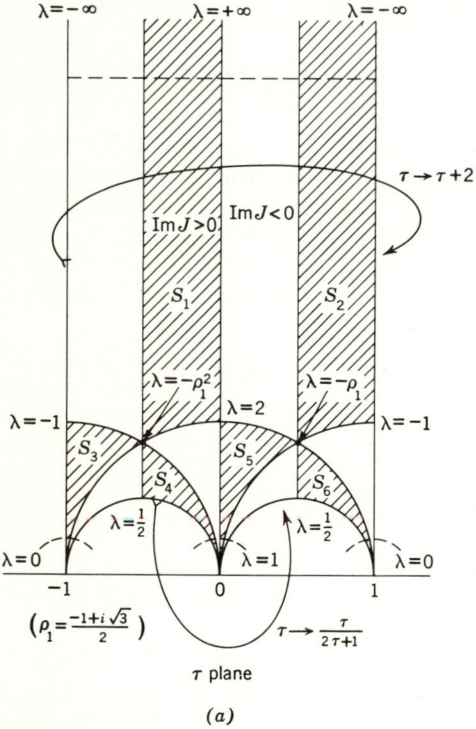

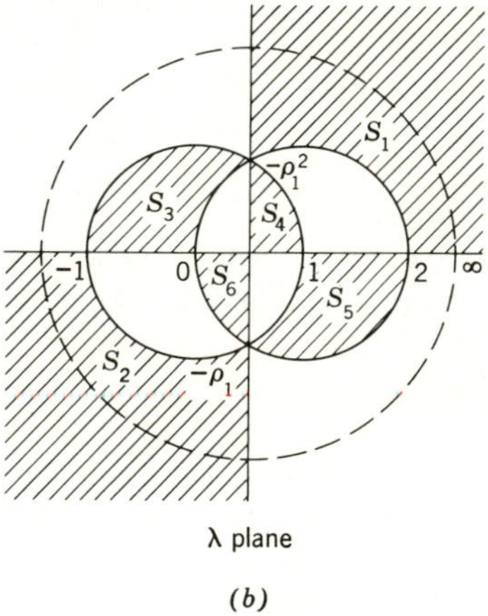

Fig. 6-1 The fundamental domain $\mathfrak{M}^*$ for λ showing (a) the τ plane with identifying operations. Dotted line and circle are explained in Exercise 7; (b) the image $\lambda(\mathfrak{M}^*)$; the images of S_k are explained in Exercise 8.

Now it happens that the classical theory of elliptic functions actually produces a function $\lambda(\tau)$ which is meromorphic over $\mathfrak{M}^*$ and which takes every value on the λ sphere exactly once for some $\tau \in \mathfrak{M}^*$. Such a function is seen in Exercise 4 to be

$$\lambda(\tau) = \frac{\wp(\omega_1) - \wp(\omega_2)}{\wp(\omega_1 + \omega_2) - \wp(\omega_2)} \qquad \tau = \frac{\omega_2}{\omega_1} \tag{6-32}$$

where $\wp(u)$ is taken relative to the usual period module

$$\mathbf{\Omega} = \{2\omega_1 m + 2\omega_2 n\}$$

Since $\lambda(\tau) = (e_1 - e_2)/(e_3 - e_2)$, one of the four cross ratios of the roots (see Lemma 4-3 and Exercise 2, Sec. 4-5), we can combine these results with Lemma 2-3 (see Sec. 2-6) to obtain the relation

$$J(\tau) = \frac{4}{27} \frac{(\lambda^2 - \lambda + 1)^3}{\lambda^2(\lambda - 1)^2} \tag{6-33}$$

This relation (6-33) is not surprising since each value of J (generally) determines *six* values of τ in $\mathfrak{M}^*$, hence six values of $\lambda(\tau)$, one in each replica of $\mathfrak{M}$. Actually, it is seen (Exercises 5 to 8) that the $\lambda(\tau)$ defined by (6-32) satisfies the mapping properties depicted in Fig. 6-1*a* and *b*.

We thus have amply illustrated the desirability of a concept of Riemann manifold which is not bound to n-sheeted surfaces but is sufficiently broad to include geometric configurations identified under groups.

This is as far as we extend the abstraction process in this text.

EXERCISES

1 Uniqueness part of Lemma 6-2: (*a*) First show that for every $\tau \in \mathfrak{M}$, it follows that $|c\tau + d| \geq 1$ for every pair of integers (c,d) not both zero. Indeed, draw the circles $|c\tau + d| = 1$, and show they lie within some circle of type $|\tau + d| = 1$, if $c > 1$. (*b*) Show that the inequality in $|c\tau + d| \geq 1$ becomes equality for $\tau \in \mathfrak{M}$ only if $c = 0$, $d = 1$ or $c = 1$, $d = 0$ (the lower boundary of $\mathfrak{M}$). (*c*) Show that if $\tau_1 \in \mathfrak{M}$ and $\tau_2 \in \mathfrak{M}$ and if (6-20) holds, then

$$\operatorname{Im} \tau_1 = \operatorname{Im} \frac{\tau_2}{|c\tau_2 + d|^2}$$

and thus $\operatorname{Im} \tau_1 \leq \operatorname{Im} \tau_2$ ($\leq \operatorname{Im} \tau_1$ by the same argument). Consider the possibilities $\tau_1 = \tau_2 + d$ or $\tau_1 = -1/\tau_2$ if $\tau_1 \in \mathfrak{M}$ and $\tau_2 \in \mathfrak{M}$.

2 (*Lagrange's reduction method*, 1773) Show that if $\tau \in \mathfrak{M}$ and $\tau = \omega_2/\omega_1$, then $\Omega = 2\omega_1$ is a nonzero vector in $\mathbf{\Omega}$ of minimal length ($|\Omega|$), and $\Omega = 2\omega_2$ is a (nonzero) vector in $\mathbf{\Omega}$ of minimal length which is not colinear with $2\omega_1$. (Actually, $2\omega_1$ can be imagined to be 1 and $2\omega_2$ can be imagined to be τ.) Show that there is ambiguity in choice of $2\omega_1$ and $2\omega_2$ (ignoring sign) when τ satisfies an equality in the definition (6-19) of $\mathfrak{M}$. [Actually, the cases $\tau = i$ and $(-1 + i\sqrt{3})/2$ are of special interest.] Conversely, for given $\mathbf{\Omega}$, form a basis of minimal vectors and show (6-19).

3 Show that if [in (6-31)] $S(\tau) \equiv S_k(\tau) \pmod 2$, we can write $S(\tau) = S^*(S_k(\tau))$ where $S^* \in \mathcal{G}^*$. For instance, when $k = 2$, then (6-31) implies $S(\tau) = (a\tau + b)/(c\tau + d) \equiv (\tau + 1)/(0\tau + 1)$; that is, a, b, d are odd and c is even. Thus we write

$S(\tau) = S^*(S_2(\tau)) = S^*(\tau + 1)$, and we obtain $S(\tau)$ in the form $(a\tau + b)/(c\tau + d) = [a(\tau + 1) + (b - a)]/[c(\tau + 1) + (d - c)]$, where a and $a - c$ are odd while $b - a$ and c are even.

4 Show that $\lambda(\tau)$ in (6-32) is invariant under a change of basis where $\omega_2' \equiv \omega_2$ and $\omega_1' \equiv \omega_1 \pmod{2\Omega}$. [This leads to $\mathcal{G}^*$ in (6-29).]

5 Show that according to (6-32), $\lambda(\tau) \to \infty$ as $\tau \to i\infty$. [*Hint:* Let $2\omega_1 = 1$ and $2\omega_2 = \tau$ and show that in $\wp(\omega_i) = \Sigma \cdots$, only the terms $\Omega_{m,n}$ with $n = 0$ are dominant.] Thus, $e_1 - e_2 \to \Sigma(\frac{1}{2} + m)^{-2} - \Sigma(\tau/2 + m)^{-2} \to \pi^2[1 - \csc^2(\pi\tau/2)]$, and $e_3 - e_2 \to \pi^2 [\sec^2(\pi\tau/2) - \csc^2(\pi\tau/2)]$ as $\tau \to i\infty$.

6 From the preceding exercise, identify the root of (6-33) as $\tau \to i\infty$ to be approximately $\lambda(\tau) = [27\, J(\tau)/4]^{1/2}$. Thus, $\lambda(\tau)$ behaves like $\exp \pi i\tau$ as $\tau \to \infty$.

7 From the preceding exercise, show $\lambda(\tau)$ assumes each value of λ once in $\mathfrak{M}^*$. [*Hint:* Consider $\int d \log\, [\lambda(\tau) - \lambda_0]/2\pi i$ about the boundary of $\mathfrak{M}^*$ as in Sec. 4-6. Isolate $\tau = 0, -1, 1, \infty$ this time.]

8 Verify the labeling of Fig. 6-1a and b (particularly the value of λ at the vertices). Note that the transformations $\lambda \to \lambda,\ 1 - \lambda,\ 1/\lambda,\ 1 - 1/\lambda,\ 1/(1 - \lambda),\ \lambda/(1 - \lambda)$ effectively identify the images of S_i in Fig. 6-1b. They are the six values of the cross ratio λ under permutation. (See Exercise 3, Sec. 2-6.)

PART III
DERIVATION OF EXISTENCE THEOREMS

RETURN TO REAL VARIABLES

Now we should have reached a degree of serene confidence that situations involving Riemann manifolds over the z sphere can be expressed in terms of a single *complex* variable z. From the point of view of algebra, this is much more convenient than writing $z = x + iy$ and using the *real* variables x and y.

Yet when we try to set up the basic existence theorem for complex manifolds, we see that we must return to the use of x and y, and indeed, we must also break up complex functions (in fact differentials) into real and imaginary parts.

We develop motivations by topology before confirming the classical motivations by applied mathematics. Although this may seem flagrantly defiant of history, it is more realistic from the point of view of modern mathematics to use this order of presentation.

CHAPTER 7
TOPOLOGICAL CONSIDERATIONS

In Sec. 5-2, we saw how Riemann manifolds naturally divide themselves into conformal equivalence classes. Two manifolds which are conformally equivalent are necessarily topologically equivalent, but the converse is not true, as we shall make amply clear.† Still, a first step would consist of a *topological* classification of all finitely triangulated Riemann manifolds according to certain canonical models.

In effect, such a manifold is characterized by the "genus" p, or the number of "handles" and the number of boundary segments n which it possesses. For the analyst, however, the geometry of the model is not as significant as the algebraic structure known as "homology" [of 1-chains particularly (compare Sec. 1-6)]. From this structure, a first attempt is made at an existence theorem for differentials prescribed on a manifold.

THE TWO CANONICAL MODELS

7-1 Orientability

We return to the concept of a triangulated Riemann surface $\mathfrak{M}$ over the z sphere. Its points P consist of interior points of triangles Δ_i (in the u_i plane) and boundary points of such triangles which are *shared* by two or more triangles in the manner of (5-18), by a mapping over the z sphere. We define a region $\mathfrak{R}$ ($\subseteq \mathfrak{M}$) as a connected open set (in $\mathfrak{M}$) (with suitable modifications of the definition, for the point at ∞ of the z sphere). We define a curve by (piecewise smooth) parametrizations, $u_i(s)$ joining where the different triangles Δ_i join. The regions and curves on $\mathfrak{M}$, of course, generally extend over several of the triangles Δ_i.

No assumption is made that $\mathfrak{M}$ is compact, but the

† In Theorem (6-3), all period parallelograms (tori) are not conformally equivalent. We shall see this for different annuli in Exercise 11, Sec. 9-2, and for related plane regions in Theorem A-4.

boundary of $\mathfrak{M}$ (consisting of the "unshared" boundary points of Δ_i) is assumed to consist wholly of closed sides of Δ_i, so that no isolated boundary points† occur.

The *analytic* nature of the joining of triangles is actually important topologically for one property, orientability, which is expressed in the following result.

THEOREM 7-1 (*Orientability*) *The triangles of the Riemann manifold and their boundary segments*

$$\partial\Delta_i = \mathfrak{L}_{i1} + \mathfrak{L}_{i2} + \mathfrak{L}_{i3} \qquad \partial\Delta_I = \mathfrak{L}_{I1} + \mathfrak{L}_{I2} + \mathfrak{L}_{I3} \tag{7-1}$$

are oriented in each u_i and u_I sphere so that any arcs which are identified have canceling orientations. Thus, if $|\mathfrak{L}_{ij}|$ is identified with $|\mathfrak{L}_{IJ}|$ as an arc $|\mathfrak{L}^*|$ connecting P and Q on $\mathfrak{M}$, then one of the segments, say $\mathfrak{L}_{ij}$, will go from P to Q while the other segment, say $\mathfrak{L}_{IJ}$, will go from Q to P when the identification is made with $\mathfrak{L}^*$.

The proof is based on the observation (made in Theorem 3-7, Sec. 3-4) that the orientation of a boundary is preserved by *analytic* mappings which identify $\mathfrak{L}_{ij}$ and $\mathfrak{L}_{IJ}$ [see (5-18)]. But in these mappings, Δ_i and Δ_I are made to lie on opposite sides of the common image $\mathfrak{L}_{(iI)}$ of $\mathfrak{L}_{ij}$ and $\mathfrak{L}_{IJ}$ [see (5-19)], and clearly, since one triangle is on the left of $\mathfrak{L}_{(iI)}$ and the other is on the right, only one of the sides $\mathfrak{L}_{ij}$ and $\mathfrak{L}_{IJ}$ can have the same orientation as $\mathfrak{L}_{(iI)}$.

This result is better appreciated if we remember how we excluded the Möbius strip (Exercise 1), where the identifying mappings are not always analytic (but are possibly conjugate-analytic).

COROLLARY 1 *A triangulated Riemann manifold $\mathfrak{M}$ with a closed curve $\mathfrak{C}$ can be subdivided so that a right- and left-hand neighborhood of $\mathfrak{C}$ can be formed from some of the triangles Δ_i, and these neighborhoods are disjoint except for where they meet at $\mathfrak{C}$.*

Again, let us recall that the triangles Δ_i lie scattered over many different u_i planes, so that the desired neighborhood triangles would be similarly scattered. We first subdivide the triangles so that $\mathfrak{C}$ is now included in the triangulation (as part of $\partial\Delta_i$). We next consider the triangles Δ_i which adjoin $\mathfrak{C}$ by a point or a side (of Δ_i). They can be classified as lying on the left or right, according to the orientation of $\mathfrak{C}$. (We might have to refine the triangulation of $\mathfrak{M}$ even further, so that a triangle Δ_i cannot "wind its way around" so as to adjoin $\mathfrak{C}$ on both the left and right, but such detailed matters will be left to the reader.)

† The exclusion of isolated boundary points will hold for the rest of the text until Appendix B.

LEMMA 7-1 *A finite manifold $\mathfrak{M}$ can be equivalently triangulated so that any finite number of preassigned interior points do not lie on the boundary of any of its triangles.*

A formal proof of this result is annoying but not difficult. We recall that the matching of sides of Δ_i and Δ_I is accomplished by *analytic* functions, and therefore such mappings are valid in *neighborhoods* (not just on the arcs where $\partial\Delta_i$ and $\partial\Delta_I$ are identified). This lemma, then, is in effect like moving the boundary $|\partial\Delta_i|$ and $|\partial\Delta_I|$ a little bit, so that some point formerly on the boundary is now interior to one of the triangles. We did this in the theory of elliptic functions to keep zeros and poles off the boundary of the period parallelogram (see Lemmas 4-1 and 4-2 of Sec. 4-4).

We next apply the orientability concept to the evaluation of residues. We define the residue of a differential $w(\zeta)\,d\zeta$ within a curve $\mathcal{C}$ as

$$\frac{(\mathcal{C}, w(\zeta)\,d\zeta)}{2\pi i} = \int_{\mathcal{C}} \frac{w(\zeta)\,d\zeta}{2\pi i} \tag{7-2}$$

(assuming no singularities of w lie on $\mathcal{C}$). By definition of the differential $w(\zeta)\,d\zeta$, this is independent of the triangulation of $\mathfrak{M}$. We define the *residue at P* as

$$\frac{(\mathcal{C}, w(\zeta)\,d\zeta)}{2\pi i} = \operatorname{res} w(\zeta_P)\,d\zeta \tag{7-3}$$

where $\mathcal{C}$ is a closed curve sufficiently close to P (so that it contains no other singularities of w). We also define the *total residue* of $w(\zeta)\,d\zeta$ on $\mathfrak{M}$ (assumed to be finitely triangulated) as

$$\sum_i \frac{(\partial\Delta_i, w(\zeta)\,d\zeta)}{2\pi i} = \text{total res } w(\zeta)\,d\zeta \text{ on } \mathfrak{M} \tag{7-4}$$

where i is summed over all triangles of $\mathfrak{M}$. [Again it must be shown that this value is independent of triangulation (see Exercise 2 below).] Because of the independence of triangulation, we can always assume no singularities lie on the path of integration.

THEOREM 7-2 *The total residue of a meromorphic differential on a compact manifold is* 0.

This is so, because in (7-4), the segments of $\partial\Delta_i$ will cancel in pairs by Theorem 7-1.

COROLLARY 1 *A nonconstant meromorphic function w on a compact Riemann manifold $\mathfrak{M}$ takes each value a (including ∞) the same number of times, namely, the so-called order of w.*

The proof follows the pattern of Corollary 1 to Theorem 1-9 (Sec. 1-8) or Lemma 4-1 (Sec. 4-4). Consider the differential $d \log [w(\zeta) - a]/2\pi i$ whose total residue by Theorem 7-2 is $0 =$ zeros minus poles for the function $w(\zeta) - a$. (See Exercises 3 and 4.)

EXERCISES

1 Consider the Möbius strip as a unit square with sides identified through a 180° twist as follows: Let Δ_1 be the triangle $P_{11} = 0$, $P_{12} = i$, $P_{13} = 1 + i$; for Δ_2, take $P_{21} = 0$, $P_{22} = 1 + i$, $P_{23} = 1$. Then let the sides $(P_{11}P_{13})$ and $(P_{21}P_{22})$ be identified by the mapping $z = u_1 = u_2$ (identity). Let the sides (with the 180° twist) $(P_{23}P_{22})$ and $(P_{11}P_{12})$ be identified by $z = u_2 = 1 + i + \bar{u}_1$. Then show why the triangles Δ_1, Δ_2 *cannot* be oriented so that the images of the sides cancel under identification.

2 Show that the total residue (7-4) is independent of the triangulation of $\mathfrak{M}$. First, show that a refinement of the indicated triangulation will not affect the result.

3 Complete the proof of Corollary 1 to Theorem 7-2. In particular, prove that the number of points P where $w(P) = a$ is finite.

4 Show that the multiplicity of $w(P) = a$ at some root P on $\mathfrak{M}$ is independent of the parameter ζ under conformal equivalence of representations of $\mathfrak{M}$. (See Sec. 5-2.)

5 State and prove Liouville's theorem on a compact Riemann manifold.

7-2 *Canonical Subdivisions*

We proceed intuitively by deriving a canonical (or standardized) model of a finite Riemann manifold $\mathfrak{M}$ equivalent to $\mathfrak{M}$ under continuous (rather than analytic) transformations.

We now designate the Riemann manifold $\mathfrak{M}$ as a *Jordan manifold* if every closed curve $\mathfrak{C}$ in $\mathfrak{M}$ subdivides $\mathfrak{M}$ into two disjoint manifolds separated by $\mathfrak{C}$. Clearly the plane and sphere are such manifolds (Sec. 1-4), and hence so is every subregion of the plane or sphere (see Exercise 1). We shall ultimately show (conversely) *that every finitely triangulated Riemann manifold which is a Jordan manifold is topologically equivalent to a sphere or a plane region* (of finite connectivity). This is presented as Theorem 7-3 and Exercise 2, but we assume it for intuitive purposes now, in order to derive our models.

RECTANGULAR MODEL

Our first step is to show that we can reduce $\mathfrak{M}$, by the deletion of an even number of closed curves ($2p$), to another manifold ($\mathfrak{M}^p$) which is a Jordan manifold.

If $\mathfrak{M}$ is already a Jordan manifold, set $\mathfrak{M} = \mathfrak{M}^0$ (symbolically, $p = 0$).

Otherwise, a simple closed curve $\mathfrak{Q}_1$ exists in $\mathfrak{M}$ which fails to subdivide $\mathfrak{M}$ into two (disjoint) submanifolds. This means another curve $\mathfrak{Q}_1'$ joins both the right- and left-hand neighborhood of $\mathfrak{Q}_1$, say, at some point P_1; and we can conveniently represent $\mathfrak{Q}_1'$ as a simple closed curve crossing

$\mathcal{Q}_1$ only at P_1. We call $\mathcal{Q}_1$ and $\mathcal{Q}_1'$ *conjugate* curves and we note that their roles are interchangeable; indeed, $\mathcal{Q}_1$ is a curve joining the right- and left-hand neighborhoods of $\mathcal{Q}_1'$ at P_1. Call $\mathfrak{M}^1$ the manifold which remains if arcs $|\mathcal{Q}_1|$ and $|\mathcal{Q}_1'|$ are removed from $\mathfrak{M}$. If $\mathfrak{M}^1$ is a Jordan manifold, we stop here ($p = 1$).

Otherwise, again, a simple closed curve $\mathcal{Q}_2$ exists in $\mathfrak{M}^1$ not intersecting $\mathcal{Q}_1$ and $\mathcal{Q}_1'$ and such that another curve, $\mathcal{Q}_2'$ in $\mathfrak{M}^1$, joins the right- and left-hand side of $\mathcal{Q}_2$, crossing only at point P_2 of $\mathcal{Q}_2$. We remove $\mathcal{Q}_2$ and $\mathcal{Q}_2'$ from $\mathfrak{M}^1$ and form $\mathfrak{M}^2$. If $\mathfrak{M}^2$ is not a Jordan manifold, it is clear how we proceed further. Inductively, we remove a succession of $2p$ curves called *crosscuts*†

$$\mathcal{Q}_1, \mathcal{Q}_1', \ldots, \mathcal{Q}_p, \mathcal{Q}_p' \tag{7-5}$$

which are conjugate in p pairs, only $\mathcal{Q}_t$ and $\mathcal{Q}_t'$ intersecting one another in a crossing at P_t. The remaining manifold [after removal of points on the $2p$ curves (7-1)] is called $\mathfrak{M}^p$. Ultimately, we presume $\mathfrak{M}^p$ to be a Jordan manifold; then p is the genus‡ of $\mathfrak{M}$. In Fig. 7-1*a* we see how the genus can be interpreted intuitively as the number of handles placed on a sphere.

Let us imagine we have a rubber-sheet model of $\mathfrak{M}$ as in Fig. 7-1*a*. We cut it along all crosscuts (7-5) (in other words, we delete the crosscuts

† The term *crosscut* is used to denote any curve introduced for the canonical subdivision. (This is broader than the older usage of the term to denote cuts which have end points on the boundary.)

‡ We shall ultimately see from Theorem 7-4 (see Sec. 7-3) that p is finite and is dependent only on $\mathfrak{M}$ and not on the many possible ways we could draw the curves (7-5).

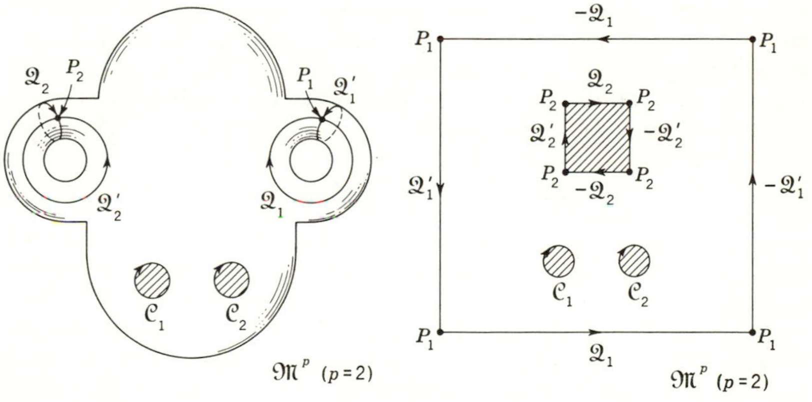

(*a*) (*b*)

Fig. 7-1 Rectangular model $\mathfrak{M}^p$ ($p = 2$, $n = 2$).

from $\mathfrak{M}$) to obtain $\mathfrak{M}^p$, and open up the rubber-sheet model $\mathfrak{M}^p$. We then obtain our *rectangular model* of $\mathfrak{M}$ as the $(p + n - 1)$-tuply-connected region bounded by p sets of *four* (not two) crosscuts $\mathcal{Q}_1$, $\mathcal{Q}_1'$, $-\mathcal{Q}_1$, $-\mathcal{Q}_1'$, etc., as well as n boundary curves. (See Fig. 7-1*b*.) If we consider $\partial\mathfrak{M}^p$, we find that the sides $\mathcal{Q}_1$ and $-\mathcal{Q}_1$ have opposing orientations (as shown). In the special case of a plane region, $p = 0$ so that there are no crosscut boundaries; if there also are no curve boundaries ($n = 0$), of course, the rectangular model is the whole sphere.

Thus, the boundary of $\mathfrak{M}^p$ is p (≥ 0) rectangles and n (≥ 0) closed curves, where the rectangles are oriented so that opposing sides cancel. The boundary in Fig. 7-1*b*, indeed, is labeled to show this cancellation by following the periphery in such a fashion that the region $\mathfrak{M}^2$ lies on the inside (see arrows). Each of the crosscuts is traversed twice but in opposite directions, of course.

The process of finding these crosscuts is called the "*canonical subdivision*" (or "canonical dissection") of the manifold $\mathfrak{M}$. We next consider another such system, which is even more standard (i.e., more "canonical"), as it leads to a *regular* polygon.

THE POLYGONAL MODEL

To obtain a second model of $\mathfrak{M}$, let us connect all points P_i (where the crosscuts $\mathcal{Q}_i$ and $\mathcal{Q}_i'$ intersect) to a common point P by means of some simple curves $\mathcal{R}_i$ in $\mathfrak{M}$ which do not intersect the crosscuts or one another (see Fig. 7-2*a*). We then stretch the curves $\mathcal{Q}_i$, $\mathcal{Q}_i'$ (as though they were rubber strings) along $\mathcal{R}_i$ by retracting their intersections P_i to reach P. Needless to say, we shall imagine that no new points of intersection are introduced as these crosscuts are retracted to P.

We let $\mathcal{Q}_i$ and $\mathcal{Q}_i'$ also denote (for simplicity) the new crosscuts through P (see Fig. 7-2*b*). We also take one point S_j on each $\mathcal{C}_j$ and connect it to P by a simple curve $\mathcal{K}_j$ in $\mathfrak{M}$ as shown. The $2n$ curves $\mathcal{K}_j$, $\mathcal{C}_j$ are imagined to be "close" to one another and (of course) to be nonintersecting with one another and with any other curve previously mentioned.

If we delete from $\mathfrak{M}$ the arcs

$$|\mathcal{Q}_i|,\ |\mathcal{Q}_i'|,\ |\mathcal{K}_j| \qquad i = 1, 2, \ldots, p;\ j = 1, 2, \ldots, n \tag{7-6}$$

we obtain a simply-connected Jordan domain $\mathfrak{M}^*$ (see Fig. 7-2*b*) which would reasonably seem equivalent to a polygon whose boundary consists of the $4p + 3n$ curves $\pm\mathcal{Q}_i$, $\pm\mathcal{Q}_i'$; $\pm\mathcal{K}_j$, $\mathcal{C}_j$. The boundary of $\mathfrak{M}^*$ would seem to contain only the $(n + 1)$ vertices P, S_j, but when we dissociate the arcs (7-6), the point P has $4p + n$ images for the arcs (7-6), and each S_j has two images making a total of $4p + 3n$ vertices on $\mathfrak{M}^*$ again. This process is depicted (at least intuitively) in Fig. 7-3*a* and *b*. (Compare Exercise 3.)

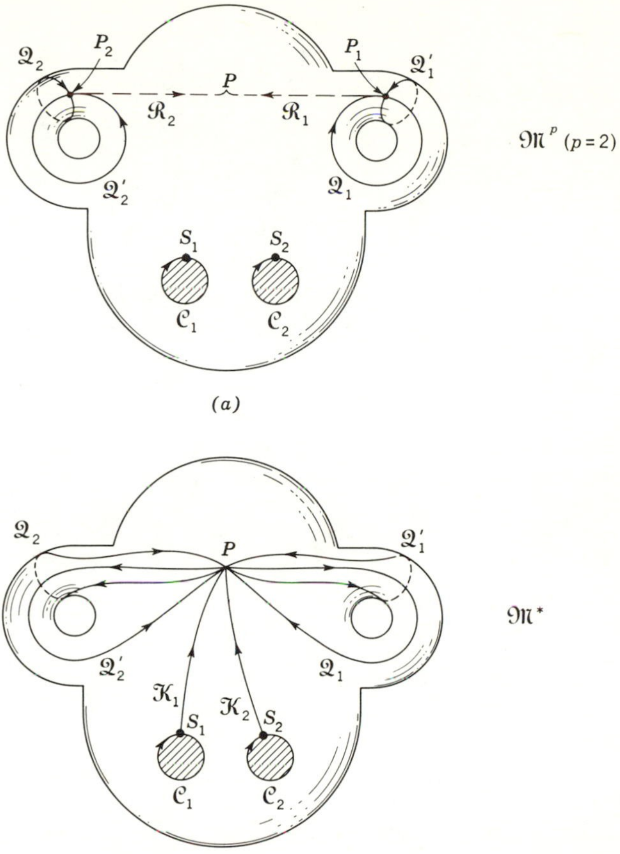

Fig. 7-2 The retraction of the crosscuts to a common intersection. In Fig. 7-2*b*, the arcs $|\mathcal{K}_i|$ are deleted from $\mathfrak{M}^p$ to yield $\mathfrak{M}^*$.

We now can have a *regular* $(4p + 3n)$-gon as the polygonal model $\mathfrak{M}^*$. If we traverse the boundary of $\mathfrak{M}^*$ in "positive" fashion (area to the left), we find a succession of sides on $|\partial\mathfrak{M}^*|$, namely

(7-7) $$\mathcal{Q}_i,\ \mathcal{Q}'_i,\ -\mathcal{Q}_i,\ -\mathcal{Q}'_i,\ \cdots \qquad \mathcal{K}_j,\ \mathcal{C}_j,\ -\ \mathcal{K}_j$$

where i goes from 1 to p *and then* j goes from 1 to n. In the sense of a chain,

$$\partial\mathfrak{M}^* = \Sigma\mathcal{C}_j \qquad j = 1, 2, \ldots, n$$

since all other sides cancel. If $p = 1$ and $n = 0$, $\mathfrak{M}$ is the torus and

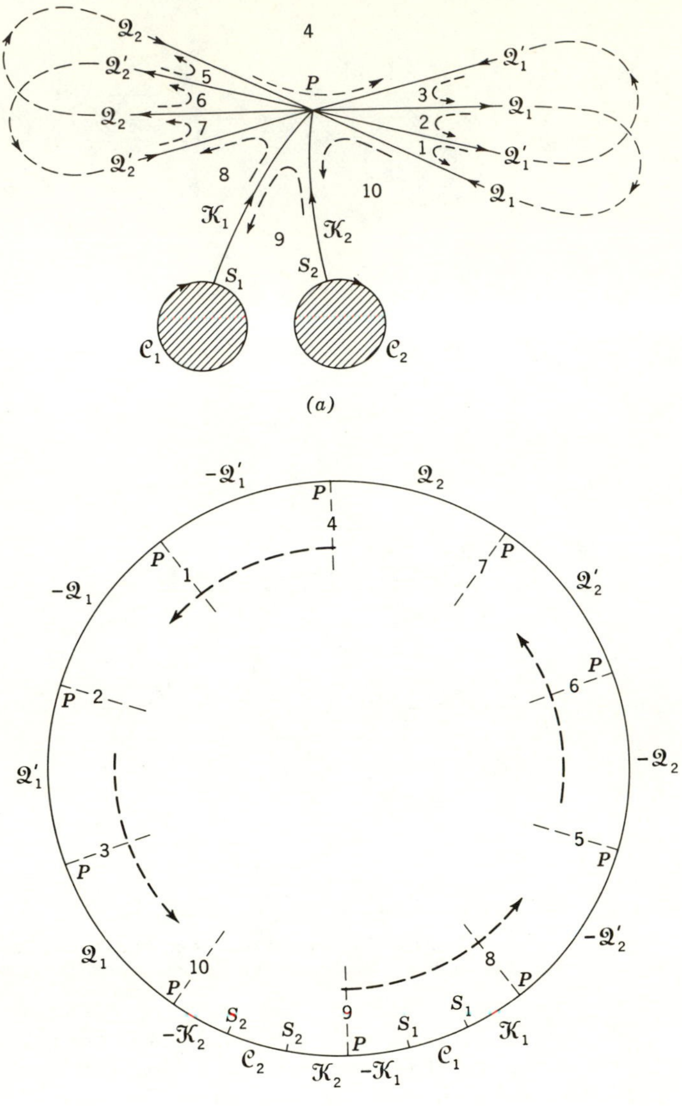

Fig. 7-3 The boundary of the polygon $\mathfrak{M}^*$ ($p = 2$, $n = 2$). (a) The neighborhood of P in Fig. 7-2b slightly enlarged; (b) the polygon $\mathfrak{M}^*$ opened up flat ($4p + 3n = 14$ sides).

there is no difference between the rectangular and polygonal models. If $p = 0$, then $\mathfrak{M}^p$ or $\mathfrak{M}^*$ is the sphere with n holes, and naturally either model is almost trivial.

We are now *asserting* the following result which is intuitive for the present.

THEOREM 7-3 *A finitely triangulated Riemann manifold $\mathfrak{M}$ can be transformed by deletion of $2p$ arcs into a rectangular model $\mathfrak{M}^p$ and by the deletion of $2p + n$ arcs into a polygonal model $\mathfrak{M}^*$, as shown in Figs.* 7-1*b and Fig.* 7-3*b, with due allowance being made for the trivial cases $p = 0$, $n = 0$, etc. In a topological sense, either model is determined completely by the values p and n.*

We have no reason to assert that every other set of crosscuts will lead to the same values of p (and n), but this is shown in the next section. A proof of Theorem 7-3 is reserved for Sec. 7-4.

EXERCISES

1 Show that if a Riemann manifold $\mathfrak{M}$ is a Jordan manifold, then so is every submanifold of $\mathfrak{M}$.

2 Show that if Theorem 7-3 is valid, it necessarily follows that every (finitely triangulated) Jordan manifold is equivalent topologically to an n-tuply-connected region of the plane. (*Hint:* Consider the models where $p \neq 0$.)

3 Verify the assertion that Fig. 7-3*a* and *b* are equivalent by showing that the 10 sectors about P in Fig. 7-3*a* correspond to the 10 sectors shown in Fig. 7-3*b*. Here the dotted lines in either figure show the boundary $\partial\mathfrak{M}^*$; note that Sector 1 is oriented by the dotted lines as *opposed* to $\mathfrak{Q}_1$ and $\mathfrak{Q}_i'$, hence its location between $-\mathfrak{Q}_1'$ and $-\mathfrak{Q}_1'$.

4 Let $p > 0$, $n = 0$ so that Fig. 7-3*a* and *b* refer to a $4p$-gon. Explain the numbering of the sectors in general fashion. (First try $p = 3$.)

7-3 The Euler-Poincaré Theorem

Consider now the polygonal model (accepting Theorem 7-3 for the present). Clearly, n is uniquely defined as the number of boundary continua, which is certainly a topological invariant. The fact that p is a topological invariant follows from this result:

THEOREM 7-4 (*Euler-Poincaré*) *Consider a Riemann manifold $\mathfrak{M}$ represented as a collection of curvilinear polygons. Consider its closure $\overline{\mathfrak{M}}$ so that the boundary sides are included in $\overline{\mathfrak{M}}$ (but exclude from the hypothesis the case of isolated boundary points, as usual). Then let F be the number of polygons in $\overline{\mathfrak{M}}$, let E be the number of edges in $\overline{\mathfrak{M}}$, and let V be the number of vertices in $\overline{\mathfrak{M}}$ (after we identify common edges and vertices of polygons). Then the so-called "alternating sum" has value as follows:*

$$F - E + V = 2 - 2p - n \tag{7-8}$$

PROOF It is easier to prove the result in a slightly more general context than stated here. Consider on $\mathfrak{M}$ a *connected* system of E (>0) curve segments and V *vertices* (including the curves and vertices in $|\partial\mathfrak{M}|$), that is, a *graph*. The one restriction on the graph is that the (open) faces thus determined must be topologically disks (simply-connected) and the (open) edges must be topologically segments (never closed curves). Thus, the complement in the sphere of a single line segment ($E = 1$), connecting two points ($V = 2$), is such a face ($F = 1$) which checks ($p = n = 0$).

We shall show generally that in such a system $F - E + V$ depends on $\mathfrak{M}$ alone.

We must show *any other* polygonal subdivision yields the same value of $F - E + V$. We show this in several steps:

(*a*) If we introduce a vertex on an edge already present in a subdivision, we thereby increase V by 1 and E by 1, leaving $F - E + V$ unchanged. The same happens if we introduce a vertex in the interior of a face and join it to another vertex already present.

(*b*) If we introduce an edge (joining two vertices already present), we thereby increase E by 1 and F by 1, leaving $F - E + V$ unchanged.

(*c*) Thus, if we have two different subdivisions of $\mathfrak{M}$, we can superimpose the two subdivisions to produce a so-called "common refinement" by using the operations of (*a*) and (*b*) on either original subdivision.

Thus, $F - E + V$ has the *same value for any two subdivisions* of $\mathfrak{M}$ (as it has for the common refinement of any two subdivisions).

Now let us consider the polygonal model as a $(4p + 3n)$-gon (see Fig. 7-3*b*). The cuts $\mathcal{Q}_i$, $\mathcal{Q}_i'$; $\mathcal{C}_j$, $\mathcal{K}_j$ produce a single polygon although the sides are identified so that on $\mathfrak{M}$ the tally of vertices and edges is subject to such identifications. Thus

$F = 1$	(the polygon)
$E = 2p + 2n$	($2p$ crosscuts $\mathcal{Q}_i$, $\mathcal{Q}_i'$ and n edges $\mathcal{K}_j$ with n edges $\mathcal{C}_j$)
$V = 1 + n$	(common vertex P and vertices on each $\mathcal{C}_j$)

Hence the alternating sum assumes the value in (7-8). Q.E.D.

In particular, for a sphere, $n = p = 0$ so that $F - E + V = 2$ (Euler's formula).

The formula (7-8) can be applied in most elegant and abstract fashion to one compact Riemann manifold, namely, the Riemann surface of an algebraic function.

COROLLARY 1 (*Riemann*) *Let a compact Riemann manifold* $\mathfrak{M}$ *have* m *sheets and total order of ramification*

$$b = \sum_{i=1}^{s} (e_i - 1) \qquad \text{(summed over all branch points)}$$

where e_i *is the exponent of uniformization at each of, say,* s *branch points. Then*

$$p = 1 - m + \tfrac{1}{2}b \tag{7-9}$$

PROOF Let us consider $\mathfrak{M}$ extended over the Riemann sphere $\mathfrak{R}$, and let us project each branch point onto $\mathfrak{R}$. We then perform a subdivision of $\mathfrak{R}$ into polygons so that these branch points are present as vertices. Thus the subdivision of $\mathfrak{R}$ has f faces, e edges, and v vertices where $f - e + v = 2$ (the genus of $\mathfrak{R}$ being 0), according to (7-8). If we project this subdivision onto the m sheets of $\mathfrak{R}$, clearly we obtain a subdivision of $\mathfrak{R}$ which has

(7-10) mf faces and me edges but $mv - b$ vertices

since b points are "lost" to the subdivision on $\mathfrak{R}$. Hence, the Euler-Poincaré theorem, as applied to $\mathfrak{M}$ (not $\mathfrak{R}$), yields

$$mf - me + mv - b = 2 - 2p$$

and the left-hand side becomes $2m - b$. Q.E.D.

Note that the *whole* structure of the Riemann surface $\mathfrak{M}$ is not relevant here. We need consider each branch point only "locally" to determine the e_i and b, and to thereby "see $\mathfrak{M}$ globally!".

Perhaps the most widely used case is the function

$$w^2 = (z - \beta_1) \cdots (z - \beta_{2k}) \qquad \beta_i \text{ distinct} \tag{7-11}$$

Here the Riemann surface for w (over the z sphere) has two sheets $(=m)$ and $2k$ ramification points (of total order $2k = b$) for each of $\beta_1, \ldots, \beta_{2k}$ and thus $p = k - 1$. The Riemann surface for (7-11) is called "elliptic" if $k = 2$ and "hyperelliptic" if $k > 2$. We now see it is possible to obtain an algebraic Riemann surface of genus p for any $p \geq 0$. Thus any *topological* structure is achieved by the illustration (7-11) but unfortunately not any *conformal*† structure.

EXERCISES

1 Consider $w^2 = P(z)$, where $P(z)$ is a polynomial of odd degree g and has g distinct roots; show the corresponding Riemann surface has genus $(g - 1)/2$.

† It is shown in Exercise 7 of Sec. 14-1, that the Riemann surface for $z^4 + w^4 = 1$ is *not* conformally equivalent to a hyperelliptic surface although the correct genus $(= 3)$ is achieved if $k = 4$.

2 Assume that the equation (with positive integral exponents)

$$w^t = (z - \beta_1)^{u_1} \cdots (z - \beta_k)^{u_k} \qquad t, u_j > 0 \text{ for all } j \tag{7-12}$$

is reduced and β_j are distinct. Then let the fractions $u_j/t = g_j/h_j$ (in lowest terms) and let $(u_1 + \cdots + u_k)/t = g_0/h_0$. Show that the singularities over $z = \beta_j$ each contribute $(h_j - 1)t/h_j$ to the order of ramification and thus

$$p = 1 - t + \left(\frac{t}{2}\right)\left[\sum_{j=0}^{k}\left(1 - \frac{1}{h_j}\right)\right]$$

3 Prove that the equation for w defined in Exercise 2 is reduced unless a factor (>1) of t divides *each* u_j. [*Hint:* Otherwise some subset numbering, say, v $(<t)$ of the t formal root values $w_1, \ldots, w_t$ in (7-12), would belong to a rational equation and, say, the (symmetric) product $\phi = w_1 \cdots w_v$ $(v < t)$ would be rational. But then

$$\phi = (z - \beta_1)^{u_1 v/t} \cdots (z - \beta_k)^{u_k v/t}$$

would be a rational function for some $v < t$. Complete the proof.

4 Find the genus of the folium of Descartes $z^3 + w^3 - 3azw = 0$ from the information in Sec. 5-4, Exercise 2. (Check by Exercises 4 and 8, Sec. 5-4.)

5 (*a*) Show that the function $z^m + w^m = 1$ has genus $(m - 1)(m - 2)/2$. (*b*) What about $z^m + w^n = 1$?

6 Show that formula (7-8) for the noncompact case follows from the compact case $(n = 0)$, by filling in each hole with a polygon.

7 Show that for a compact manifold with $p = 0$ or 1, we can triangulate in such a fashion that there is "complete economy" of vertices (*every* two vertices determine an edge). [*Hint:* Apply $2E = 3F$ and $V(V - 1)/2 = E$ to (7-8).] (The answers are $F = 4$, $E = 6$, $V = 4$ or $F = 14$, $E = 21$, $V = 7$, respectively. Draw the triangulations of each.)

8 Show that the rectangular model (p rectangles and n boundary components) has genus p by showing a decomposition which satisfies Theorem 7-4. (*Hint:* Connect all rectangles and boundary curves to a common interior point.) (Why must we draw connecting curves here and not for the polygonal model?)

7-4 Proof of Models

We now finally verify Theorem 7-3 which asserts that certain models are topologically equivalent to an arbitrary Riemann manifold. At one time (say, prior to 1900), such a result would have been regarded as intuitively obvious from the earlier discussion, and indeed we shall only sketch a proof, relying heavily on the reader's acceptance of minor details. We are dealing with algebraic tools which bear little resemblance to complex variable theory of the hard classical type but which nevertheless have been increasingly absorbed by modern complex analysis.

If we consider the triangles Δ_i (in the u_i plane) $(1 \le i \le F)$, we have certain relations given by an $F \times F$ matrix of so-called "incidence numbers" e_{ij}, where $e_{ij} = 1$ if Δ_i adjoins Δ_j on a common side and $e_{ij} = 0$

otherwise. This information will be referred to generally as the *simplicial structure*† of $\overline{\mathfrak{M}}$ (the closure of $\mathfrak{M}$).

Our purpose is to think of the polygonal and rectangular models as triangulated by closed (ordinary) straight triangles

$$\delta_1, \ . \ . \ . \ , \delta_F$$

which have the same simplicial structure as the triangles

$$\Delta_1, \ . \ . \ . \ , \Delta_F$$

which constitute $\mathfrak{M}$. In the case of the triangles Δ_i in the u_i planes, we were concerned with *actual mappings* of adjacent triangles, but in the case of the triangles δ_i, we are concerned wholly with *incidence relations* of sides. (Clearly, the vertices fall in place naturally from the simplicial structure.) We now borrow a result from topology: *A finitely triangulated Riemann manifold is determined topologically by its simplicial structure.*

The proof of this last result is a matter of tedious details which are referred to the literature. The idea is to imagine that bicontinuous mappings stretch each triangle Δ_i onto δ_i (like a rubber sheet) but in such a fashion that the relations of adjacent triangles (say, Δ_i and Δ_I) are *preserved at the boundary* of the triangles (δ_i and δ_I).

LEMMA 7-2 *The manifold $\mathfrak{M}$ has a closure $\overline{\mathfrak{M}}$ with the same simplicial structure as some convex polygon Π in the plane with sides identified in pairs except for n closed sequences $\mathcal{C}_j$ ($j = 1, 2, \ . \ . \ . \ , n$) with the following properties (for noncompact $\mathfrak{M}$):*

(a) $\mathcal{C}_j$ consists of a simple closed curve in $\partial\overline{\mathfrak{M}}$ which connects a point of $\partial\overline{\mathfrak{M}}$, say S_j, to itself.

(b) $\mathcal{C}_j$ is preceded and followed on the boundary of Π by matching sides $\mathcal{K}_j'$ and $-\mathcal{K}_i'$ which connect S_j to some point P_j interior to $\mathfrak{M}$.

PROOF Assume $\mathfrak{M}$ is compact at first. Start with any curvilinear triangle Δ_1 and let it correspond to an ordinary equilateral triangle δ_1. Then if another triangle of $\mathfrak{M}$, say Δ_2, adjoins Δ_1, we represent it as an isosceles triangle δ_2 with base adjoining Δ_1 in appropriate fashion and with base angle $60°/2$. Generally, to form a polygon Π, we keep adjoining δ_m to a previous set of triangles $\delta_1, \ . \ . \ . \ , \delta_{m-1}$ in some connected fashion simulating the joining of Δ_m to $\Delta_1, \ . \ . \ . \ , \Delta_{m-1}$ in $\mathfrak{M}$. Thus if Δ_m joins Δ_r on a certain side as base, then δ_m joins δ_r on the same side but is represented as an isosceles triangle with base angles $60°/2^{m-1}$. Thus no angle of the polygon will exceed $\Sigma 60°/2^{m-1} = 120°$.

If $\mathfrak{M}$ is not compact, we begin with a triangle which does not have a side on the boundary. Eventually, we obtain a triangle, say δ_1, which

† Such $F \times F$ matrices at one time were the core of old-fashioned combinatorial topology or analysis situs. We refer to the literature for the manipulative techniques.

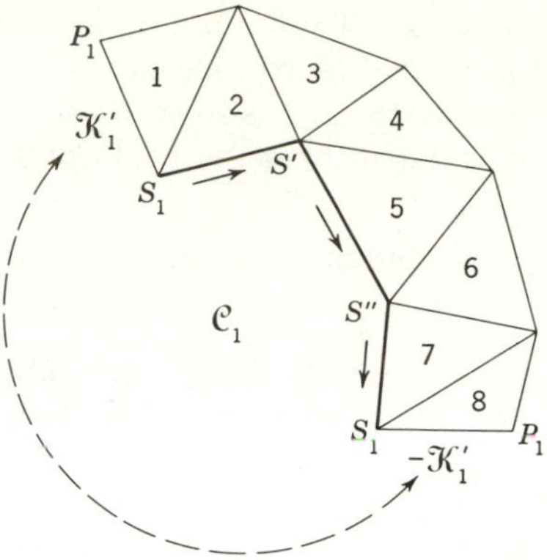

Fig. 7-4 The curve $\mathcal{C}_1$ (of $\partial\overline{\mathfrak{M}}$) is drawn in heavy outline. (In the construction used to prove Lemma 7-2, the triangles would become increasingly obtuse. We ignore this in the diagram, of course, for purposes of legibility.)

has a vertex, say S_1, *on the boundary.* (See Fig. 7-4.) Then some triangle, say δ_2, with S_1 as a vertex will lie on the left of the boundary (in the usual orientation of the boundary sides of the δ_i). In this manner we get to S', the second vertex (of δ_2), such that S_1S' belongs to $\partial\mathfrak{M}$. We treat S' the same way as S_1, thus obtaining a sequence of sides forming a polygonal curve

$$\mathcal{C}_1 = S_1S'S'' \cdots S_1 \tag{7-13}$$

which can end only in the same point S_1 [say, with δ_8 (see Fig. 7-4)]. The sequence (7-13), therefore, cannot "close up" without coming to a side (see $-\mathcal{K}_1'$ in Fig. 7-4) which matches a previous side ($\mathcal{K}_1' = P_1S_1$).

Q.E.D.

Our objective now is to show that with clever retriangulation, the polygon Π can be reassembled to any one of the canonical models we desire.

SIMPLE PERIPHERAL SEQUENCE

We are now considering certain polygons such as the recent example Π and the polygonal model $\mathfrak{M}^*$ (of Sec. 7-2). They have *peripheral sequences* of sides with the property *that certain sides pair off (negatively), but otherwise the sides constitute simple polygonal "curves" which never intersect one another except at the boundary vertices of* Π. Call this the *simple boundary property* of the polygon Π.

We shall denote a peripheral sequence [as (7-7) in Sec. 7-2] by

$$\mathbf{S} = \mathcal{A}, \mathcal{B}, \mathcal{C}, \ldots, \mathcal{K}, \mathcal{L} \tag{7-14}$$

This sequence can be written in reverse order

$$-\mathbf{S} = -\mathcal{L}, -\mathcal{K}, \ldots, -\mathcal{B}, -\mathcal{A} \tag{7-15}$$

and it can be permuted cyclically

$$\mathbf{S} = \mathcal{B}, \mathcal{C}, \ldots, \mathcal{K}, \mathcal{L}, \mathcal{A} \tag{7-16}$$

We also introduce abbreviations $\mathbf{Q}$, $\mathbf{K}$, etc., which are regarded as collections of sides. They are often called "words." In particular, using the previous symbols, we form the words

$$\mathbf{Q}_i = \mathcal{Q}_i, \mathcal{Q}'_i, -\mathcal{Q}_i, -\mathcal{Q}'_i \tag{7-17a}$$
$$\mathbf{K}_j = \mathcal{K}_j, \mathcal{C}_j, -\mathcal{K}_j \tag{7-17b}$$

Thus, when $p > 0$ $(n \geq 0)$,

$$\mathbf{S}^* = \mathbf{Q}_1, \mathbf{Q}_2, \ldots, \mathbf{Q}_p, \mathbf{K}_1, \ldots, \mathbf{K}_n \tag{7-18a}$$

serves as a peripheral sequence (with simple boundary property) for the canonical polygon. The case $p = 0$ is taken care of by the following result:

LEMMA 7-3 *If a polygon* Π *has peripheral sequence*

$$\mathbf{S}^* = \mathbf{T}, -\mathbf{T}, \mathbf{K}_1, \ldots, \mathbf{K}_n \qquad n \geq 0 \tag{7-18b}$$

(with simple boundary property) where $\mathbf{T} = \mathcal{A}, \mathcal{B}, \ldots, \mathcal{L}$ *denotes some sequence of sides, then* Π *is topologically equivalent to a sphere* ($p = 0$) *with* n *boundary contours.*

The proof is shown in Fig. 7-5 (with $n = 1$). We can imagine that the polygon Π (in Fig. 7-5*a*) is made of sheet rubber and it is twisted so that the curve $\mathcal{C}$ is "engulfed inside" by the matching of $\mathcal{K}_1$ and $-\mathcal{K}_1$ (as in Fig. 7-5*b*). We then get two symmetric halves bounded by $\mathbf{T} = \mathcal{A}, \mathcal{B}, \ldots, \mathcal{L}$ and $-\mathbf{T} = -\mathcal{L}, \ldots, -\mathcal{B}, -\mathcal{A}$ which are (topologically) the hemispheres, of course. Therefore a detailed proof would consist of decomposing Fig. 7-5*a* and *b* into triangles and showing both have the same simplicial structure. [This is done in the next result, Lemma 7-4.]

In the meantime, we see that our problem is the following: Lemma 7-2 yields the peripheral sequence for Π

$$\mathbf{S} = \mathbf{A}_1, \mathbf{K}'_1, \mathbf{A}_2, \mathbf{K}'_2, \ldots, \mathbf{A}_n, \mathbf{K}'_n \tag{7-19}$$

where $\mathbf{K}'_j = \mathcal{K}'_j, \mathcal{C}_j, -\mathcal{K}'_j$ and the $\mathbf{A}_j$ collectively have canceling pairs of sides, always forming a simple boundary for Π. We must reduce (7-19) to (7-18*a*) or (7-18*b*) by retriangulation. This will establish the polygonal

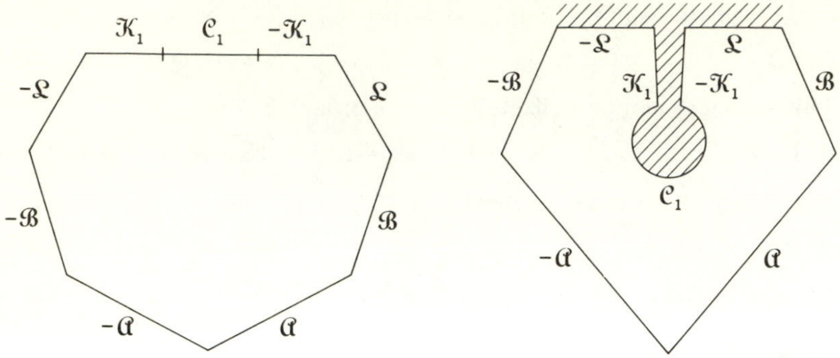

(a) (b)

Fig. 7-5

model (see Lemmas 7-4 to 7-8). The rectangular model is an easy corollary (see Lemma 7-9).

LEMMA 7-4 *If* $\mathbf{S} = \mathbf{T}, \mathcal{A}, -\mathcal{A}, \mathbf{U}$ *is given, then adjacent negative sides can be canceled unless we are reduced to the form* (7-18*b*) *for a sphere.*

For proof, we examine Fig. 7-6. We suppose $\mathcal{A}$ goes from Q to P as shown but is preceded by a side from S to Q and $-\mathcal{A}$ is followed by a side from Q to R. If E is a point on RS, we just match triangles as shown. (The wavy line in Fig. 7-6 indicates a collection of unspecified sides.)

LEMMA 7-5 *If* $\mathbf{S} = \mathbf{X}, \mathcal{L}, \mathcal{M}, \mathbf{Y}, -\mathcal{M}$, *we can regard* $\mathcal{L}, \mathcal{M}$ *as a polygonal curve* $\mathcal{D}$ *for which* $\mathbf{S} = \mathbf{X}, \mathcal{D}, \mathbf{Y}, -\mathcal{D}, \mathcal{L}$ (*always preserving the simple boundary property*).

For proof, we examine Fig. 7-7. We form the triangle $\mathcal{L}, \mathcal{M}, -\mathcal{D}$ and cut it off in (*a*), then paste it back in (*b*).

Of course, symbolically, these last two lemmas seem trivial; they seem to involve cancellations of adjacent letters, but the important feature is that the "word type" cancellation leaves us with a peripheral sequence with simple boundary property. (We do not always bother to repeat this fact.) Thus, Lemma 7-5 really tells us that the peripheral sequence reduction problem is paralleled by the formal operation. Thus, the replacement of $\mathcal{L}, \mathcal{M}$ by $\mathcal{D}$ is geometrically justifiable and is not just a formal operation! [Again we do not always repeat the fact that any polygon is topologically equivalent to a convex polygon (so that interior diagonals can always be drawn to correspond to curves lying in $\mathcal{M}$).]

LEMMA 7-6 *The peripheral sequence* (7-15) *can be reduced so that only one vertex appears except for different vertices* S_j *on each of the sides represented by* $\mathcal{C}_j$ *in the noncompact case.*

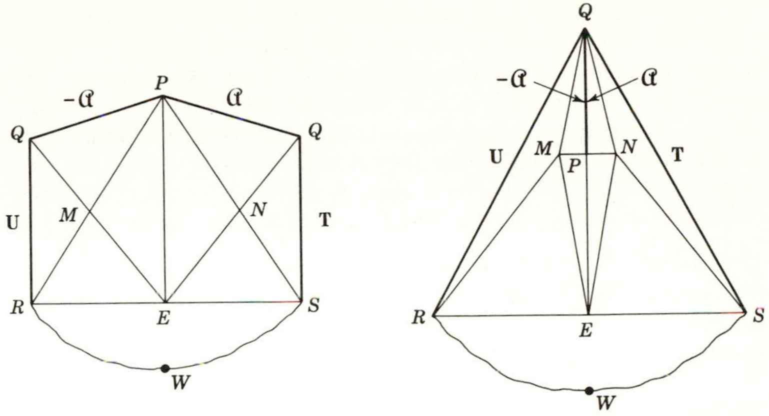

(*a*) (*b*)

Fig. 7-6 Here **T** (and **U**) denote a collection of sides from W to Q via S (and R, respectively).

For proof note that Lemma 7-5 shows how the occurrence of any one vertex (say Q) can be eliminated (at the expense of reintroducing another P). We therefore reduce the peripheral sequence (7-15) so that P is the only vertex even on the *word* $\mathbf{K}_j'$ of (7-19), which is now $\mathbf{K}_j$ of (7-17*b*). Also, $\mathcal{K}_j$ goes from $\mathcal{C}_j$ to P.

LEMMA 7-7 *Under the conditions of Lemma* 7-6 (*one vertex*) *the peripheral sequence* Π *can be reduced to the form where* (*ignoring* $\mathbf{K}_j$) *between any* $\mathfrak{A}$ *and* $-\mathfrak{A}$ *there must exist some* $\mathfrak{B}$ *whose negative* $-\mathfrak{B}$ *is not also included between* $\mathfrak{A}$ *and* $-\mathfrak{A}$.

(*a*) (*b*)

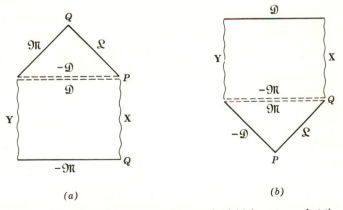

Fig. 7-7 Note that the point Q at the summit of (*a*) is suppressed at the expense of P reintroduced at the bottom of (*b*).

For proof note that otherwise (aside from $\mathbf{K}_j$) the subsequence between $\mathcal{A}$ and $-\mathcal{A}$ would contain only matched pairs $\mathcal{B}$, $-\mathcal{B}$, $\mathcal{C}$, $-\mathcal{C}$, etc. (in some rearranged order). Thus if a diagonal $\mathcal{D}$ is drawn between a (matching) midpoint on $\mathcal{A}$ and $-\mathcal{A}$, it would be a closed curve dividing the polygon into two disjoint polygons (since there is no way to go from one side of $\mathcal{D}$ to the other through the matched sides $\mathcal{B}$, $-\mathcal{B}$, etc.). Hence, on both sides of $\mathcal{D}$ the polygon has different end points. This is a contradiction to the hypothesis that Lemma 7-6 has already been applied.

LEMMA 7-8 *Let* $\mathbf{S}$ *contain* $\mathcal{A}$, $\mathcal{B}$, $-\mathcal{A}$, $-\mathcal{B}$ *with some sides in between; then an equivalent triangulation produces* $\mathcal{Q}$, $\mathcal{Q}'$, $-\mathcal{Q}$, $-\mathcal{Q}'$ *as successive sides without increasing the number of sides.*

For proof consider

$$\mathbf{S} = \mathcal{A}, \mathbf{X}, \mathcal{B}, \mathbf{Y}, -\mathcal{A}, \mathbf{Z}, -\mathcal{B}, \mathbf{W} \tag{7-20a}$$

Then if $\mathcal{Q}' = \mathbf{Z}, -\mathcal{B}, \mathbf{W}$ and if $\mathcal{Q} = -\mathbf{X}, -\mathbf{W}, -\mathcal{A}$, then we can show by an actual substitution and cancellation,

$$\mathbf{S} = \mathcal{A}, \mathbf{X}, \mathbf{W}, -\mathcal{Q}', \mathbf{Z}, \mathbf{Y}, -\mathcal{A}, \mathcal{Q}' \tag{7-20b}$$

$$\mathbf{S} = -\mathcal{Q}, -\mathcal{Q}', \mathbf{Z}, \mathbf{Y}, \mathbf{W}, \mathbf{X}, \mathcal{Q}, \mathcal{Q}' \tag{7-20c}$$

which leads to the desired result by cyclic rearrangement. The use of (7-20*b*) eliminates $\mathcal{B}$ and (7-20*c*) eliminates $\mathcal{A}$ by a replacement process reminiscent of Lemma 7-5.

Now we note that (7-20*c*) has the same gaps as (7-20*b*), but they are collected in a sequence. Hence the repeated application of Lemma 7-8 should produce the sequence something such as

$$\mathbf{S} = \mathbf{Q}_1, \mathbf{K}_1, \mathbf{K}_2, \mathbf{Q}_2, \mathbf{K}_2, \ldots \tag{7-21a}$$

which resembles (7-18*a*) except that the $\mathbf{K}_i$ are not together, or else something such as

$$\mathbf{S} = \mathcal{A}, \mathbf{K}_1, \mathcal{B}, \mathbf{K}_2, \mathbf{K}_3, \ldots, -\mathcal{B}, -\mathcal{A} \tag{7-21b}$$

which resembles (7-18*b*) except that the $\mathbf{K}_i$ are not together.

For the final simplification, write

$$\mathbf{S} = \mathcal{K}_1, \mathcal{C}_1, -\mathcal{K}_1, \mathbf{X}, \mathcal{K}_2, \mathcal{C}_2, -\mathcal{K}_2, \mathbf{Y} \tag{7-22}$$

Then if $\mathbf{X}, \mathcal{K}_2 = \mathcal{K}_2'$, we have

$$\mathbf{S} = \mathcal{K}_1, \mathcal{C}_1, -\mathcal{K}_1, \mathcal{K}_2', \mathcal{C}_2, -\mathcal{K}_2', \mathbf{X}, \mathbf{Y} \tag{7-23}$$

and by repetition of this step we can bring the $\mathbf{K}_1, \mathbf{K}_2, \ldots$ together in (7-21*a*) and (7-21*b*) without disturbing the other terms.

This takes care of the polygonal model.

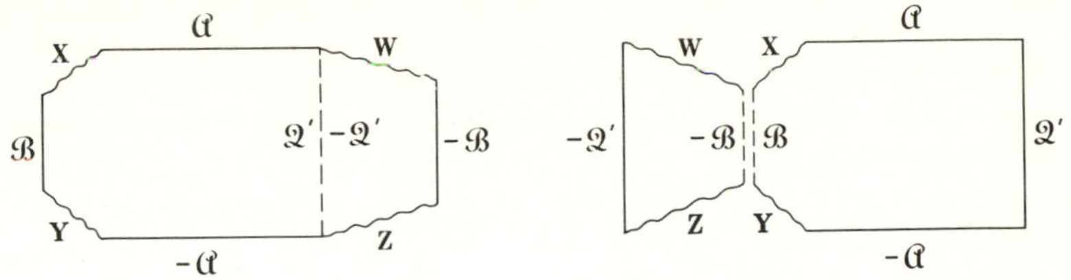

Fig. 7-8 For Exercise 3.

LEMMA 7-9 *The polygonal model is equivalent to the rectangular model.*

For proof, let us start with the rectangular model. We reassemble it according to Lemma 7-2 as a polygon Π which is topologically equivalent to one of the canonical polygons by a retriangulation. This retriangulation process is clearly *reversible,* and this fact establishes the lemma.

EXERCISES

1 Show that with the incidence matrices described at the beginning of the section $\sum_{j=1}^{F} e_{ij} = 3$ if Δ_i is not part of the boundary of $\mathfrak{M}$.

2 Simplify the peripheral sequence $\mathfrak{A}$, $\mathfrak{B}$, $\mathfrak{C}$, $-\mathfrak{B}$, $-\mathfrak{A}$, $-\mathfrak{C}$, assuming all terms are curves going from P to P.

3 Justify the transition from (7-20*a*) to (7-20*b*) (see Fig. 7-8) and (7-20*b*) to (7-20*c*) by a diagram showing how the sides $\mathfrak{Q}'$ and $\mathfrak{Q}$ are representable as diagonals.

4 The rectangular model for $p = 2$, $n = 0$ can be made into a polygon by joining the common point P_1 of $\mathfrak{Q}_1$, $\mathfrak{Q}_1'$ to the common point P_2 of $\mathfrak{Q}_2$, $\mathfrak{Q}_2'$ by an arc $\mathfrak{A}$. The peripheral sequence is then

$$S = \mathfrak{Q}_1, \mathfrak{Q}_1', -\mathfrak{Q}_1, -\mathfrak{Q}_1', \mathfrak{A}, \mathfrak{Q}_2, \mathfrak{Q}_2', -\mathfrak{Q}_2, -\mathfrak{Q}_2', -\mathfrak{A}$$

Transform this into the canonical form (7-18*a*) by Lemmas 7-4 to 7-8.

5 Show how the polygonal model of Sec. 7-2 can also be represented topologically as a $4p$-gon with n holes.

HOMOLOGY AND ABELIAN DIFFERENTIALS

7-5 Boundaries and Cycles

The various canonical models of a Riemann manifold $\mathfrak{M}$ are conceived independently of the complex variable applications of Sec. 7-1, such as Theorem 7-2 on the total residue. The significance of the model can be understood as well from certain line integral results in *real* variable theory. We shall never pursue the matter beyond the grossly intuitive level, but we shall mainly draw an analogy between the curves drawn on a manifold and the integrands of such line integrals. We must now define a seemingly ostentatious formalism, which the reader will recognize as a generalization of that of Sec. 1-6.

CHAINS

We begin by defining *t chains* as finite linear combinations of regions $\mathcal{R}_i$ ($t = 2$), simple curves $\mathcal{G}_i$ ($t = 1$), and points P_i ($t = 0$) with integral coefficients m_i

(7-24) $$\Sigma^{(t)} \begin{cases} \sigma^{(2)} = \Sigma m_i \mathcal{R}_i \\ \sigma^{(1)} = \Sigma m_i \mathcal{G}_i \\ \sigma^{(0)} = \Sigma m_i P_i \end{cases}$$

Here $\Sigma^{(t)}$ is the linear space of chains $\sigma^{(t)}$ closed under addition and subtraction (of chains with the same t). Here there is a natural (positive) orientation of regions and points (but not arcs); thus we can say $-\mathcal{R}_i$ is not a region but a 2-chain, etc. In adding chains we recognize a natural simplification process; we can break up $\mathcal{R}_i$ and $\mathcal{G}_i$ into nonoverlapping segments so that the coefficients of like segments are collected in (7-24) with cancellations. Thus we can have 0 representing a *null chain* (for any t). For certain purposes of completeness we shall need to say formally,

(7-25) $$\sigma^{(t)} = 0 \qquad \text{for } t > 2,\ t < 0$$

To further simplify matters, we assume $\partial\mathfrak{M}$ (the collection of the unmatched sides of the triangulation) is piecewise analytic (in the u_i and ζ_i parameter planes) and that all parts of $\mathcal{G}_i$ and $\partial\mathcal{R}_i$ are piecewise analytic curves.†

BOUNDARIES

We now define the boundary operator ∂. The most critical case is $\partial\mathcal{R}_i$, where the orientation (of the left-hand bordering) takes care of matters as usual, thus $\partial\mathcal{R}_i = \Sigma$ (certain closed curves). For completeness, we define in natural fashion $\partial\mathcal{G}_i = P_i' - P_i$ as a 0-chain where $\mathcal{G}_i$ is a curve segment from P_i to P_i'. If $\mathcal{G}_i$ is closed, then $P_i' = P_i$ and $\partial\mathcal{G}_i = 0$. Formally, $\partial\sigma^{(0)} = 0$ in agreement with (7-25) since the boundary is of negative dimension. Call the set of $\partial\sigma^{(t)}$ the "boundaries" B^{t-1}. Thus

(7-26) $$B^t \begin{cases} t = 1 & \partial\sigma^{(2)} = \Sigma m_i\, \partial\mathcal{R}_i \\ t = 0 & \partial\sigma^{(1)} = \Sigma m_i\, \partial\mathcal{G}_i = \Sigma m_i(P_i' - P_i) \\ t = -1 & \partial\sigma^{(0)} = 0 \qquad \text{(formally)} \end{cases}$$

It is now clear that in each case

(7-27) $$\partial\partial\sigma^{(t)} = 0$$

† It will be seen in Sec. 11-2 that the problem of extending analytic functions to the boundary of a region is essentially the same when the boundary is analytic as when the boundary is merely smooth. For this reason an assumption of analyticity is harmless.

The only nontrivial case naturally is $\partial\partial\mathcal{R}_i$; but here $\partial\mathcal{R}_i$ is a collection of simple closed curves $\mathcal{C}$ and $\partial\mathcal{C} = 0$. Here, formally, B^t consists only of 0 if $t \geq 2$ or $t \leq -1$.

The boundary operator is *distributive*, $\partial(\sigma_1 + \sigma_2) = \partial\sigma_1 + \partial\sigma_2$. Thus it is clear the boundaries (as do chains) form a linear space under addition. If $\sigma_1{}^{(t)} = \partial\sigma_1{}^{(t+1)}$ and $\sigma_2{}^{(t)} = \partial\sigma_2{}^{(t+1)}$, then $\sigma_1{}^{(t)} \pm \sigma_2{}^{(t)}$ is also a boundary (of $\sigma_1{}^{(t+1)} \pm \sigma_2{}^{(t+1)}$) trivially.

A very important result is given in Fig. 7-3, illustrating the general finite Riemann manifold $\mathfrak{M}$, namely

(7-28)
$$\partial\mathfrak{M} = \mathcal{C}_1 + \cdots + \mathcal{C}_n$$

Thus $\partial\mathfrak{M} = 0$ if $\mathfrak{M}$ is compact. This result follows from noting that $\partial\mathfrak{M}$ is the sum of the (signed) peripheral sequence [as given in (7-7)]. Thus, the concept of "point-set" boundary $|\partial\mathfrak{M}| = \Sigma\, |\mathcal{C}_j|$ is essentially in agreement with the formal definition as an operator on chains.

CYCLES AND HOMOLOGY

Actually, $\partial\sigma^{(t)}$ can equal 0 without our having $\sigma^{(t)} = \partial\sigma^{(t+1)}$ for some $\sigma^{(t+1)}$ [or without our using (7-27)]. In general we call $\sigma^{(t)}$ a *cycle* (denoted by Z^t) exactly when $\partial\sigma^{(t)} = 0$. Thus all boundaries are cycles but not conversely. For example, a closed curve $\mathcal{C}$ is always a cycle, but it need not be a boundary in a non-Jordan manifold (such as a torus) or in an n-tuply-connected region where $\mathcal{C}$ is homologous to one of several boundary curves. More trivially, any *single* point P is a cycle ($\partial P = 0$) but is not the boundary of a 1-chain. Clearly cycles also form a linear space under addition.

Generally, it is easy to identify cycles

(7-29)
$$Z^t = \begin{cases} \sigma^{(2)} = m\mathfrak{M} \ (\mathfrak{M} \text{ compact) and 0 otherwise} \\ \sigma^{(1)} = \text{sum of 1-chains of closed curves} \\ \sigma^{(0)} = \text{all 0-chains} \end{cases}$$

The difficult case ($t = 2$) is part of Theorem 7-5 and the other cases are intuitively clear. Of course, Z^t is formally 0 for $t < 0$ and $t > 2$.

We say that two chains $\sigma_1{}^{(t)}$ and $\sigma_2{}^{(t)}$ are *homologous* if their difference is a boundary, symbolically

$$\sigma_1{}^{(t)} \sim \sigma_2{}^{(t)} \qquad \text{or} \qquad \sigma_1{}^{(t)} - \sigma_2{}^{(t)} = \partial\sigma^{(t+1)}$$

for some $\sigma^{(t+1)}$. We say that any two homologous chains belong to the same *homology class.*

The theory of homology usually is confined to homologous *cycles,* where the main result of the canonical model is expressed in the following results:

THEOREM 7-5 *Let $\mathfrak{M}$ be a finite manifold of genus p and with n boundary continua. Then any 1-cycle $\mathcal{Z}$ on $\mathfrak{M}$ is represented by the homology class given by the 1-chain*

$$\mathcal{H} = \sum_{i=1}^{p} m_i \mathcal{Q}_i + \sum_{i=1}^{p} m_i' \mathcal{Q}_i' + \sum_{j=1}^{n-1} g_j \mathcal{C}_j \tag{7-30}$$

where $\mathcal{Q}_i$, $\mathcal{Q}_i'$ are 2p fixed crosscuts and $\mathcal{C}_j$ are $(n - 1)$ boundary continua and the coefficients are integers. Furthermore, the homology classes represented by (7-30) are independent†; i.e., for a given cycle $\mathcal{Z}$, only one choice of m_i, m_i', g_j can produce an $\mathcal{H}$ which is homologous with $\mathcal{Z}$. [If $n = 0$ or 1 the elements $\mathcal{C}_j$ are of course ignored in (7-30), and if $p = 0$ in addition, the only homology class is 0.]

The proof is regarded as an intuitive procedure here, although the details are below the level of difficulty of, say, the Jordan curve theorem. First of all, we break up $\mathcal{Z}$ into a sum of simple closed curves and show each such curve is homologous with some $\mathcal{H}$ as represented in (7-30). We thus take $\mathcal{Z}$ to be a simple closed curve. Using the canonical model of a $(4p + 3)$-gon (see Fig. 7-3), we consider several possibilities.

(*a*) If $\mathcal{Z}$ is entirely interior to the polygon, it is clearly homologous to zero by the Jordan curve theorem.

(*b*) If $\mathcal{Z}$ is not entirely interior to the polygon, $\mathcal{Z}$ can be deformed to the boundary of the polygon so that it represents a succession of sides beginning and ending with two images of the *common* vertex P. Such a succession consists of

$$\pm\mathcal{Q}_i, \ \pm\mathcal{Q}_i', \ \pm\mathcal{K}_j, \ \mathcal{C}_j$$

but any $\mathcal{K}_j$ will be canceled by a $-\mathcal{K}_j$ and $\sum_{1}^{n} \mathcal{C}_j = \partial\mathfrak{M} \sim 0$.

Thus $\mathcal{Z} \sim \mathcal{H}$ as shown in (7-30).

It remains to see that the homology classes in (7-30) are independent; i.e., for no choice of m_i, m_i', g_j could we have $\mathcal{H} \sim 0$ or $\mathcal{H} = \partial\sigma^{(2)}$ unless all m_i, m_i', g_j are zero. This, again, is easy to see intuitively. If we break up any arbitrary $\sigma^{(2)}$ into nonoverlapping regions $\sigma^{(2)} = \sum_{i=1}^{t} k_i \mathcal{R}_i$, then unless all k_i are equal (indeed unless $\sigma^{(2)} = k\mathfrak{M}$), a contradiction is obtained.

For example, if $k_1 \neq k_2$ and if $\mathcal{R}_1$ and $\mathcal{R}_2$ join on a common boundary $\mathcal{C}$, then $\partial\sigma^{(2)} = \partial(k_1\mathcal{R}_1 + k_2\mathcal{R}_2 + \cdots) = (k_2 - k_1)\mathcal{C} + \cdots \neq 0$. Thus $\partial\sigma^{(2)} = 0$ is impossible unless $\sigma^{(2)} = k\mathfrak{M}$. Thus the only case of $\mathcal{H} \sim 0$ is

† The *unique* representability of $\mathcal{Z}$ by $\mathcal{H}$ is expressed in the statement that in (7-29) $\mathcal{Q}_i$, $\mathcal{Q}_i'$ and $\mathcal{C}_j$ form a *homology basis*.

$\mathcal{H} = k\partial\mathfrak{M}$ and $\mathcal{H} \sim 0$ leads to $\mathcal{C}_1 + \cdots + \mathcal{C}_n = 0$ *only*. (See Exercise 2.) But $\mathcal{C}_n$ is absent in (7-30), hence $\mathcal{H} \sim 0$ only if all m_i, m_i', g_j vanish. This proves Theorem 7-5.

If we let $p = 0$, $n = 1$, we are back to the Jordan curve theorem (see Sec. 1-4).

COROLLARY 1 *In a simply-connected region of the plane, all cycles are boundaries.*

In this proof, the main advantage of the canonical model was that the problem was transformed from a manifold extending over several sheets to a manifold located in the plane, where intuition is more reliable! A more rigorous treatment is a matter of "foundations" largely. In fact, the most commonly accepted procedure is to regard the basic components $\mathcal{R}_i$, $\mathcal{G}_i$, P_i of (7-24) as chosen from a finite collection of triangles (simplexes) and their sides and vertices. Hence the factor of geometrical intuition is transferred to the identification of the sides of adjoining triangles, etc., and the proof of Theorem 7-5 seems more straightforward. Such matters are referred to the literature.

The homology classes (7-30) are represented† by an integral vector with arbitrary components

$$\mathcal{H}^1 \cong (m_1, \ldots, m_p, m_1', \ldots, m_p', g_1, \ldots, g_{n-1}) \tag{7-31}$$

of dimension $2p$ if $n = 0$ ($\mathfrak{M}$ compact), of $2p + n - 1$ if $n > 0$ ($\mathfrak{M}$ noncompact). It is left as Exercise 1 to show that the 0-homology classes are represented by mP_0 where P_0 is any fixed point of $\mathfrak{M}$, or

$$\mathcal{H}^0 \cong (m) \tag{7-32}$$

which represents $\mathcal{H}^0$ as an integral vector of dimension 1. It has likewise been shown (by Theorem 7-5) that $\mathcal{H}^2 = k\mathfrak{M}$ if $\mathfrak{M}$ is compact, 0 if $\mathfrak{M}$ is noncompact, or

$$\mathcal{H}^2 \cong (k) \text{ if } n = 0 \qquad \text{or } (0) \text{ if } n > 0 \tag{7-33}$$

Summarizing, we call the dimensionality of the homology classes $\mathcal{H}^t$ the *Betti numbers* β^t and we find

(7-34*a*)	$\beta^0 = 1$	
(7-34*b*)	$\beta^1 = 2p$ ($\mathfrak{M}$ compact)	or $2p + n - 1$ ($\mathfrak{M}$ noncompact)
(7-34*c*)	$\beta^2 = 1$ ($\mathfrak{M}$ compact)	or 0 ($\mathfrak{M}$ noncompact)

In particular for a plane simply-connected region ($p = 0$, $n = 0$ or 1), there is no homology class except 0, or every cycle (e.g., closed curve) is the boundary of a 2-chain. This is the Jordan curve theorem again.

† The symbol $\cong$ is used to denote structural isomorphism.

EXERCISES

1 Show that the 0-chain $\sum_{i=1}^{t} m_i P_i$ is homologous with $\left(\sum_{i=1}^{t} m_i\right) P_0$ for any P_0, by joining each P_i to P_0 by a curve $\mathcal{G}_i$ and considering $\partial(\Sigma m_i \mathcal{G}_i)$. Also show $mP_0 \sim 0$ only when $m = 0$, by considering when $\partial\sigma^1 = mP_0$ is possible (geometrically).

2 From the proof of Theorem 7-5, point out why $\mathcal{C}_1 + \cdots + \mathcal{C}_n \sim 0$ is essentially the only homology relation among the $\mathcal{C}_j$.

7-6 Complex Existence Theorem

We are now prepared to interpret the homology theory of cycles in terms of a theorem of *complex* analysis. The *unsatisfactory* nature of this theorem provides a turning point in the development, inasmuch as we must return to *real* analysis in the next chapter.

LINE INTEGRALS

We have defined the meromorphic differential on $\mathfrak{M}$, $f(\zeta)\,d\zeta$ (see Sec. 5-2), and from it, we define $\int f(\zeta)\,d\zeta$ over a segment of path $\mathcal{G}$, which is assumed to contain no singularity of $f(\zeta)$. It is clearly no more difficult to think of the differential $f(\zeta)\,d\zeta$ over a path of integration consisting of a 1-chain, $\sigma^{(1)} = \Sigma m_i \mathcal{G}_i$. This is written

$$(\sigma^{(1)}, f(\zeta)\,d\zeta) = \int_{\sigma^{(1)}} f(\zeta)\,d\zeta = \Sigma m_i \int_{\mathcal{G}_i} f(\zeta)\,d\zeta \tag{7-35}$$

[according to (7-24)]. The format (σ,ω) is appropriate because we have a function (like the scalar product) which is *bilinear* (linear in σ and ω separately).

The significance of Theorem 7-5 is now seen in the light of *analysis:*

THEOREM 7-6 *Let $f(\zeta)\,d\zeta$ be a holomorphic differential on the finite Riemann manifold $\mathfrak{M}$, then if $\mathcal{Z}$ is a 1-cycle,*

$$(\mathcal{Z}, f(\zeta)\,d\zeta) = \sum_{i=1}^{p} m_i(\mathcal{Q}_i, f(\zeta)\,d\zeta) + \sum_{i=1}^{p} m_i'(\mathcal{Q}_i', f(\zeta)\,d\zeta) + \sum_{j=1}^{n-1} g_j(\mathcal{C}_j, f(\zeta)\,d\zeta) \tag{7-36}$$

where $\mathcal{Z}$ belongs to the homology class represented by

$$\mathcal{Z} \sim \sum_{i=1}^{p} m_i\mathcal{Q}_i + \sum_{i=1}^{p} m_i'\mathcal{Q}_i' + \sum_{j=1}^{n-1} g_j\mathcal{C}_j$$

The proof consists of the recognition of the fact that if $\mathcal{Z} = \partial\sigma^{(2)}$, then

$$(\partial\sigma^{(2)}, f(\zeta)\,d\zeta) = 0 \tag{7-37}$$

This is merely an adaptation of the Cauchy integral theorem, since $\partial\sigma^{(2)}$ is a linear (integral) combination of $\partial\mathcal{R}_i$ where each region $\mathcal{R}_i$ lies wholly in one of the parameter planes. Q.E.D.

The form $(\mathcal{Z}, f(\zeta)\,d\zeta)$ helps us to consummate the analogy between chains and differentials. It is trivial that 1-chains can be constructed homologous to any element represented by the integral vector (7-31). Yet it is *not* true that a differential $f(\zeta)\,d\zeta$ can be constructed which produces any $2p + n - 1$ preassigned values ($2p$ for $\mathfrak{M}$ compact)†

$$(\mathcal{Q}_i, f(\zeta)\,d\zeta),\ (\mathcal{Q}'_i, f(\zeta)\,d\zeta),\ (\mathcal{C}_j, f(\zeta)\,d\zeta) \qquad i = 1, 2, \ldots, p;\ j = 1, 2, \ldots, n-1 \tag{7-38}$$

Here we interpret $(\mathcal{C}_j, f(\zeta)\,d\zeta)$ as the value of the indicated integral over any curve $\mathcal{C}'$ interior to $\mathfrak{M}$, homologous to $\mathcal{C}_j$, and sufficiently close to $\mathcal{C}_j$. Thus we avoid having to know whether or not the differential $f(\zeta)\,d\zeta$ is defined on the boundary $|\partial\mathfrak{M}|$ in any sense.

The best we can do with the values (7-38) is to preassign their *real* parts, and this will be seen to lead to two major existence theorems which return us to *real* analysis.

ABELIAN DIFFERENTIALS (AND INTEGRALS)

Let us consider a meromorphic differential on a Riemann manifold $\mathfrak{M}$, namely $f(\zeta)\,d\zeta$. Let us assume that when $\mathfrak{M}$ is noncompact, then $f(\zeta)$ can be extended continuously to $\partial\mathfrak{M}$.

We next call such a meromorphic differential *abelian* if, in addition,

$$\operatorname{Im} f(\zeta)\,d\zeta = 0 \qquad \text{on } |\partial\mathfrak{M}| = |\mathcal{C}_1| \cup \cdots \cup |\mathcal{C}_n| \tag{7-39}$$

This means that if any curve segment in $|\partial\mathfrak{M}|$ is parametrized by $\zeta = \zeta(s)$ for s real, then $f(\zeta)\zeta'(s)$ is purely real. If $\mathfrak{M}$ is compact, of course, no further restriction is needed, and every meromorphic differential is (vacuously) abelian. The justification for a definition such as (7-39) came originally from hydrodynamics and will be discussed in more detail in Chap. 9.

We say an abelian differential is of the *first kind* if it is analytic; i.e., if $f(\zeta)$ is everywhere finite (in the local parameter). We say an abelian differential is of the *second kind* if $f(\zeta)\,d\zeta$ has residue 0 at all poles (if any). We call an abelian differential one of the *third kind* if $f(\zeta)$ has (possibly) poles with nonzero residues. The categories are defined to be inclusive (every differential of the first kind is necessarily one of the second kind, etc.). It may be noted that these designations (actually the work of Riemann) serve to generalize Legendre's classification of elliptic integrals (compare Sec. 4-2).

When we are dealing with a primitive $\int f(\zeta)\,d\zeta = F(\zeta)$, we call $F(\zeta)$ an "abelian integral" (of first, second, or third kind). Clearly $F(\zeta)$ need not be *single-valued;* and, as in the case of elliptic integrals, $F(\zeta)$ can be

† The unassignability of arbitrary values to (7-38), indeed, is a consequence of Riemann's period relation (Sec. 8-5) which connects these values.

multivalued† with periods (7-38), as well as periods showing logarithmic behavior about a pole when we deal with integrals of the third kind. (See Exercise 2.)

COMPLEX EXISTENCE THEOREM *Let a finite number of singular parts of an abelian differential on a finite Riemann manifold*

$$f_i(\zeta)\,d\zeta = \left(\frac{c_{-1}^{(i)}}{\zeta - \zeta_i} + \frac{c_{-2}^{(i)}}{(\zeta - \zeta_i)^2} + \cdots + \frac{c_{-m_i}^{(i)}}{(\zeta - \zeta_i)^{m_i}}\right) d\zeta \tag{7-40}$$

be prescribed for ζ *near singularities* $\zeta = \zeta_i$ $(i = 1, 2, \ldots, s)$. *Let the real parts of the periods on the* $2p$ *crosscuts [and the possible* $(n-1)$ *periods on* $\partial\mathfrak{M}$*] be prescribed:*

$$\begin{gathered}\operatorname{Re}\,(\mathfrak{Q}_i, f(\zeta)\,d\zeta),\ \operatorname{Re}\,(\mathfrak{Q}_i', f(\zeta)\,d\zeta),\ (\mathfrak{C}_j, f(\zeta)\,d\zeta)\ (=\text{real})\\ i = 1, 2, \ldots, p;\ j = 1, 2, \ldots, n-1\end{gathered} \tag{7-41}$$

Then a unique abelian differential $f(\zeta)\,d\zeta$ *exists with singularities* $f_i(\zeta)$ *and prescribed periods, provided only that the total residue of all the singularities*

$$\sum_{i=1}^{s} c_{-1}^{(i)} = \text{pure imaginary} \tag{7-42a}$$

when $\mathfrak{M}$ *is noncompact; when* $\mathfrak{M}$ *is compact we require*

$$\sum_{i=1}^{s} c_{-1}^{(i)} = 0 \tag{7-42b}$$

For now, let us merely note that condition (7-42*a*) is necessary, since (7-42*b*) follows from Theorem 7-2 (Sec. 7-1) concerning the total residue.

LEMMA 7-10 *If an abelian differential* $f(\zeta)\,d\zeta$ *exists on a noncompact Riemann manifold with* $\partial\mathfrak{M} = \mathfrak{C}_1 + \cdots + \mathfrak{C}_n$, *then its singularities and periods are related by*

$$2\pi i \sum_{i=1}^{s} c_{-1}^{(i)} = \sum_{j=1}^{n} (\mathfrak{C}_j, f(\zeta)\,d\zeta)(=\text{real}) \tag{7-43}$$

PROOF This is clear from the fact that analogously with Theorem 7-2 (Sec. 7-1),

$$2\pi i\,(\text{total res}\,f(\zeta)\,d\zeta) = (\partial\mathfrak{M}, f(\zeta)\,d\zeta) \tag{7-44}$$

which leads directly back to (7-43). Q.E.D.

Since $f(\zeta)\,d\zeta$ is real on $\mathfrak{C}_j$ (by definition of an abelian differential), (7-42*a*) is necessary. If (7-42*a*) is satisfied, the residues $c_{-1}^{(i)}$ can all be

† We show in Appendix B that certain infinite manifolds ("covering surfaces" of $\mathfrak{M}$) are the natural domain of existence of abelian *integrals* (while abelian *differentials* are, of course, definable *strictly on* $\mathfrak{M}$ since they are not multivalued).

arbitrary since the *one unprescribed value* $(\mathcal{C}_n, f(\zeta)\, d\zeta)$ will take whatever (real) value is necessary to satisfy (7-43).

Nevertheless, the condition (7-41) is unsatisfactory since it relates to real parts (even when $\mathfrak{M}$ is compact), although the problem is posed in complex variables (and differentials). We resolve this difficulty in the next chapter and ultimately replace the complex existence theorem by means of two real existence theorems.

EXERCISES

1 Verify the complex existence theorem (but ignore uniqueness) for elliptic integrals of the first kind by letting u be the usual parameter of the period parallelogram, so that ku is an abelian integral of the first kind (or $k\, du$ is an abelian differential) for complex k. Show that its periods are $(\mathcal{Q}_1, k\, du) = 2\omega_1 k$, $(\mathcal{Q}_1', k\, du) = 2\omega_2 k$ with suitable designations. Is it possible to prescribe Re $(2\omega_1 k)$, Re $(2\omega_2 k)$ uniquely by choice of complex k? (*Hint:* Make sure to use the fact that ω_2/ω_1 is not real.)

2 Consider how the complex existence theorem applies to Theorem 4-3 on the existence of an elliptic function (abelian *differential*) with prescribed singularities.

3 Show that the classification of abelian differentials is independent of conformal mapping of the Riemann manifold $\mathfrak{M}$ (including $\partial\mathfrak{M}$) onto another manifold.

4 Prove the following result: Consider $(\mathcal{Z}, f(\zeta)\, d\zeta)$ where $\mathcal{Z}$ is a general 1-cycle and $f(\zeta)\, d\zeta$ is a general (abelian) differential of first kind. Show that if $(\mathcal{Z}, f(\zeta)\, d\zeta) = 0$ for all cycles $\mathcal{Z}$ and a fixed differential of the first kind $f(\zeta)\, d\zeta$, then such a differential is identically zero. Show that conversely if $(\mathcal{Z}, f(\zeta)\, d\zeta) = 0$ for all differentials of the first kind and a fixed 1-cycle $\mathcal{Z}$, then such a 1-cycle is homologous to 0. (*Hint:* Use the existence or uniqueness part of the complex existence theorem.)

5 Verify the abelian differential $i\, dz/z$ for the annulus $\mathcal{R}$: $r_1 < |z| < r_2$. What are the periods?

CHAPTER 8
HARMONIC DIFFERENTIALS

Our main objective is (and will remain) the proving of Riemann's fundamental theorem (to the effect that to every preassigned compact Riemann manifold an algebraic function exists which "determines" such a manifold as a domain of definition; see Sec. 6-2). This theorem might seem to provide no scope for real variable theory since its statement is entirely in terms of complex variables [*complex* planes for the triangulation of the Riemann manifold, *complex* algebraic functions like $w = w(z)$, etc.]. Strangely enough, no proof of this theorem has yet been found which avoids the critical use of real variable theory in some form. Indeed, some inkling of this fact is found in the complex existence theorem (Sec. 7-6) which involves the use of the *real* part of certain periods.

We begin by considering real functions (of real variables).

REAL DIFFERENTIALS

8-1 Cohomology

Rather than deal with *multivalued* functions, we find it easier in principle to speak of *single-valued* differentials. To take the easiest case, let θ be the angle subtended at the origin, in cartesian coordinates. Thus

$$\theta = \arctan \frac{y}{x} \tag{8-1}$$

We note that θ is indeterminate in the sense that if $(x,y) \neq (0,0)$, the ratio y/x corresponds to infinitely many values of θ differing by a multiple of π. If we differentiate, however, we obtain

$$d\theta = \frac{x\,dy}{x^2 + y^2} - \frac{y\,dx}{x^2 + y^2} \tag{8-2}$$

which is a *single-valued differential* as we view the coefficients of dx and dy. We explain this terminology by saying that despite our indeterminate knowledge of θ, we still know $d\theta$ precisely. Thus strictly speaking, $d\theta$ is not the differential of anything definite. We should really not write $d\theta$ but

$$\omega = \frac{x\,dy}{x^2 + y^2} - \frac{y\,dx}{x^2 + y^2} \tag{8-3}$$

as the primary entity, and then we should define θ as a local integral of ω

$$\theta = \int \omega \tag{8-4}$$

Then θ is single-valued in a region $\mathcal{R}$ which does not surround the origin. Now θ is defined, but only from ω.

To generalize these ideas further, let us define (analogously with t chains) a set of "differentials," called t forms $\Omega^{(t)}$, written formally as

$$\Omega^{(t)} \begin{cases} \omega^{(2)} = A(x,y)\,dx\,dy & t = 2 \\ \omega^{(1)} = A(x,y)\,dx + B(x,y)\,dy & t = 1 \\ \omega^{(0)} = A(x,y) & t = 0 \end{cases} \tag{8-5}$$

where the real functions $A(x,y)$, $B(x,y)$ are *single-valued* in a region $\mathcal{R}$ of the xy plane and sufficiently smooth (as many derivatives exist as will be required).†

We *formally* define the differential operator d as follows:

$$B_{t+1} = d\Omega^{(t)} \begin{cases} d\omega^{(2)} = 0 \qquad \text{(identically)} \\ d\omega^{(1)} = \left(\dfrac{-\partial A}{\partial y} + \dfrac{\partial B}{\partial x}\right) dx\,dy \\ d\omega^{(0)} = \dfrac{\partial A}{\partial x}\,dx + \dfrac{\partial A}{\partial y}\,dy \end{cases} \tag{8-6}$$

By analogy with chains (see Sec. 7-5) the set $\Omega^{(t)}$ is called t *cochains.* They are closed under multiplication by real constants and addition (or the t forms are a linear space over the reals for each t). The differential operator d is called a *coboundary* operator and the aggregate of $d\omega^{(t)}$ are called the $(t + 1)$ *coboundaries,* hence the notation B_{t+1}. To carry the analogy further, those $\omega^{(t)}$ for which the forms are *closed,* or

$$Z_t\colon\ d\omega^{(t)} = 0 \tag{8-7}$$

are called *cocycles* and are denoted by Z_t. As before, the cocycles and coboundaries are a linear space over the reals (and the unmentioned dimensions $t > 2$, $t < 0$ always correspond only to 0-forms). A form is called "exact" (meaning "an exact differential") if it is a coboundary.

† We can restrict A and B to have t orders of continuous derivatives in the domain of definition; but in the important cases A and B will be analytic in x and y individually, so that the exact restriction does not materially affect the theory.

Actually, the designation of the differential (form) is quite general and is extendable to any number of dimensions. It is based on formal operations known as the "exterior differential calculus," "Grassmann algebra," etc. Basically, we have a noncommutative associative algebra of differentials (and formal products) over coefficients consisting of functions, such that typically

$$dx\,dy = -dy\,dx$$

This has the intuitive sense that an "area integral" for a region in the xy plane is the negative of the value of the area of the same region regarded as lying in the yx plane (with orientation reversed). From this anticommutation, $dx\,dx = -dx\,dx = 0$ (if we set $x = y$), etc. Thus if we write

$$\text{(8-8}a\text{)}\qquad dA = \frac{\partial A}{\partial x}\,dx + \frac{\partial A}{\partial y}\,dy \qquad dB = \frac{\partial B}{\partial x}\,dx + \frac{\partial B}{\partial y}\,dy$$

and

$$\text{(8-8}b\text{)}\qquad \begin{aligned} d\omega^{(1)} &= d(A\,dx + B\,dy) = dA\,dx + dB\,dy \\ &= \frac{\partial A}{\partial y}\,dy\,dx + \frac{\partial B}{\partial x}\,dx\,dy \end{aligned}$$

then we can easily conclude (8-6) by the formal rules of anticommutation. (Indeed, in two-dimensional space, no triple product of *different* differentials can occur in the two-dimensional plane, so $d\omega^{(2)} = 0$ always.)

From (8-8*a*) we obtain only two nonzero terms in $dA\,dB$ (namely, the terms in $dx\,dy$ and $dy\,dx$)

$$\text{(8-8}c\text{)}\qquad \begin{aligned} dA\,dB &= \left(\frac{\partial A}{\partial x}\frac{\partial B}{\partial y} - \frac{\partial B}{\partial x}\frac{\partial A}{\partial y}\right) dx\,dy \\ &= \frac{\partial(A,B)}{\partial(x,y)}\,dx\,dy \end{aligned}$$

The role of the Jacobian in changing variables of integration can also be justified by this type of calculus.

The role of the differentials dx, dy might seem more special than dA, dB unless, for example,

$$d(A\,dx) = dA\,dx + A\,d(dx)$$

Actually, this is consistent with (8-8*b*) from the rules

$$d(dx) = 0 \qquad d(dy) = 0$$

The important result is that consequently, under such rules,

$$\text{(8-9)}\qquad d(d\omega^{(t)}) = 0$$

The only nontrivial case of (8-9) occurs when $t = 0$, $\omega^{(0)} = A(x,y)$. Then

$$d(d\omega^{(0)}) = d\left(\frac{\partial A}{\partial x}\,dx + \frac{\partial A}{\partial y}\,dy\right)$$
$$= \left(-\frac{\partial}{\partial y}\frac{\partial A}{\partial x} + \frac{\partial}{\partial x}\frac{\partial A}{\partial y}\right)dx\,dy$$

by (8-6). Thus $d(d\omega^{(0)}) = 0$ by "interchangeability of cross derivatives."

We can now define cohomology classes of differentials by saying two t cocycles (closed forms) are *cohomologous*

$$\omega_1^{(t)} \sim \omega_2^{(t)}$$

exactly when

$$\omega_1^{(t)} - \omega_2^{(t)} = d\omega_0^{(t-1)}$$

for some $(t - 1)$ form $\omega_0^{(t-1)}$, defined in the *whole region* $\mathcal{R}$. Thus the coboundaries are the cocycles cohomologous with 0. We therefore arrive at t *cohomology classes* H_t of cocycles (closed forms) Z_t.

Going back to our elementary example (8-3) again (namely, $\omega = d\theta$), we know ω is a 1-cocycle. It is cohomologous to 0 if we take a region of definition $\mathcal{R}$ which does not surround the origin.

The above formalism is most seriously involved in extending these definitions of $\omega^{(t)}$ over a Riemann manifold $\mathfrak{M}$. This is done by defining $\omega^{(t)}$ over each triangle of the manifold. Then, where the triangles adjoin one another, a mapping function (5-5) is defined which joins the edges. We must make the differentials coherent, as in Sec. 5-2.

Let us suppose that (x,y) and (x',y') are for simplicity the two variables (the u planes of Sec. 5-1) and that they are joined by an analytic mapping

$$x' = \phi(x,y) \qquad y' = \psi(x,y) \tag{8-10}$$

Then the condition that $\omega^{(t)}$ be defined over the manifold is that (where the adjoining triangles have overlapping mapping functions)

$$\begin{aligned} &A(x',y')\,dx'\,dy' = A(\phi(x,y),\psi(x,y))\,d\phi\,d\psi \\ &A(x',y')\,dx' + B(x',y')\,dy' \\ &\qquad\qquad = A(\phi(x,y),\psi(x,y))\,d\phi + B(\phi(x,y),\psi(x,y))\,d\psi \\ &A(x',y') = A(\phi(x,y),\psi(x,y)) \end{aligned} \tag{8-11}$$

These conditions would, of course, have to be expanded according to the method of (8-8*a*) to (8-8*c*), $d\phi = (\partial\phi/\partial x)\,dx + (\partial\phi/\partial y)\,dy$, etc. Thus each statement (8-11) has the form

$$\omega_0^{(t)}(\text{in } x',y') = \omega_1^{(t)}(\text{in } x,y)$$

We now must verify that the cohomology theory is independent of coordinates so that the mode of definition of the manifold $\mathfrak{M}$ is not part of the theory. (This independence was almost trivial in the case of the homology theory; e.g., the boundary of a region will transform into the boundary of the image *by continuity* when a region is transformed in biunique fashion.) Here we must verify somewhat laboriously that under the transformation (8-10), if on change from (x',y') to (x,y) we find

$$\omega_0^{(t)} \text{ transforms into } \omega_1^{(t)}$$

then

$$d\omega_0^{(t)} \text{ transforms into } d\omega_1^{(t)}$$

This is left as Exercise 1.

Thus we have a cohomology theory on Riemann manifolds $\mathfrak{M}$ referred to the cartesian components (x,y) of the local coordinates $u = x + iy$. (Yet this theory is independent of coordinates.) Any statement concerning real differentials extends to the complex differentials (usually 1-forms)

$$dG = d\psi + i\,d\psi^* = \frac{\partial\psi}{\partial x}\,dx + \frac{\partial\psi}{\partial y}\,dy + i\left(\frac{\partial\psi^*}{\partial x}\,dx + \frac{\partial\psi^*}{\partial y}\,dy\right)$$

in an obvious way. Thus, $d \log z$ $(z = x + iy)$ is a coboundary in a region $\mathfrak{R}$ $(= \mathfrak{M})$ which does not encircle the origin, etc. The transformation of 1-forms by (8-11) is seen to be consistent with the chain rule $dG/du = (dG/dw)(dw/du)$ of calculus, so that we are in later practice almost unaware of presence of the formalism we are expounding here. (We defer the verification of this until Exercise 8, Sec. 8-3, when we reintroduce complex variables.)

SEVERAL DIMENSIONS

The method of differential forms is most effective in three-dimensional space, where it can be seen to contain "all of vector analysis." We define

$$\begin{aligned}
\omega^{(0)} &= f(x,y,z) \\
\omega^{(1)} &= A(x,y,z)\,dx + B(x,y,z)\,dy + C(x,y,z)\,dz \\
\omega^{(2)} &= A(x,y,z)\,dy\,dz + B(x,y,z)\,dz\,dx + C(x,y,z)\,dx\,dy \\
\omega^{(3)} &= f(x,y,z)\,dx\,dy\,dz
\end{aligned} \tag{8-12}$$

In the second and third cases the vector $[A,B,C]$ can be imagined as a vector field over (x,y,z) space.

We follow the usual associative and anticommutation rules to obtain, typically,

$$\begin{aligned}
dx\,dy\,dz &= (dx\,dy)\,dz = -dy\,dx\,dz = dy\,dz\,dx \\
dx\,dy\,dx &= dy\,dx\,dx = dy(dx\,dx) = 0 \\
d(A\,dx) &= dA\,dx,\ d(A\,dy\,dz) = dA\,dy\,dz, \text{ etc.}
\end{aligned}$$

where, of course, dA is $\partial A/\partial x\, dx + \partial A/\partial y\, dy + \partial A/\partial z\, dz$. Thus

$$
\begin{aligned}
d\omega^{(0)} &= \frac{\partial f}{\partial x}\,dx + \frac{\partial f}{\partial y}\,dy + \frac{\partial f}{\partial z}\,dz && \text{gradient } f\\
d\omega^{(1)} &= \left(\frac{\partial C}{\partial y} - \frac{\partial B}{\partial z}\right)dy\,dz + \left(\frac{\partial A}{\partial z} - \frac{\partial C}{\partial x}\right)dz\,dx\\
&\qquad + \left(\frac{\partial B}{\partial x} - \frac{\partial A}{\partial y}\right)dx\,dy && \text{curl } [A,B,C]\\
d\omega^{(2)} &= \left(\frac{\partial A}{\partial x} + \frac{\partial B}{\partial y} + \frac{\partial C}{\partial z}\right)dx\,dy\,dz && \text{divergence } [A,B,C]\\
d\omega^{(3)} &= 0
\end{aligned}
\tag{8-13}
$$

Actually, the usual vector functions arise from the coboundary operator on differential forms. Here the gradient or grad f is a vector whose (x,y,z) components are the coefficients of (dx,dy,dz) in $d\omega^{(0)}$; the *rotation* or curl $[A,B,C]$ is a vector function of the vector $[A,B,C]$ whose (x,y,z) components are the respective coefficients in $d\omega^{(1)}$; and the *divergence* or div $[A,B,C]$ is a scalar function of the vector $[A,B,C]$ whose value is the coefficient of $dx\,dy\,dz$ in $d\omega^{(2)}$.

The statements that the $\omega^{(t)}$ are closed ($d\omega^{(t)} = 0$), therefore, lead to properties familiar from vector analysis.

EXERCISES

1 Show that the coboundary operator d is independent of coordinates by showing that if $\omega_0^{(t+1)} = d\omega^{(t)}$ in coordinates (x,y), the relation holds on change of coordinate to (x',y') as given in (8-10). For instance, let

$$\omega_0^{(1)} = dA(x,y) = \frac{\partial A}{\partial x}\,dx + \frac{\partial A}{\partial y}\,dy$$

while

$$\frac{\partial A}{\partial x} = \frac{\partial A}{\partial x'}\frac{\partial \phi}{\partial x} + \frac{\partial A}{\partial y'}\frac{\partial \psi}{\partial x}, \text{ etc.}$$

Thus

$$
\begin{aligned}
\omega_0^{(1)} &= \frac{\partial A}{\partial x'}\left(\frac{\partial \phi}{\partial x}\,dx + \frac{\partial \phi}{\partial y}\,dy\right) + \cdots = \frac{\partial A}{\partial x'}\,dx' + \cdots\\
&= d(A(x',y'))
\end{aligned}
$$

Do likewise for the other cases of $d\omega^{(t)}$ in (8-6).

2 To extend the formalism to n variables, we define the t forms $(0 \le t \le n)$

$$\omega^{(t)} = \sum_I A_{i_1}{}^{(t)}{}_{\cdots i_t}\,dx_{i_1}\cdots dx_{i_t} = \sum_I A_I{}^{(t)}\,dX_I \tag{8-14}$$

Here $I = (i_1, \ldots, i_t)$ is an ordered subset of t subscripts. Show (*a*) that there can be at most ${}_nC_t$ different terms; that is, $dX_I = 0$ if there is a repetition in I. Likewise, define a u form

$$\omega^{(u)} = \sum_J A_J{}^{(u)}\,dX_J \qquad \text{where } J = (j_1, \ldots, j_u)$$

Then we can define $\omega^{(t)}\omega^{(u)}$ formally as $\Sigma\Sigma A_I{}^{(t)}A_J{}^{(u)}\,dX_I\,dX_J$, and

$$d\omega^{(t)} = \sum_I \sum_k \frac{\partial A_I{}^{(t)}}{\partial x_k}\,dx_k\,dX_I$$

Show (*b*) that

$$dx_k\,dX_I = (-1)^t\,dX_I\,dx_k \tag{8-15}$$

and show (*c*) that

$$d(\omega^{(t)}\omega^{(u)}) = (d\omega^{(t)})\omega^{(u)} + (-1)^t\omega^{(t)}\,d\omega^{(u)} \tag{8-16}$$

3 Verify (independently of Exercise 2*c*) for forms ϕ and $\omega^{(1)}$

$$d(\phi\omega^{(1)}) = d\phi\,\omega^{(1)} + \phi\,d\omega^{(1)}$$

4 Justify the sign in $\omega^{(t)}\omega^{(u)} = (-1)^{tu}\omega^{(u)}\omega^{(t)}$.

8-2 Stokes' Theorem

The formalism just developed for homology and cohomology is actually a very *concrete* model of a development in mathematics which is still under process of generalization. The formalism, however, is best motivated by connections with the so-called "Stokes' theorem."

We start with, say, a simply-connected region $\mathcal{R}$. Let $\omega^{(1)}$ be a 1-form in $\mathcal{R}$, then by Gauss' theorem

$$\int_{\partial\mathcal{R}} A(x,y)\,dx + B(x,y)\,dy = \iint_{\mathcal{R}} \left(-\frac{\partial A}{\partial y} + \frac{\partial B}{\partial x}\right) dx\,dy \tag{8-17a}$$

The proof of this result is an elementary matter involving integration by parts.†

We can interpret it symbolically as

$$\int_{\partial\mathcal{R}} \omega^{(1)} = \iint_{\mathcal{R}} d\omega^{(1)} \tag{8-17b}$$

Next we consider the fact that Gauss' theorem is applicable to any region $\mathcal{R}$ in a manifold $\mathfrak{M}$, possibly multiply-connected and of genus > 1 (as is $\mathfrak{M}$). From the *orientability* of $\mathfrak{M}$, we can subdivide $\mathcal{R}$ into (simply-connected) triangles such that $\int\omega^{(1)}$ cancels on adjacent sides, and thus we obtain (8-17*b*) in fullest generality (as applied to $\mathfrak{M}$, not $\mathcal{R}$).

We go one step farther and redefine the symbol $(\sigma^{(t)},\omega^{(t)})$ for integration of a t form over a t chain on $\mathfrak{M}$ [using the definitions of $\sigma^{(t)}$ in (7-24)

† We refer to texts on advanced calculus for this proof. The result (8-17*a*) is often called Gauss' theorem, but actually the theorems of Gauss, Green, Stokes, etc., are all united by the formalism of differentials, and Stokes' theorem is now accepted as the generic name for these results.

and the definition of $\omega^{(t)}$ in (8-5)]. Specifically,

$$\text{(8-18)}\qquad \begin{aligned} (\sigma^{(2)},\omega^{(2)}) &= \sum m_i \iint_{\mathcal{R}_i} A(x,y)\,dx\,dy \\ (\sigma^{(1)},\omega^{(1)}) &= \sum m_i \int_{\mathcal{G}_i} A(x,y)\,dx + B(x,y)\,dy \\ (\sigma^{(0)},\omega^{(0)}) &= \sum m_i A(x_i,y_i) \end{aligned}$$

Here, of course, $\sigma^{(t)}$ is referred to the local coordinates in each triangle so that each $\mathcal{R}_i$ or $\mathcal{G}_i$ lies in a single triangle [and indeed (x_i,y_i) is the local coordinate of the point P_i].

Then Gauss' theorem in its full glory is renamed Stokes' theorem:

THEOREM 8-1

$$\text{(8-19)}\qquad (\partial\sigma^{(t+1)},\omega^{(t)}) = (\sigma^{(t+1)},d\omega^{(t)})$$

This theorem is used here only in case $t = 1$ [the usual Gauss' theorem (8-17*b*)]. When $t = 0$, it is derivable from the so-called "fundamental theorem of calculus": If $\sigma^{(1)} = \mathcal{G}$, the curve from P to P', then $\partial\mathcal{G} = P' - P$ and (8-19) becomes

$$\omega^{(0)}(P') - \omega^{(0)}(P) = \int_{\mathcal{G}} d\omega^{(0)}$$

A very useful form of Theorem 8-1 occurs if $t = 1$ and

$$\omega^{(1)} = \phi_1\,d\phi_2,\ d\omega^{(1)} = d\phi_1\,d\phi_2$$

But $d\phi_1\,d\phi_2 = [(\partial\phi_1/\partial x)(\partial\phi_2/\partial y) - (\partial\phi_1/\partial y)(\partial\phi_2/\partial x)]\,dx\,dy$. Therefore, for a manifold (or region) $\mathfrak{M}$

$$\text{(8-20}a\text{)}\qquad (\mathfrak{M},\ d\phi_1\,d\phi_2) = (\partial\mathfrak{M},\ \phi_1\,d\phi_2)$$

or in more conventional notation, we have a statement that the Jacobian (on the left) is related to the area (as a line integral on the right):

$$\text{(8-20}b\text{)}\qquad \iint_{\mathfrak{M}} \left(\frac{\partial\phi_1}{\partial x}\frac{\partial\phi_2}{\partial y} - \frac{\partial\phi_1}{\partial y}\frac{\partial\phi_2}{\partial x}\right) dx\,dy = \int_{\partial\mathfrak{M}} \phi_1\,d\phi_2$$

SIMPLY-CONNECTED REGION

In the basic advanced calculus course the most important result of Gauss' theorem is as follows:

LEMMA 8-1 *If $\mathcal{R}$ is a simply-connected region, then the 1-form on $\mathcal{R}$, namely $\omega^{(1)} = A\,dx + B\,dy$, is an exact differential or*

$$\text{(8-21}a\text{)}\qquad \omega^{(1)} = df \quad \textit{for some } f(x,y);\ \text{i.e.,}\quad A = \frac{\partial f}{\partial x} \qquad B = \frac{\partial f}{\partial y}$$

if and only if the 1-*form* $\omega^{(1)}$ *is closed or*

$$d\omega^{(1)} = 0; \text{ i.e.,} \qquad \frac{\partial A}{\partial y} = \frac{\partial B}{\partial y} \tag{8-21b}$$

The "if" part of the proof consists of defining $\omega^{(0)} = \int \omega^{(1)}$ or

$$f(x,y) = \int_{(x_0,y_0)}^{(x,y)} A\,dx + B\,dy$$

over some arbitrary path $\mathcal{K}$ (in $\mathcal{R}$) from (x_0,y_0) to (x,y). We must then see that $f(x,y)$ is independent of $\mathcal{K}$, by showing that any two such paths, say $\mathcal{K}_1$ and $\mathcal{K}_2$, lead to the same result. This amounts to showing $\int_{\mathcal{K}_1} = \int_{\mathcal{K}_2}$ or

$$\int_{\mathcal{K}_1 - \mathcal{K}_2} \omega^{(1)} = 0$$

Now $\mathcal{K}_1 - \mathcal{K}_2$ is a closed path, hence a cycle. But in a simply-connected region (by Corollary 1 to Theorem 7-5, Sec. 7-5), all 1-cycles are boundaries; in other words, $\mathcal{K}_1 - \mathcal{K}_2 = \partial\sigma^{(2)}$ for some 2-chain $\sigma^{(2)}$. Thus

$$\int_{\mathcal{K}_1 - \mathcal{K}_2} \omega^{(1)} = \int_{\partial\sigma^{(2)}} \omega^{(1)} = \iint_{\sigma^{(2)}} d\omega^{(1)} = 0$$

Once $f(x,y)$ is seen to be definable independently of path, we see that $\partial f/\partial x = A$, $\partial f/\partial y = B$ (see Exercise 1).

The "only if" part of the proof is a trivial identity on cross derivatives, $\partial A/\partial y = \partial B/\partial x (= \partial^2 f/\partial x \partial y)$, or symbolically $d(df) = 0$.

LEMMA 8-2 *If* $\mathcal{R}_0$ *is not necessarily simply-connected, every* 1-*form which is closed is locally exact (i.e., exact in a sufficiently small neighborhood of every interior point of* $\mathcal{R}_0$).

For proof, we merely take $\mathcal{R}$ to be the sufficiently small but simply-connected neighborhood used in Lemma 8-1.

We now have a suitable generalization of Lemma 8-1.

LEMMA 8-3 *If* $\omega^{(1)}$ *is a* 1-*cocycle on* $\mathfrak{M}$ $(d\omega^{(1)} = 0)$, *then* $\omega^{(1)}$ *is exact on* $\mathfrak{M}$ $(\omega^{(1)} =$ *a coboundary,* $d\omega^{(0)})$ *if and only if for the* $2p + n - 1$ (or $2p$) *curves* $\mathcal{Q}_i$, $\mathcal{Q}_i'$, $\mathcal{C}_j$, *we have*

$$0 = (\mathcal{Q}_i,\omega^{(1)}) = (\mathcal{Q}_i',\omega^{(1)}) = (\mathcal{C}_j,\omega^{(1)}) \qquad 1 \le i \le p, 1 \le j \le n-1 \tag{8-22}$$

(*omitting* $\mathcal{C}_j$ *if* $n = 0$ *or* 1). *In particular, if* $\mathfrak{M}$ *is simply-connected* $(p = 0, n = 0$ or $1)$, *then every* 1-*cocyle is a* 1-*coboundary.*

For proof we first note that the "only if" part again is trivial. If $\omega^{(1)} = d\omega^{(0)}$, then $(\mathcal{Q}_1,d\omega^{(0)}) = (\partial\mathcal{Q}_1,\omega^{(0)}) = (0,\omega^{(0)}) = 0$, etc., by Stokes' theorem.

To prove the "if" part note that any closed curve $\mathcal{C}$ satisfies

$$\mathcal{C} \sim m_1\mathcal{Q} + \cdots + m_1'\mathcal{Q}_1' + \cdots + g_1\mathcal{C}_1 + \cdots$$

by Theorem 7-5. Hence, any $(\mathcal{C},\omega^{(1)}) = 0$.

Thus, as in Lemma 8-1, we can define $\omega^{(0)} = \int\omega^{(1)}$ over any path $\mathcal{K}$ from a fixed point on $\mathfrak{M}$ to a variable point, completing the proof.

To properly illustrate the analogy between homology and cohomology, let us return to the annulus $\mathcal{A}$ lying between the two concentric circles $\mathcal{C}_1$, $\mathcal{C}_2$ where $\mathcal{C}_j$ is $|z| = r_j$ $(r_1 < r_2)$ and the orientation gives $\partial\mathcal{A} = \mathcal{C}_1 + \mathcal{C}_2$, as usual. We have just seen that in $\mathcal{A}$ all 1-cycles $\mathcal{Z}$ satisfy the homology

$$\mathcal{Z} \sim n\mathcal{C}_1$$

where n is an integer, actually the winding number $N(\mathcal{Z};0)$. It also follows quite analogously that in $\mathcal{A}$ all 1-cocycles ω^1 satisfy the cohomology to $\rho\, d\theta$ of Sec. 8-1; for example,

$$\omega^{(1)} \sim \frac{\rho(x\,dy - y\,dx)}{x^2 + y^2}$$

where the real number ρ is now found from the period $\int\omega^{(1)}/2\pi$, according to Lemma 8-3. Indeed, Exercise 6 makes an obvious extension to a multiply-connected region.

In any case, we see that for a plane region the dimension of the 1-homology classes of 1-cycles (over integral coefficients) equals the 1-cohomology classes of 1-forms (over real coefficients). This is true in greater generality (compare Exercise 2). We shall state what we need of this analogy in more concrete analytic form.

THEOREM 8-2 *The 1-cohomology classes on $\mathfrak{M}$ are determined uniquely by the assignment of $2p + n - 1$ real numbers ($2p$ if $\mathfrak{M}$ is compact) which serve as the periods for the closed 1-forms $\omega^{(1)}$:*

$$(8\text{-}23) \qquad (\mathcal{Q}_i,\omega^{(1)}) = h_i \qquad (\mathcal{Q}_i',\omega^{(1)}) = h_i', \qquad (\mathcal{C}_j,\omega^{(1)}) = k_j$$

for $1 \le i \le p$, $1 \le j \le n - 1$ $(n \ge 2)$.

Our proof is highly "overpowered." The complex existence theorem (of Sec. 7-6) assures us of the existence of a *unique abelian (complex) differential whose real part takes these values.* Because of this, a study will next be undertaken concerning the use of special differentials which can fulfill such period conditions† for the Riemann manifold.

ALGEBRAIC INTERPRETATION

We shall eventually use the formalism of cohomology for the *interpretation* of results in analysis. It is therefore relevant to mention that

† The general theory in n dimensions is often referred to as the "Poincaré–de Rham theory."

in a more abstract vein cohomology is defined not by the t form $\omega^{(t)}$, but by the "pairing" $(\sigma^{(t)},\omega^{(t)})$. This pairing provides a mapping of the t chains $\sigma^{(t)}$ into the real numbers, namely the values of the integrals $(\sigma^{(t)},\omega^{(t)})$; and this mapping is *bilinear* (in $\sigma^{(t)}$ and $\omega^{(t)}$). Hence for some purposes of the algebraist, the forms $\omega^{(t)}$ do not need to exist but only the set of values $(\sigma^{(t)},\omega^{(t)})$ which for each $\omega^{(t)}$ provide a homomorphism (mapping) into the reals. For example, if the t forms correspond to homomorphisms, the addition of t forms corresponds abstractly to addition of homomorphisms.

For us it is necessary that such differentials $\omega^{(t)}$ actually exist! (For $t = 1$, they are the "objects" of the Riemann existence theorem.) Indeed, by Theorem 8-2, all homomorphisms are accounted for in this way.

EXERCISES

1 Show that if $f = \int_{\mathcal{K}(x_0,y_0)}^{(x,y)} A\,dx + B\,dy$ is independent of path $\partial f/\partial x = A$ [by choosing a path $\mathcal{K}$ which approaches (x,y) horizontally; that is, $\mathcal{K}$ can be the succession of two-line segments first from (x_0,y_0) to (x_0,y) and then from (x_0,y) to (x,y)]. Likewise, show $\partial f/\partial y = B$.

2 Prove that the t cohomology classes H_t of a Riemann manifold $\mathfrak{M}$ are β_t-dimensional linear spaces over the reals where [analogously with (7-34*a*) to (7-34*c*)]

(8-24*a*) $$\beta_0 = 1$$
(8-24*b*) $$\beta_1 = 2p + n - 1 \;(\mathfrak{M}\text{ noncompact}) \qquad \text{or } 2p \;(\mathfrak{M}\text{ compact})$$

Show (8-24*a*) by showing that the 0-cocycles are constants and the 0-coboundaries are (formally) 0. Show (8-24*b*) follows from Theorem 8-2. (The result for H_2 is more complicated, and we must refer to works on differential forms in the Bibliography.)

3 Consider Stokes' theorem for the coboundaries (8-13). (Derive the gradient theorem, the curl-type Stokes' theorem, and the divergence theorem.)

4 Show $A\,dx + B\,dy + C\,dz = \omega^{(1)}$ is closed exactly when $\omega^{(1)} = df(x,y,z)$ *locally* (for f single-valued in any small neighborhood of the region of space where $\omega^{(1)}$ is defined).

5 Prove (8-20*b*) directly by integration by parts.

6 Prove Theorem 8-2 directly for the case where $\mathfrak{M}$ is a region ($p = 0$, $n \geq 2$) lying interior to $\mathcal{C}_n$ and exterior to $-\mathcal{C}_j$ ($1 \leq j \leq n - 1$), where $-\mathcal{C}_j$ is the boundary of a disk $\mathcal{R}_j$. Here let $z_j = x_j + iy_j$ be any point in $\mathcal{R}_j$, and let us define $d \arg (z - z_j)$ or

$$\omega_j^{(1)} = \frac{(x - x_j)\,dy - (y - y_j)\,dx}{|z - z_j|^2}$$

Then show any closed 1-form in $\mathfrak{M}$ is cohomologous to a linear combination $\sum_{1}^{n-1} \omega_j^{(1)}\rho_j$ with real constants. [Here $\omega_j^{(1)}$ $(j = 1, \ldots, n - 1)$ constitutes a "cohomology basis."]

8-3 Conjugate Forms

We now consider real 1-forms which are the real parts of complex differentials. Let us start with the 1-form

(8-25a)
$$\omega = A\,dx + B\,dy$$

and formally define the *conjugate*

(8-25b)
$$\omega^* = -B\,dx + A\,dy$$

A connection with complex analysis seems manifestly clear from

(8-26)
$$\omega + i\omega^* = (A - iB)(dx + i\,dy)$$

Thus from the point of view of complex analysis, we say

$$\frac{\omega + i\omega^*}{dz} \to \text{limit as } dz \to 0$$

and in other words, the definition of ω^* was chosen for its connection with the concept of a derivative (analyticity).

Another connection with i comes from $(\omega^*)^* = -\omega$ (just as $i^2 = -1$). If $\omega = df = \partial f/\partial x\,dx + \partial f/\partial y\,dy$, then

(8-27a)
$$(df)^* = -\frac{\partial f}{\partial y}\,dx + \frac{\partial f}{\partial x}\,dy$$

Now $d(df)^*$ is not always zero, indeed

(8-27b)
$$d(df)^* = \Delta f\,dx\,dy$$

where

(8-27c)
$$\Delta f = \frac{\partial^2 f}{\partial x^2} + \frac{\partial^2 f}{\partial y^2}$$

CAUCHY-RIEMANN EQUATIONS

We call the 1-form ω *harmonic*, if ω and ω^* are both closed. This means

(8-28a)
$$\frac{\partial A}{\partial y} = \frac{\partial B}{\partial x} \qquad (\omega \text{ closed})$$
$$\frac{\partial B}{\partial y} = -\frac{\partial A}{\partial x} \qquad (\omega^* \text{ closed})$$

These are called the *Cauchy-Riemann* equations on $A - iB$. Usually they are expressed in terms of $u(x,y) + iv(x,y)$ as

(8-28b)
$$\frac{\partial u}{\partial x} = \frac{\partial v}{\partial y}$$
$$\frac{\partial v}{\partial x} = -\frac{\partial u}{\partial y}$$

THEOREM 8-3 (*Cauchy-Riemann*) *A harmonic differential* (1-*form*) *is precisely the real part of an* (*analytic*) *complex differential.*

The proof is usually given in the form of Exercises 1 and 2, which show that $u + iv$ satisfy the Cauchy-Riemann equations (8-28*b*) exactly when $u + iv$ is *locally* an analytic function of $x + iy$. Of course, an analytic function $w(z) = u + iv$ is the local derivative of something [namely, of $\int w(z)\, dz$].

COROLLARY 1 *The harmonic differentials* ω *are* (*locally*) *precisely the differentials* df *of solutions* f *to Laplace's equation,* $\Delta f = 0$.

COROLLARY 2 *If* ω *is a harmonic differential, so is* ω^*. (*See Exercise* 7.)

If the harmonic differential ω is exact in some region $\mathcal{R}$, we write $\omega = d\phi$ and call ϕ *a harmonic function* in $\mathcal{R}$ (which may be a Riemann manifold). Very often the term *harmonic integral* is used for the first primitive $\int d\phi$ of a harmonic differential, but a harmonic integral (unlike a harmonic function) is then multivalued. Indeed, the real (or imaginary) part of an abelian integral is a special case of a harmonic integral, as we soon see. (This is illustrated in Exercise 4 for the torus.)

We shall yield to classical terminology by writing all harmonic differentials as $d\psi$ *even though* ψ *need not be single-valued* (*except in the neighborhood of a point*). This avoids the use of extra symbols (such as $\omega^{(1)}$ for $d\psi$). Thus, given ψ (or even $d\psi$), then whether or not ψ^* is single-valued, we can still write $d\psi^*$ as though ψ^* were uniquely defined (see Exercise 3). Of course, since $(d\psi)^*$ is closed, we are taking advantage of a notational convenience,

$$(d\psi)^* = d(\psi^*) \qquad \text{(written } d\psi^*\text{)} \tag{8-29}$$

(This trick of notation holds only because $d\psi$ is harmonic.)

THEOREM 8-4 *A harmonic differential, integral, or function defined in one coordinate system remains harmonic in a new coordinate system derived from a conformal* (*biunique analytic*) *mapping.*

The proof consists of the observation that an analytic function also preserves analyticity under a biunique conformal mapping (by Theorem 3-6). By Theorem 8-3 this invariance property extends to harmonic differentials, etc. Thus, symbolically, if

$$\phi_1(x',y') + i\phi_2(x',y') = x + iy \tag{8-30}$$

is an analytic function and if $\psi(x,y)$ is a harmonic function, then $\psi(\phi_1(x',y'),\phi_2(x',y'))$ is a harmonic function in x' and y'. We must speak of a *biunique* mapping only so that we can speak of $\psi(x,y)$ being defined in a region $\mathcal{R}'$ (in the $x'y'$ plane) which is a one-to-one image of $\mathcal{R}$ (in the

xy plane) under (8-30). Clearly, the use of local uniformizing parameters extends these results to Riemann manifolds.

Thus harmonic *functions* or *integrals* can be regarded as values $\psi(P)$ pinned onto various points P of a Riemann manifold $\mathfrak{M}$ so that if $\mathfrak{M}$ is transformed conformally, ψ takes the same value on the image of P as $\psi(P)$. For harmonic *differentials* the relationship is harder to visualize, so that we do find it easier to talk about ψ than its differential $\omega^{(1)}$, even if ψ is not single-valued. The Cauchy-Riemann equations are then (for $\psi + i\psi^*$)

(8-31) $$\frac{\partial \psi}{\partial x} = \frac{\partial \psi^*}{\partial y} \qquad \frac{\partial \psi}{\partial y} = -\frac{\partial \psi^*}{\partial x}$$

SEVERAL DIMENSIONS

The concept of a harmonic differential is easier to generalize to n dimensions than the concept of a complex variable. The important property is that [from (8-25*a*) and (8-25*b*)],

(8-32*a*) $$\omega\omega^* = (A^2 + B^2)\, dx\, dy$$

In essence, (8-32*a*) is sufficient as a defining property.

For example, in three dimensions, the conjugates pair off as follows

(8-33*a*) $$\omega_0 = A(x,y,z)$$
$$\omega_0^* = A(x,y,z)\, dx\, dy\, dz$$
(8-32*b*) $$\omega_0\omega_0^* = A^2(x,y,z)\, dx\, dy\, dz$$
(8-33*b*) $$\omega_1 = A\, dx + B\, dy + C\, dz$$
$$\omega_1^* = A\, dy\, dz + B\, dz\, dx + C\, dx\, dy$$
(8-32*c*) $$\omega_1\omega_1^* = (A^2 + B^2 + C^2)\, dx\, dy\, dz$$

More generally, the process of conjugation is discussed in Exercise 10. For now, let us say that generally, ω_i is *harmonic* exactly when both $d\omega_i$ and $d\omega_i^*$ vanish.

Thus in (8-33*a*) the harmonic forms are (only)

(8-34*a*) $$A = \text{const}$$

In (8-33*b*), however, the differentials ω_1 (or ω_1^*) are harmonic exactly when $\omega_1 = df(x,y,z)$ (see Exercise 4, Sec. 8-2) and accordingly,

$$d\left(\frac{\partial f}{\partial x}\, dy\, dz + \frac{\partial f}{\partial y}\, dz\, dx + \frac{\partial f}{\partial z}\, dx\, dy\right) = 0$$

This leads, on expansion, to

(8-34*b*) $$\Delta f = \frac{\partial^2 f}{\partial x^2} + \frac{\partial^2 f}{\partial y^2} + \frac{\partial^2 f}{\partial z^2} = 0$$

the Laplace equation in three variables.

It is now not surprising that the Laplace equation is the basic analytic tool of extending complex analysis to any number of dimensions. Various boundary-value problems arise, however, in regard to it. They are called "Dirichlet problems" if they are related to the prescription of conditions on the boundary of the region of definition (among other prescribed data). We examine two such problems in the next two sections which supersede the complex existence theorem.

DIRICHLET INTEGRAL

To understand the relationships (8-32*a*) to (8-32*c*) in terms of differential forms, we need certain basic identities which lead to them. By Exercise 3 of Sec. 8-1 above, for functions ϕ_1, ϕ_2 (not necessarily harmonic),

$$d\phi_1\,(d\phi_2)^* = -\phi_1\,d(d\phi_2)^* + d(\phi_1(d\phi_2)^*) \tag{8-35}$$

Thus, by Stokes' theorem,

$$(\mathfrak{M}, d\phi_1\,(d\phi_2)^*) = -\,(\mathfrak{M}, \phi_1\,d(d\phi_2)^*) + (\partial\mathfrak{M}, \phi_1(d\phi_2)^*) \tag{8-36}$$

We now rewrite the terms in (8-36) using more conventional symbols:

$$D_{\mathfrak{M}}[\phi_1,\phi_2] = \iint\limits_{\mathfrak{M}} \left(\frac{\partial\phi_1}{\partial x}\frac{\partial\phi_2}{\partial x} + \frac{\partial\phi_1}{\partial y}\frac{\partial\phi_2}{\partial y}\right) dx\,dy \tag{8-37a}$$

But by (8-27*a*),

$$d\phi_1\,(d\phi_2)^* = \left(\frac{\partial\phi_1}{\partial x}\frac{\partial\phi_2}{\partial x} + \frac{\partial\phi_1}{\partial y}\frac{\partial\phi_2}{\partial y}\right) dx\,dy$$

Thus

$$D_{\mathfrak{M}}[\phi_1,\phi_2] = (\mathfrak{M}, d\phi_1\,(d\phi_2)^*) \tag{8-37b}$$

This expression is the *bilinear form of the Dirichlet integral*

$$D_{\mathfrak{M}}[\phi] = D_{\mathfrak{M}}[\phi,\phi] = \iint\limits_{\mathfrak{M}} \left[\left(\frac{\partial\phi}{\partial x}\right)^2 + \left(\frac{\partial\phi}{\partial y}\right)^2\right] dx\,dy \tag{8-38}$$

Also, of course, by (8-27*a*) and (8-27*b*)

$$d(d\phi_2)^* = \Delta\phi_2\,dx\,dy \tag{8-39}$$

Thus by applying (8-36), we finally obtain

$$D_{\mathfrak{M}}[\phi_1,\phi_2] = -\iint\limits_{\mathfrak{M}} \phi_1\,\Delta\phi_2\,dx\,dy + \int_{\partial\mathfrak{M}} \phi_1(d\phi_2)^* \tag{8-40}$$

where†

$$\int_{\partial\mathfrak{M}} \phi_1\,(d\phi_2)^* = \int_{\partial\mathfrak{M}} -\phi_1\frac{\partial\phi_2}{\partial y}\,dx + \phi_1\frac{\partial\phi_2}{\partial x}\,dy$$

† In Sec. 9-1, we see how to write $\int_{\partial\mathfrak{M}} \phi_1(d\phi_2)^*$ as $-\int_{\partial\mathfrak{M}} \phi_1(\partial\phi_2/\partial n)\,ds$, where s is arc length on $|\partial\mathfrak{M}|$ and n is inner normal to $|\partial\mathfrak{M}|$.

Returning to *harmonic* functions ϕ, we find from (8-40)

$$D_{\mathfrak{M}}[\phi] = \int_{\partial\mathfrak{M}} \phi \, d\phi^* \tag{8-41}$$

(We do not require single-valuedness for ϕ^*, but only for ϕ.)

The relation (8-41) is quite deep. It relates the self-energy of a charge distribution $D_{\mathfrak{M}}[\phi]$ with the boundary data on $\partial\mathfrak{M}$. This relation provides the core of an historically famous development of minimal principles (i.e., the principle that harmonic functions will minimize self-energy). We return to this matter in Appendix A.

EXERCISES

1 Consider $w = u(x,y) + iv(x,y)$ as $w = f(z)$ for $z = x + iy$. (*a*) Show that the Cauchy-Riemann equations on $u + iv$ come from the formal operations $\partial f/\partial x = \partial f/i\,\partial y\,[= f'(z)]$. (*b*) Assuming two orders of continuous derivatives and the Cauchy-Riemann equations, show that the Taylor series yields

$$f(x + h, y + k) = f(x,y) + (h + ik)\left(\frac{\partial u}{\partial x} + i\frac{\partial y}{\partial x}\right) + \text{error}$$

where the error is quadratic in h and k [$<$ constant $(h^2 + k^2)$]. (*c*) From this result prove that the Cauchy-Riemann equations on $u + iv$ imply analyticity of $u + iv$ in terms of $x + iy$ (assuming continuity of derivatives shown).

2 Use Morera's theorem to justify the difficult part of Theorem 8-3 by a direct construction. Show that if the Cauchy-Riemann equations are valid, then $F(z) = \int^z \omega + i\omega^*$ is analytic in z *in the neighborhood* of any z $(\in\mathfrak{M})$.

3 Consider cases of harmonic differentials where $d\psi$, $d\psi^*$ are both, neither, or mixed single-valued in the region of definition, by using $\alpha \log z$ for different complex constants α.

4 Consider the period parallelogram structure $\mathfrak{M}$ given by vectors $2\omega_1$, $2\omega_2$ (at, say, the origin) in the $x + iy = z$ plane. (*a*) Show that x is a harmonic integral, but not a single-valued function, on $\mathfrak{M}$. (*b*) Show that $\phi_1 = \text{Re}\,(z/2\omega_1)$, $\phi_2 = \text{Re}\,(z/2\omega_2)$ are harmonic integrals for which

$$(\mathfrak{Q}_1,d\phi_1) = 1 \qquad (\mathfrak{Q}_1,d\phi_2) = \text{Re}\,\frac{\omega^1}{\omega_2}$$

$$(\mathfrak{Q}_1',d\phi_1) = \text{Re}\,\frac{\omega_2}{\omega_1} \qquad (\mathfrak{Q}_1',d\phi_2) = 1$$

where $\mathfrak{Q}_1$ is the side (0 to $2\omega_1$) and $\mathfrak{Q}_1'$ is the side (0 to $2\omega_2$) on the parallelogram $\mathfrak{M}$. (*c*) Show that any real harmonic integral $\alpha x + \beta y = \phi$ can be expressed in the form $\lambda\phi_1 + \mu\phi_2$ by proper choice of the (real) quantities λ and μ. (*d*) Show that $(\mathfrak{Q}_1,d\phi)$ and $(\mathfrak{Q}_1',d\phi)$ can be chosen arbitrarily. Note the necessity of choosing $\omega_2/\omega_1 \neq$ real.

5 Consider a linear mapping of the xy plane onto the uv plane,

$$u = \alpha x + \beta y \qquad v = \gamma x + \delta y \qquad \alpha\delta - \beta\gamma \neq 0$$

Show this mapping is conformal at the origin precisely when $\alpha = \delta$, $\beta = -\gamma$. [*Hint:* If $\alpha = \delta$, $\beta = -\gamma$, we can write $R = (\alpha^2 + \beta^2)^{1/2}$ and $\alpha = R\cos\theta$, $\beta = R\sin\theta$, etc.

If the mapping is conformal at the origin, we can perform a rotation so that the x and y axes lie on the u and v axes, respectively; then show conformality implies $u/x = v/y = R$.]

6 Prove that if a biunique transformation with continuous partial derivatives is conformal, the Cauchy-Riemann equations are satisfied. (Consider $\alpha = \delta$, $\beta = -\gamma$ in Exercise 5.)

7 Prove Corollary 2 of Theorem 8-3 by writing $-i[\omega + i\omega^*] = \omega^* - i\omega$.

8 Show that the process of taking complex differentials $dG(z) = G'(z)\,dz$ coincides with the formal process of using the coboundary operator. Thus if $G(z) = \psi + i\psi^*$ and $G'(z) = u(x,y) + iv(x,y)$, then

$$d\psi + i\,d\psi^* = (u + iv)(dx + i\,dy)$$

(This should be an equivalent form of the Cauchy-Riemann equations.)

9 (*a*) Show for harmonic functions

$$\begin{aligned} d\phi_2\,d\phi_1^* &= d\phi_1\,d\phi_2^* = -d\phi_1^*\,d\phi_2 = -d\phi_2^*\,d\phi_1 \\ &= \left[\frac{\partial\phi_1}{\partial x}\frac{\partial\phi_2}{\partial x} + \frac{\partial\phi_1}{\partial y}\frac{\partial\phi_2}{\partial y}\right] dx\,dy \end{aligned}$$

Derive this result from $d\phi\,d\phi^* = [(\partial\phi/\partial x)^2 + (\partial\phi/\partial y)^2]\,dx\,dy$ (by taking $\phi = \phi_1 + \phi_2$). (*b*) Verify from (8-35) that if ϕ_1 and ϕ_2 are harmonic then $\phi_1(d\phi_2)^* - \phi_2(d\phi_1)^*$ is closed.

10 Define conjugates generally in n-dimensional space as follows: Let $0 \le t \le n$

$$\omega^{(t)} = \Sigma A_I^{(t)}\,dX_I$$

(see Exercise 2, Sec. 8-1) and let

$$\omega^{(t)*} = \Sigma A_I^{(t)}\,dX_J$$

where dX_J is so chosen that $dX_I\,dX_J = dX = dx_1 \cdots dx_n$, or

$$\omega^{(t)}\omega^{(t)*} = \left[\sum_I \left(A_I^{(t)}\right)^2\right] dX$$

From this definition, show $\omega^{(1)} = \Sigma A_j\,dx_j$ is harmonic precisely when $A_j = \partial f/\partial x_j$, where

$$\sum_1^n \frac{\partial^2 f}{\partial x_j{}^2} = 0$$

11 Verify that $D_{\mathfrak{M}}[\psi]$ is invariant under conformal mapping of $\mathfrak{M}$. (*a*) Do this first by using (8-41) and noting that the harmonic character of ψ is unchanged under conformal mapping. (*b*) Verify this directly by letting $x + iy = w(u + iv)$ and showing

$$\frac{\partial\psi}{\partial x} - i\frac{\partial\psi}{\partial y} = \frac{\partial(\psi + i\psi^*)}{\partial x} = \frac{d(\psi + i\psi^*)}{dz}$$

$$\left(\frac{\partial\psi}{\partial x}\right)^2 + \left(\frac{\partial\psi}{\partial y}\right)^2 = \left|\frac{d(\psi + i\psi^*)}{dz}\right|^2$$

(*c*) Also show that $du\,dv/dx\,dy$ (= ratio of area elements) $= \partial(u,v)/\partial(x,y) = |dw/dz|^2$.

12 In n dimensions show $[\omega^{(t)*}]^* = \omega^{(t)}(-1)^{(n-1)t}$.

DIRICHLET PROBLEMS

8-4 The Two Existence Theorems

We are now ready to make an obvious extension of the concept of *complex* meromorphic and abelian differentials (see Sec. 7-6) to *real* harmonic differentials.

We say that a real differential $d\psi$ is *meromorphic* on $\mathfrak{M}$ if it is the real part of a complex meromorphic differential, $f(\zeta)\,d\zeta$ defined on $\mathfrak{M}$,

$$d\psi = A(x,y)\,dx + B(x,y)\,dy = \operatorname{Re}\,[f(\zeta)\,d\zeta] \tag{8-42a}$$

The conjugate differential [see (8-25a) and (8-25b)] is also meromorphic

$$d\psi^* = -B(x,y)\,dx + A(x,y)\,dy = \operatorname{Im}\,[f(\zeta)\,d\zeta] = \operatorname{Re}\,[-if(\zeta)\,d\zeta] \tag{8-42b}$$

In practice, the parameters $\zeta = x + iy$ will refer to suitable planes in which the triangles Δ are given with coherent neighborhood structures.

The real differential $d\psi$ is called *abelian* (of first, second, or third kind) if $f(\zeta)\,d\zeta$, occurring in (8-42a), is *abelian* (of first, second, or third kind, respectively), in the sense of Sec. 7-6. Thus $d\psi^* = 0$ on $|\partial\mathfrak{M}|$. This means that if $x(s),y(s)$ represents a parametrization of, say, $\mathfrak{C}_j$ (a component of $\partial\mathfrak{M}$), then $d\psi^*$ vanishes or

$$\frac{d\psi^*}{ds} = -B(x(s),y(s))x'(s) + A(x(s),y(s))y'(s) = 0 \tag{8-43}$$

In writing (8-43) we make demands of continuity† on $A(x,y)$ and $B(x,y)$ on $|\partial\mathfrak{M}|$ as well as $\mathfrak{M}$. Therefore, we shall speak of our real abelian differential $d\psi$ as defined on the closure $\overline{\mathfrak{M}}$ (rather than on $\mathfrak{M}$).

The problem of determining $d\psi$ from boundary data (and periods) has been referred to as a "Dirichlet problem." We are concerned primarily with two such problems whose solution is asserted in the following theorems:‡

First Dirichlet Problem and Existence Theorem for Abelian Differentials *An abelian differential $d\psi$ on $\overline{\mathfrak{M}}$ is (uniquely) determined by the following data:*

First the abelian property is required,

$$d\psi^* = 0 \qquad \textit{along each } \mathfrak{C}_j,\ j = 1, 2, \ldots, n \tag{8-44a}$$

† We also make the assumption that on $|\partial\mathfrak{M}|$, the tangent $x'(s),y'(s)$ is continuous. But in Sec. 11-3 we show that any manifold can be defined by a triangulation which is analytic near any preassigned point of $|\partial\mathfrak{M}|$ and that the abelian differentials are necessarily analytic there, so that $f(\zeta)\,d\zeta$ can be extended locally beyond $|\partial\mathfrak{M}|$.

‡ Refer to Table 1 at the end of the text from now on through Chap. 13.

and next the periods are assigned

$$(8\text{-}44b)\qquad \begin{array}{lll} (\mathcal{Q}_i,d\psi) = h_i & (\mathcal{Q}'_i,d\psi) = h'_i & i = 1, 2, \ldots, p, \text{ if } p > 0 \\ (\mathcal{C}_j,d\psi) = k_j & j = 1, 2, \ldots, n-1, \text{ if } n > 1 & \end{array}$$

for real constants h_i, h'_i, k_j; and finally the singularities of $f_t(\zeta)\, d\zeta$ are imposed by the conditions

$$(8\text{-}44c)\qquad d\psi - \mathrm{Re}\,[f_t(\zeta)\, d\zeta] = d\psi_t \qquad (= \textit{regular differential near } \zeta = \zeta_t)$$

Here the prescribed meromorphic functions $f_t(\zeta)$ near $\zeta = \zeta_t$ on $\mathfrak{M}$ ($t = 1, 2, \ldots, s$) are subject to

$$(8\text{-}44d)\qquad \begin{array}{ll} \displaystyle\sum_{t=1}^{s} \mathrm{res}\,[f_t(\zeta_t)\, d\zeta] = 0 & (\mathfrak{M} \textit{ compact}, n = 0) \\ \mathrm{Re} \displaystyle\sum_{t=1}^{s} \mathrm{res}\,[f_t(\zeta_t)\, d\zeta] = 0 & (\mathfrak{M} \textit{ noncompact}, n > 0) \end{array}$$

It should be quite clear that the real version implies the complex version of the first existence theorem (see Sec. 7-6), particularly since from (8-44*c*), it follows that

$$(8\text{-}45)\qquad d\psi^* - \mathrm{Im}\,[f_t(\zeta)\, d\zeta] = d\psi_t^*$$

(and that $d\psi_t^*$ is necessarily a regular differential as is $d\psi_t$). The real version is neater since only real periods (8-44*b*) enter to begin with.

To prove this first existence theorem (real version), however, we have to rework the problem so as to refer only to $d\psi$ [and not also to $d\psi^*$ as in (8-44*a*)].

SECOND DIRICHLET PROBLEM AND EXISTENCE THEOREM *A meromorphic differential $d\psi$ on $\mathfrak{M}$ is (uniquely) determined by the following data:*

First the single-valuedness of ψ at the boundary $|\partial\mathfrak{M}|$ is asserted

$$(8\text{-}46a)\qquad (\mathcal{C}_j,d\psi) = 0 \qquad j = 1, 2, \ldots, n, \text{ if } n \geq 1$$

and next the periods on the canonical crosscuts are prescribed

$$(8\text{-}46b)\qquad (\mathcal{Q}_i,d\psi) = h_i \qquad (\mathcal{Q}'_i,d\psi) = h'_i \qquad i = 1, 2, \ldots, p, \text{ if } p > 0$$

for real constants h_i and h'_i; and also the singularities of $f_t(\zeta)\, d\zeta$ are imposed by

$$(8\text{-}46c)\qquad d\psi - \mathrm{Re}\,[f_t(\zeta)\, d\zeta] = d\psi_t \qquad (= \textit{regular differential near } \zeta = \zeta_t)$$

Here the prescribed meromorphic functions $f_t(\zeta)$ near $\zeta = \zeta_t$ on $\mathfrak{M}$ ($t = 1, 2, \ldots, s$) are subject to

$$(8\text{-}46d)\qquad \begin{array}{ll} \displaystyle\sum_{t=1}^{s} \mathrm{res}\,[f_t(\zeta_t)\, d\zeta] = 0 & (\mathfrak{M} \textit{ compact}, n = 0) \\ \mathrm{Im} \displaystyle\sum_{t=1}^{s} \mathrm{res}\,[f_t(\zeta_t)\, d\zeta] = 0 & (\mathfrak{M} \textit{ noncompact}, n > 0) \end{array}$$

Also, if we perform a canonical dissection $\mathfrak{M}^p$ on $\mathfrak{M}$, then ψ is not only single-valued in $\mathfrak{M}^p$ but ψ must have prescribed continuous† *boundary values on $|\partial\mathfrak{M}|$. This latter condition means that if $\mathcal{C}_j$ is parametrized (on some triangle Δ in the u plane) by $u = u(s)$ (s real) and if a continuous function $g(s)$ is prescribed, then if P_0 corresponds to $u(s_0)$ (on Δ), we have the limit*

(8-46*e*) $$\lim_{P \to P_0} \psi(P) = g(s_0)$$

We make no requirement that the limit (8-46*e*) be uniform along $\mathcal{C}$ in any way. The problem in which (8-46*e*) *alone* provides nonzero data ($p = 0$) is called the "ordinary" (or "pure") Dirichlet problem usually on a simply- or multiply-connected region ($n \geq 1$).

We must interpret $(\mathcal{C}_j, d\psi)$ in (8-46*a*) to mean $(\mathcal{C}_j', d\psi)$ for any homologous closed curve $\mathcal{C}_j'$ lying wholly interior to $\mathfrak{M}$, sufficiently close to $\mathcal{C}_j$ (and similarly oriented).

Of course, the condition (8-46*a*) means that no question of multivaluedness of ψ arises if we keep from crossing the crosscuts (or keep to $\mathfrak{M}^p$). *Hence (by Exercise 1) if each $h_i = h_i' = 0$, then a differential for the second existence theorem, $d\psi$, will be exact or ψ is single-valued on $\mathfrak{M}$ (indeed on $\bar{\mathfrak{M}}$)*, unless singularities are introduced by $f_t(\zeta)\, d\zeta$ corresponding to terms such as

(8-47*a*) $$f_t(\zeta)\, d\zeta = -\frac{ic\, d\zeta}{\zeta - \zeta_t} \qquad c \text{ real}$$

(8-47*b*) $$\psi = c \arg (\zeta - \zeta_t) + \text{harmonic function} \qquad c \text{ real}$$

These singularities are mutually canceling [as provided in (8-46*d*)] so as not to affect the single-valuedness of ψ on $\mathcal{C}_j$ [as provided in (8-46*a*)]. (See Exercise 2.)

LEMMA 8-4 *For a solution to the first existence theorem*

(8-48*a*) $$\sum_{j=1}^{n} (\mathcal{C}_j, d\psi) + 2\pi \operatorname{Im} \sum_{t=1}^{s} \operatorname{res} [f_t(\zeta_t)\, d\zeta] = 0$$

and for a solution to the second existence theorem

(8-48*b*) $$\sum_{j=1}^{n} (\mathcal{C}_j, d\psi^*) - 2\pi \operatorname{Re} \sum_{t=1}^{s} \operatorname{res} [f_t(\zeta_t)\, d\zeta] = 0$$

(See Exercise 3.)

In Sec. 9-1 we shall justify (8-44*d*), (8-46*d*), (8-48*a*), and (8-48*b*) as physical "conservation laws."

In the meantime we remark that the proof of Riemann's fundamental theorem (in Sec. 13-2) requires the first existence theorem, which in turn

† In Sec. 11-1 we modify (8-46*e*) to take into account certain discontinuous functions $g(s)$ prescribed at the boundary. For now, consider $g(s)$ continuous.

requires the second existence theorem. Thus, despite our original interest in algebraic functions and compact Riemann manifolds (Sec. 6-2), we can settle for nothing less than this highly transcendental† noncompact problem!

EXERCISES

1 Show that if $h_i = h_i' = 0$ and no terms (8-47*a*) occur in the second existence theorem, then the differential $d\psi$ is single-valued. Use the homology basis as in Lemma 8-3 (Sec. 8-2).

2 Show that the conditions (8-44*d*) and (8-46*d*) are necessitated by (8-44*a*) and (8-46*a*) in the first and second existence theorem, respectively. (Compare Exercise 3.)

3 By taking $\int_{\partial\mathfrak{M}} d\psi + i\,d\psi^*$, show (8-48*a*) and (8-48*b*). (Compare Lemma 7-10 in Sec. 7-6.)

4 Consider $d\psi = c_1\,dx + c_2\,dy$ as the general (real) abelian differential for the (torus) parallelogram determined by the vectors $(2\omega_1, 2\omega_2)$ in the $x + iy$ plane ($\omega_2/\omega_1 \neq$ real). Discuss the relation to Exercise 1 of Sec. 7-6.

5 Consider $d\psi = c\,d\theta$ as the general (real) abelian differential for the annulus $\mathfrak{R}$: $r_1 < |z| < r_2$ (in polar coordinates $z = re^{i\theta}$). Discuss the relation to Exercise 5 of Sec. 7-6.

8-5 The Two Uniqueness Proofs

In this section we shall prove that the differentials for either existence theorem are unique (if they exist).

RIEMANN'S PERIOD RELATIONS

Let us consider a Riemann manifold $\mathfrak{M}$ in terms of its polygonal model with peripheral sequence of $4p + 3n$ sides (see Sec. 7-4), . . . , $\mathcal{Q}_i, \mathcal{Q}_i', -\mathcal{Q}_i, -\mathcal{Q}_i', \ldots, \mathcal{K}_j, \mathcal{C}_j, -\mathcal{K}_j; \cdots$ $(1 \le i \le p,\ 1 \le j \le n)$. If we delete the $(2p + n)$ arcs $|\mathcal{Q}_i|, |\mathcal{Q}_i'|, |\mathcal{K}_j|$ from $\mathfrak{M}$, then we obtain a manifold $\mathfrak{M}^*$ topologically equivalent to the interior of a $(4p + 3n)$-gon. (See Sec. 7-2.) Of course, $\mathfrak{M}^*$ is simply-connected (as is its closure $\overline{\mathfrak{M}}^*$) so that if any harmonic differential dG is given on $\overline{\mathfrak{M}}^*$, then dG is exact on $\mathfrak{M}^*$ (by Lemma 8-1) and, indeed, G can be chosen as single-valued on $\mathfrak{M}^*$, and in fact on $\overline{\mathfrak{M}}^*$.

THEOREM 8-5 (*Riemann's period relation*, 1857) *Let dF and dG be closed differentials on* $\overline{\mathfrak{M}}$. *Then if we define G uniquely on* $\overline{\mathfrak{M}}^*$, *we have*

$$\text{(8-49)}\quad \int_{\partial\mathfrak{M}^*} G\,dF = \sum_{i=1}^{p} \begin{vmatrix} (\mathcal{Q}_i, dG) & (\mathcal{Q}_i, dF) \\ (\mathcal{Q}_i', dG) & (\mathcal{Q}_i', dF) \end{vmatrix} - \sum_{j=1}^{n} (\mathcal{C}_j, dG)(\mathcal{K}_j, dF) + \sum_{j=1}^{n} (\mathcal{C}_j, G\,dF)$$

† We even have to extend the second existence theorem to discontinuous boundary data (see previous footnote)!

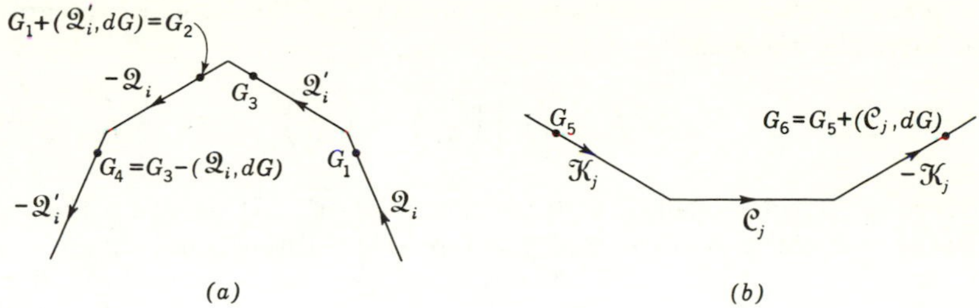

Fig. 8-1 For Riemann's period relations, (a) on $\mathbf{Q}_i$, (b) on $\mathbf{K}_j$.

PROOF We refer to Fig. 8-1 which shows how to break up $\int_{\partial\mathfrak{M}^*}$ as follows:

(8-50a) $$\int_{\mathcal{Q}_i} G_1\,dF + \int_{-\mathcal{Q}_i} G_2\,dF = \int_{\mathcal{Q}_i} G_1\,dF - \int_{\mathcal{Q}_i} [G_1 + (\mathcal{Q}_i',dG)]\,dF = -(\mathcal{Q}_i',dG)(\mathcal{Q}_i,dF)$$

where G_1 and G_2 are values of G at corresponding points of $\pm\mathcal{Q}_i$. Likewise (see Exercise 1),

(8-50b) $$\int_{\mathcal{Q}_i'} G_3\,dF + \int_{-\mathcal{Q}_i'} G_4\,dF = +(\mathcal{Q}_i,dF)(\mathcal{Q}_i,dG)$$

and

(8-50c) $$\int_{\mathcal{K}_j} G_5\,dF + \int_{-\mathcal{K}_j} G_6\,dF = -(\mathcal{C}_j,dG)(\mathcal{K}_j,dF)$$

Q.E.D.

COROLLARY 1 *Riemann's period relation is valid if dF and dG are real or complex meromorphic differentials, provided no singularity occurs on $|\partial\mathfrak{M}^*|$ and G is single-valued on $|\partial\mathfrak{M}^*|$.*

The condition that G is definable as single-valued on $|\partial\mathfrak{M}^*|$ is $(\partial\mathfrak{M}^*,dG) = 0$, which can be reduced to

(8-51) $$\sum_{\mathfrak{M}} \text{res}\; dG = 0$$

Under such circumstances, we can make an exact evaluation of $\int G\,dF$ in (8-49).

COROLLARY 2 *Let dF and dG be complex meromorphic differentials on a compact manifold $\mathfrak{M}$ such that G can be defined uniquely on $\mathfrak{M}^*$. Then*

(8-52) $$2\pi i \sum_{\mathfrak{M}} \text{res}\;(G\,dF) = \sum_{i=1}^{p} \begin{vmatrix} (\mathcal{Q}_i,dG) & (\mathcal{Q}_i,dF) \\ (\mathcal{Q}_i',dG) & (\mathcal{Q}_i',dF) \end{vmatrix}$$

At this stage we are prepared to make use of the Dirichlet integral, in particular the identity (8-41)

$$D_{\mathfrak{M}}[\phi] = \iint_{\mathfrak{M}} \left[\left(\frac{\partial \phi}{\partial x}\right)^2 + \left(\frac{\partial \phi}{\partial y}\right)^2\right] dx\, dy = \int_{\partial \mathfrak{M}} \phi\, d\phi^* \tag{8-53}$$

where ϕ is single-valued and harmonic on $\mathfrak{M}$. It is clear that if $D_{\mathfrak{M}}[\phi] = 0$, then† $\partial\phi/\partial x \equiv \partial\phi/\partial y \equiv 0$ or ϕ is constant; that is,

$$d\phi = d\phi^* = 0$$

Theorem 8-6 *The first existence theorem defines a differential $d\psi$ uniquely if it exists.*

PROOF Take any two solutions on $\overline{\mathfrak{M}}$, namely, $d\psi_1$ and $d\psi_2$, to the problem as stated in (8-44*a*) to (8-44*d*). Then if we consider

$$d\psi = d\psi_1 - d\psi_2$$

we see all periods (8-44*b*) vanish and the singularities (8-44*c*) cancel out.

Thus in Riemann's period relation (8-49) we let $dG = d\psi$, $dF = d\psi^*$. Clearly, from (8-44*b*),

$$(\mathcal{Q}_i, d\psi) = (\mathcal{Q}'_i, d\psi) = 0 \qquad i = 1, 2, \ldots, p \tag{8-54a}$$

$$(\mathcal{C}_j, d\psi) = 0 \qquad j = 1, 2, \ldots, n-1 \text{ and } n \tag{8-54b}$$

(see Lemma 8-4). Thus ψ is single-valued on $\mathfrak{M}$. Also $(\mathcal{C}_j, \psi\, d\psi^*) = 0$ by (8-44*a*). Hence, by (8-53), $\int_{\partial\mathfrak{M}^*} \psi\, d\psi^* = 0$ and with it, $D_{\mathfrak{M}}[\psi] = 0$. Therefore, $d\psi = 0$ on $\mathfrak{M}$. Q.E.D.

MAXIMUM-MINIMUM PRINCIPLE

The uniqueness part of the second existence theorem could be concluded from Riemann's period relation if we knew $d\psi$ (and therefore) $d\psi^*$ were defined on $\overline{\mathfrak{M}}$ [not just interior to $\mathfrak{M}$ (see Exercise 2)]. The uniqueness proof can be executed, however, with a simpler but equally deep device.

We recall the maximum-minimum principle as expressed in Corollary 2 to Theorem 3-5 in Sec. 3-3. A harmonic function cannot take a local maximum or a minimum at an interior point of a region $\mathcal{R}$. We also have this result for a manifold $\mathfrak{M}$ (instead of $\mathcal{R}$) since every harmonic function on $\mathfrak{M}$ is defined locally on a region $\mathcal{R}$ (for example, a parameter triangle Δ).

† Here and elsewhere implicit use is made of the result that for f continuous over a domain of integration (in any number of dimensions), then from $f \geq 0$ and $\int f = 0$ we can conclude $f = 0$ identically. Generally, we shall not bother to point out each actual use of this result.

LEMMA 8-5 *If a harmonic function $\psi(z)$ defined on a noncompact manifold $\mathfrak{M}$ has the limit*

$$\psi(z) \to 0$$

as z approaches any point z_0 of $|\partial\mathfrak{M}|$, then $\psi(z)$ is identically zero in $\mathfrak{M}$.

PROOF Assume ψ is nonconstant. Since ψ cannot have a maximum in $\mathfrak{M}$, it has a maximizing sequence z_n such that sup ψ (on $\mathfrak{M}$) = lim (z_n) (possibly ∞). But the sequence z_n has no interior limiting point; therefore, it has a subsequence converging to the boundary, say to z_0, whence sup $\psi = 0$. Likewise, inf $\psi = 0$. Thus $\psi = 0$, contrary to assumption.

Q.E.D.

THEOREM 8-7 *The second existence theorem defines a differential $d\psi$ uniquely if it exists.*

PROOF First, suppose $\mathfrak{M}$ is not compact. Then, as in Theorem 8-6, if $d\psi_1$ and $d\psi_2$ are two differentials satisfying conditions (8-46*a*) to (8-46*e*), then by setting $d\psi_1 - d\psi_2 = d\psi$ we define a harmonic *function* ψ (single-valued in $\mathfrak{M}$) satisfying Lemma 8-5.

Now if $\mathfrak{M}$ is compact, $\partial\mathfrak{M} = 0$ so that the uniqueness will be seen to follow from the fact that otherwise the above differentials $d\psi$ and $d\psi^*$ are harmonic on $\overline{\mathfrak{M}}$ as well as on $\mathfrak{M}$ (trivially)! We can then apply the Riemann period relation as though we were proving uniqueness for the first existence theorem. Q.E.D.

EXERCISES

1 Verify (8-50*b*) and (8-50*c*) from Fig. 8-1.

2 Assume that $d\psi$ and $d\psi^*$ were defined on $\overline{\mathfrak{M}}$ in the *second* existence theorem (rather than merely on $\mathfrak{M}$). Show that the uniqueness proof by Theorem 8-5 can be consummated by applying (8-46*a*) to (8-46*e*) in homogeneous form [$h_i = h_i' = 0$, $f_t(\zeta) = 0$, $g(s_0) = 0$ or $\psi = 0$ on $\mathcal{C}_j$, etc.].

3 Give an alternate proof of Theorem 8-7 for $\mathfrak{M}$ compact by using the same line of reasoning as for $\mathfrak{M}$ noncompact. (Where *is* the maximum of ψ when $\partial\mathfrak{M} = 0$?)

4 Let $\mathfrak{M}$ be a compact manifold of genus one (torus). Show that according to either existence theorem, $\mathfrak{M}$ has one complex abelian differential of the first kind du determined uniquely by $(\mathcal{Q}_1,d\psi)$ and $(\mathcal{Q}_1',d\psi)$, where $du = d\psi + i\,d\psi^*$.

5 Continuing Exercise 4, show that there is only one complex abelian differential of the first kind to within a constant factor. Assume $du_1 = d\psi_1 + i\,d\psi_1^*$ $(\neq 0)$ and $du_2 = d\psi_2 + i\,d\psi_2^*$; then show how to determine a constant $\alpha + i\beta$ so that $du = (\alpha + i\beta)\,du_1 + du_2 = d\psi + i\,d\psi^*$ satisfies the properties $(\mathcal{Q}_1,d\psi) = (\mathcal{Q}_1',d\psi) = 0$. (Thus by Exercise 4, $du \equiv 0$.) [*Hint:* We must exclude the possibility that

$$\begin{vmatrix} (\mathcal{Q}_1,d\psi_1) & (\mathcal{Q}_1,d\psi^*) \\ (\mathcal{Q}_1',d\psi_1) & (\mathcal{Q}_1',d\psi_1^*) \end{vmatrix} = 0$$

by using Riemann's period relations.] (This method is generalized for arbitrary $p > 1$ in Sec. 13-1.)

Note we do not use the existence theorems as such, but only the uniqueness portions.

CHAPTER 9

PHYSICAL INTUITION

We have seen some justification of our existence theorems in terms of period assignments based on cohomology of 1-forms (in Sec. 8-4). Although this may be sufficient motivation for the modern abstractionist, this situation hardly does justice to the tremendous impact of rational mechanics on mathematical thinking in the tradition of Cauchy and Riemann.

In particular, the boundary-value problems which distinguish the two existence theorems could have been justified mainly on the basis of models found in two-dimensional electrostatics and hydrodynamics.† We shall give some partial justifications together with some physically plausible conjectures of (correct) mapping theorems of historical theoretical interest.

9-1 Electrostatics and Hydrodynamics

Let us first consider the concept of *free* boundary. Consider a harmonic differential $d\psi$ defined on $\overline{\mathfrak{M}}$. We parametrize some component $\mathcal{C}$ of $\partial\mathfrak{M}$ by arc length s in terms of local coordinates (x,y) as follows:

$$\frac{dx}{ds} + i\frac{dy}{ds} = \cos\theta + i\sin\theta \tag{9-1a}$$

or symbolically, $dz/ds = \exp i\theta$. Then for the normal

$$\frac{dz}{dn} = \exp i\left(\theta + \frac{\pi}{2}\right)$$

or

$$\frac{dx}{dn} + i\frac{dy}{dn} = -\sin\theta + i\cos\theta \tag{9-1b}$$

since the *inner* normal dn to $\mathcal{C}$ lies $\pi/2$ radians to the "left

† Indeed no suitably broad generalization has been found for geometric function theory from one (complex) variable to two variables. This might be due to the absence of a physical model.

of ds." Thus

$$\frac{\partial\psi}{\partial s} = \frac{\partial\psi}{\partial x}\cos\theta + \frac{\partial\psi}{\partial y}\sin\theta = \frac{\partial\psi^*}{\partial y}\cos\theta - \frac{\partial\psi^*}{\partial x}\sin\theta \tag{9-2a}$$

$$\frac{\partial\psi}{\partial n} = -\frac{\partial\psi}{\partial x}\sin\theta + \frac{\partial\psi}{\partial y}\cos\theta = -\frac{\partial\psi^*}{\partial y}\sin\theta - \frac{\partial\psi^*}{\partial x}\cos\theta \tag{9-2b}$$

from the Cauchy-Riemann equations [see (8-31)]. Thus

$$\frac{\partial\psi}{\partial n} = -\frac{\partial\psi^*}{\partial s} \qquad \frac{\partial\psi}{\partial s} = \frac{\partial\psi^*}{\partial n} \tag{9-3}$$

(As a mnemonic device, x and y become s and n, respectively.)

Now if boundary data for ψ are prescribed (say, in the second existence theorem), then $\psi(s)$ is prescribed. This means that $\partial\psi/\partial s$ is known; hence $\partial\psi^*/\partial n$ is known, etc. Thus we can give a geometrical interpretation to the condition that $d\psi^* = 0$ on $\mathcal{C}$ or $\partial\psi^*/\partial s = 0$ by writing $\partial\psi/\partial n = -\partial\psi^*/\partial s = 0$. Thus on a curve $\mathcal{C}$,

$$d\psi^* = 0 \qquad \text{means} \qquad \frac{\partial\psi}{\partial n} = 0 \tag{9-4}$$

Actually the condition $\partial\psi/\partial n = 0$ is called the "floating" or "free boundary," characterized by a "zero-gradient normal" of ψ at $\mathcal{C}$.

We shall give a hydrodynamical interpretation presently for this geometrical condition. Let us merely note that the definition (outward) *flux of ψ through* $\mathcal{C}$ is given by

$$(\mathcal{C},d\psi^*) = -\int_{\mathcal{C}} \frac{\partial\psi}{\partial n}\,ds \tag{9-5}$$

over $\mathcal{C}$ any smooth curve (closed or not).

In further preparation let us assume $\mathfrak{M}$ is an n-tuply-connected region $\mathcal{R}$ in the $z = x + iy$ plane, taken to have, say, simple analytic boundary curves $\mathcal{C}_j$ making up

$$\partial\mathcal{R} = \mathcal{C}_1 + \cdots + \mathcal{C}_n \tag{9-6}$$

We now have two cases to consider:

ELECTROSTATIC CASE (SECOND EXISTENCE THEOREM)

Here we regard $\mathcal{R}$ as a conducting sheet on which an electrostatic potential ψ is defined. From elementary physics ψ is harmonic; but more important it is *intrinsically* single-valued (intuitively speaking, "voltage is defined as measured to a definite ground"). We can imagine, moreover, that the *second existence theorem* corresponds to a definite physically reasonable situation. The boundary values (8-46*e*) correspond to a "distribution of electrodes" superimposed on the boundary while the internal sources (8-46c) produce the singularities indicated. Indeed, it is the nature of electrostatic theory that Im [res $f_t(\zeta_t)$] = 0 at each ζ_t, automatically satisfying (8-46d). [Physically this means that a potential such as $\psi = \arctan [(y - y_0)/(x - x_0)]$, the multivalued part of $\log (z - z_0)$, really cannot occur.] The conditions (8-48b) will be recognized as "conservation of flux" physically speaking.

HYDRODYNAMIC CASE (FIRST EXISTENCE THEOREM)

Here we regard $\mathfrak{R}$ as a sheet over which a two-dimensional steady (irrotational, incompressible, adiabatic, inviscid, etc.) flow occurs. From elementary physics such a flow is given by a vector $(U(x,y),V(x,y))$ at a point $z \in \mathfrak{R}$, which has the property that

(9-7) $$d\psi = U\,dx + V\,dy \qquad \text{and} \qquad d\psi^* = -V\,dx + U\,dy$$

are harmonic. The function $F = \psi + i\psi^*$ is always definable locally

(9-8a) $$dF = d\psi + i\,d\psi^* = (U - iV)(dx + i\,dy)$$

Hence every flow corresponds to a meromorphic differential dF on $\mathfrak{R}$, and

(9-8b) $$\overline{F'(z)} = U + iV = W \qquad \text{(velocity vector)}$$

The *streamlines* are given by the equation

(9-8c) $$d\psi^* = 0 \qquad \text{or} \qquad \frac{dy}{dx} = \frac{V}{U}$$

The *equipotential lines* are given by

(9-8d) $$d\psi = 0 \qquad \text{or} \qquad \frac{dy}{dx} = -\frac{U}{V}$$

Thus a definite direction of flow dy/dx is determined unless $U = V = 0$ or $|F'(z)| = |W| = [U^2 + V^2]^{1/2} = 0$. If $F'(z_0) = 0$, it *always* happens that two different streamlines intersect at z_0. (See Exercise 1.) Such a value z_0 is called a "stagnation point" for obvious reasons. It is geometrically significant as a point of nonconformality of the mapping given by $F(z)$.

The Cauchy-Riemann equations here are of incidental interest because they have a convenient differential form (rather than a derivative form):

(9-9a) $$d(d\psi) = 0 \qquad \text{means} \qquad \frac{\partial U}{\partial y} = \frac{\partial V}{\partial x} \qquad \text{(irrotationality)}$$

(9-9b) $$d(d\psi^*) = 0 \qquad \text{means} \qquad \frac{\partial U}{\partial x} + \frac{\partial V}{\partial y} = 0 \qquad \text{(incompressibility)}\dagger$$

† We now see an advantage in Cauchy's definition of an analytic function (as a *differentiable* function for a region) in contrast to Lagrange's definition (in terms of the existence of *power series*). The properties of incompressibility, irrotationality, and

We now define (as before)

$$\text{Flux of } \psi \text{ through } \mathcal{C} = \int_{\mathcal{C}} d\psi^* = (\mathcal{C}, d\psi^*) \tag{9-10a}$$

$$\text{Circulation of } \psi \text{ about } \mathcal{C} = \int_{\mathcal{C}} d\psi = (\mathcal{C}, d\psi) \tag{9-10b}$$

Hence, finally, a "complex flux" is defined by the contour integral

$$\text{circ} + i \text{ flux} = (\mathcal{C}, dF) \tag{9-10c}$$

Here we can visualize several cases where dF has residue $\neq 0$ at the origin; symbolically from (9-10c), let us take the special case called the "simple pole."

$$dF = \frac{\text{circ} + i \text{ flux}}{2\pi i z}\, dz \tag{9-11a}$$

$$\psi + i\psi^* = F = \frac{\text{circ} + i \text{ flux}}{2\pi i} \log z = \frac{(\text{circ} + i \text{ flux})(\log r + i\theta)}{2\pi i} \tag{9-11b}$$

and from (1-8b), for constant values of circ and flux,

$$\bar{W} = U - iV = F' = \frac{\text{circ} + i \text{ flux}}{2\pi i z} \tag{9-12}$$

(Various cases are sketched in Fig. 9-1. See Exercise 2.)

The natural boundary-value problem in hydrodynamics is covered by the *first existence theorem*. The primary reason for this is that a boundary component $\mathcal{C}_j$ must be a streamline as it is "impenetrable." Thus $d\psi^* = 0$ on $\mathcal{C}_j$ [see (8-44a)]. The conservation laws (8-44d) and (8-48a) are the conservation of flux and circulation. It is physically plausible that subject to these obvious conservation laws, we can prescribe any number of singularities $f_t(z)$ and "stir up" any set of circulations $(\mathcal{C}_j, d\psi)$.

The harmonic differentials which arise must incidentally be *abelian* in a hydrodynamic problem where there is no flow through the boundary (although not in an electrostatic problem).

EXERCISES

1 Let a flow be given by $U - iV = F'(z)$, where $F(z)$ is analytic near a point z_0 where $F'(z_0) = 0$ with, say, a root of order m. Then show that there will be m different streamlines ($2m$ "rays") converging in z_0. (*Hint:* Consider the mapping onto a cyclic neighborhood of order m.)

2 Write out the equation of the streamlines and velocity vectors in the case of a simple pole (9-11a) and (9-11b) for circ $= 0$ (and for flux $= 0$) using rectangular coordinates.

---

continuity are physically palpable, and they require only the continuity of second derivatives. On the other hand, the existence of a power series expansion is not physically palpable as an assumption.

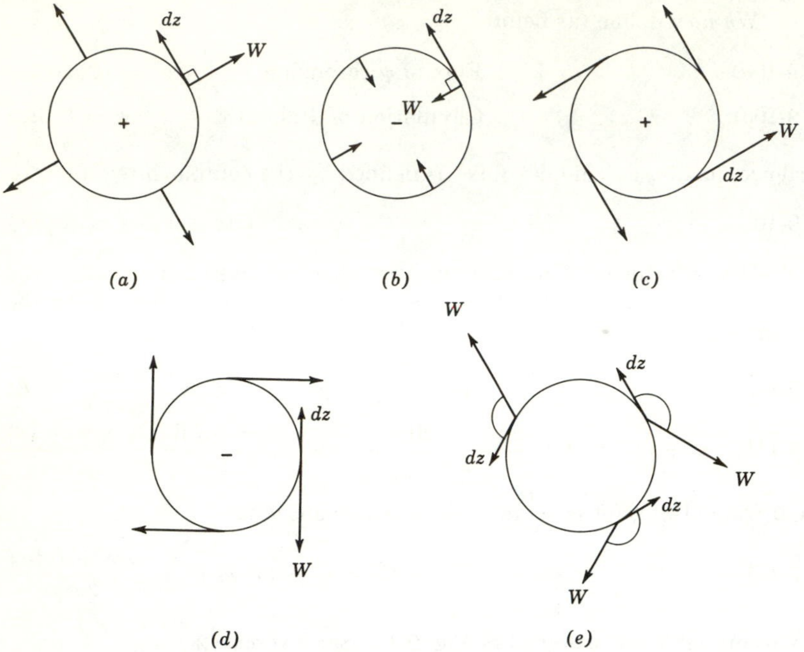

Fig. 9-1 (*a*) Positive flux (circ = 0); (*b*) negative flux (circ = 0); (*c*) positive circ (flux = 0); (*d*) negative circ (flux = 0); (*e*) flux > 0, circ < 0.

3 Consider the complex flux (9-10*c*) in the plane for rigid rotation $U = -\omega y$, $V = \omega x$ (ω = angular velocity). [Show why rigid rotation is *not* a hydrodynamic flow by referring to (9-9*a*) and (9-9*b*).]

9-2 Special Solutions

We now consider special cases of the two existence theorems based on prescribing a singularity and then prescribing all possible boundary data as zero, insofar as the consistency condition in Lemma 8-4 will permit.

We first consider the *Green's function* based on the singularity

$$\psi = \log \frac{1}{|z - z_0|} + \text{regular harm funct} \tag{9-13a}$$

at the point z_0 in $\mathcal{R}_n$, our region of connectivity n, with

$$\partial\mathcal{R}_n = \mathcal{C}_1 + \cdots + \mathcal{C}_n$$

There are two existence theorems, hence two cases.†

† In interpreting the analogy between *electrostatic* and *hydrodynamic* applications it is important to appreciate the fact that we cannot remain within one "medium." The reader must therefore be flexible in viewing the problems (and diagrams). When we take the case of *hydrodynamics*, we draw lines of flow (of the fluid) given by

GREEN'S FUNCTION

Hydrodynamic case	*Electrostatic case*
(first existence theorem and free boundary)	(second existence theorem and fixed boundary)
ψ_H^* = unknown function	ψ_E = unknown function

PRESCRIBED DATA

$d\psi_H^* = 0$ on $\mathcal{C}_j$ or $\psi_H^* = g_j$ unknown constant on $\mathcal{C}_j$; $(\mathcal{C}_j, d\psi_H) = 0$ for $j < n$.	$\psi_E = 0$ on $\mathcal{C}_j$ or $(\mathcal{C}_j, d\psi_E) = 0$ is trivially true for $j \leq n$.
Thus ψ_H^* is single-valued on all $\mathcal{C}_j$ and ψ_H is single-valued on all $\mathcal{C}_j$ except $\mathcal{C}_n$; $\psi_H = \mathrm{Re}\,[-i \log (z - z_0)] + \cdots$; consistency condition is $$(\mathcal{C}_n, d\psi_H) = 2\pi$$ (from Lemma 8-4 in Sec. 8-4).	Thus ψ_E is single-valued on all $\mathcal{C}_j$ but not ψ_E^*; $\psi_E = \mathrm{Re}\,(\log 1/(z - z_0)) + \cdots$; consistency condition refers to ψ_E^* (hence is not relevant).

When $\mathcal{R}_n$ is simply-connected ($n = 1$), there is no distinction between the hydrodynamical and electrostatic Green's function. (In particular, let us choose $g_n = 0$, so $\psi_H^* = 0$ on $\mathcal{C}_n$ to standardize all cases $n \geq 1$.)

Otherwise, the distinction is as follows: In the electrostatic case, all boundary components are grounded so that ψ_E (the unknown voltage) is single-valued, but not ψ_E^*. In the hydrodynamical case, the boundary components $\mathcal{C}_j$ $[j = 1, 2, \ldots, (n-1)]$ automatically assume exactly the right potentials g_j for which ψ_H^* and ψ_H are both single-valued.

In Fig. 9-2*a* to *d* this situation is depicted for $n = 2$. We note that the value $\psi_H^* = g_1$ on $\mathcal{C}_1$ appears to be so adjusted that *no stagnation point occurs* (compare Fig. 9-2*a* and *b* with Fig. 9-2*c* and *d*). Actually, our physical intuition might suggest that with free boundaries the variable values $g_1, \ldots, g_{n-1}$ ($g_n = 0$) will so adjust themselves as to achieve the "simplest possible configuration" of level lines (ψ_H^* = constant). This suggestion is mathematically justifiable if we observe that at a stagnation point, $F(z) = \psi + i\psi^*$ has a zero derivative, whereas Theorem 9-1 will show why $F = F(z)$ is at least locally biunique (as a mapping of z onto F).

Thus the conformal mapping result in Theorem 9-1 (*Sec.* 9-3) *will constitute a result in hydrodynamics that certain flow configurations exist which are free of stagnation points.*

ψ_H^* = constant; in the case of *electrostatics*, we draw lines of potential (orthogonal to the flow of current) given by ψ_E = constant [see (9-8*c*) and (9-8*d*)]. If we tried to draw ψ_H and ψ_E, say for uniformity, we would find the Green's functions do not check; one would have Im (log z) and the other would have Re (log z) as singularity! Ultimately, all our drawings and references can be taken as hydrodynamical, so that this contrast only applies to Sec. 9-2.

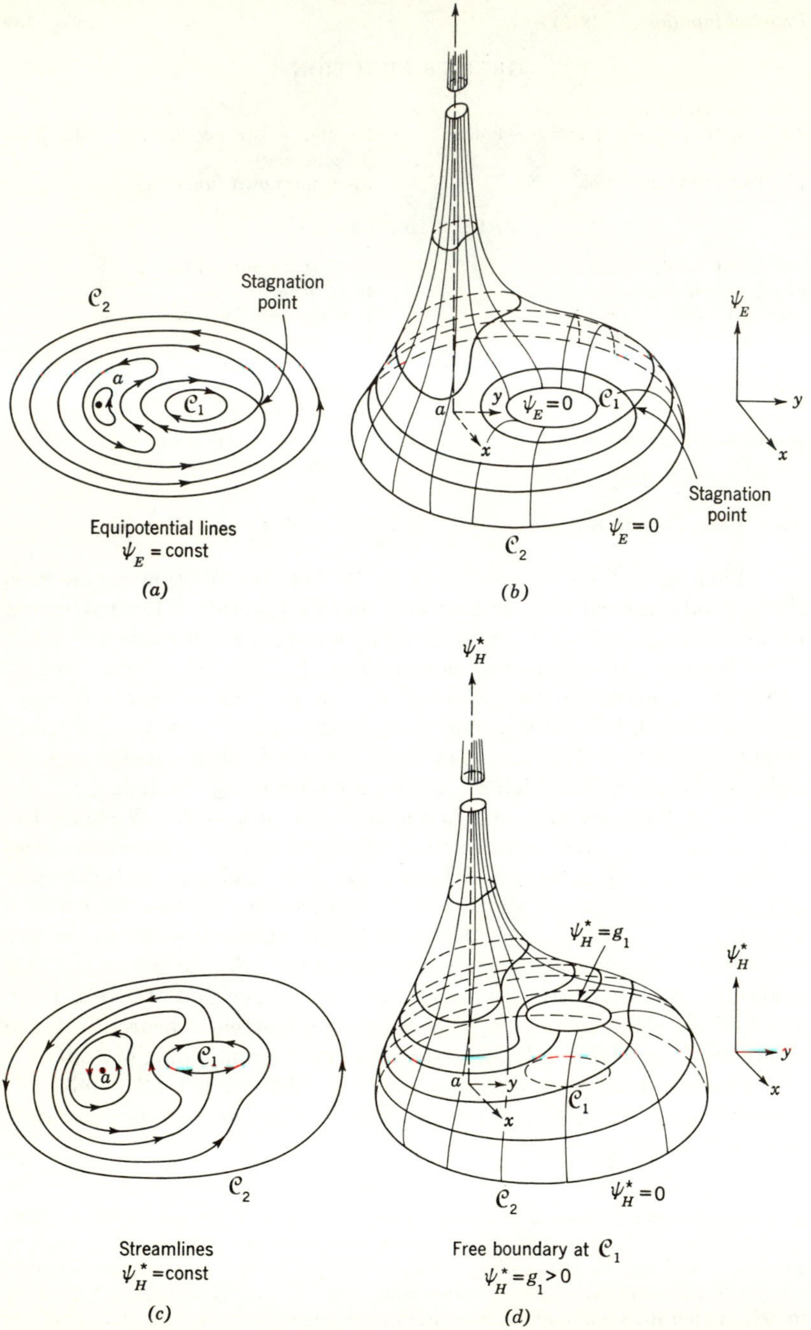

Fig. 9-2 (*a*, *b*) Electrostatic Green's function on $\mathcal{R}_2$. (*c*, *d*) Hydrodynamic Green's function on $\mathcal{R}_2$.

For further reference we define the complex (hydrodynamic) Green's function as

(9-13*b*) $$G(z;z_0) = \psi_H + i\psi_H^* = -i \log (z - z_0) + \cdots$$

DIPOLE FLOW

The simplest case of an abelian differential (strictly) of the second kind is a dipole given by the singularity

(9-14) $$f_t(z)\, dz = -\frac{\alpha\, dz}{(z - z_0)^2} \qquad \alpha \text{ complex}$$

a differential with vanishing residue at $z = z_0$. The level lines Re f_t = constant, Im f_t = constant were sketched in Fig. 2-2*c*.

Let us consider only the hydrodynamic model this time. Intuitively, it is the nature of the *hydrodynamic* case that the boundaries remain free, or ψ^* *cannot be prescribed in advance* on $\mathcal{C}_1, \ldots, \mathcal{C}_{n-1}$ (but $\psi^* = 0$ on $\mathcal{C}_n$ for standardization). The first existence theorem is applied to

(9-15*a*) $$d\psi^* = 0 \qquad \text{along each } \mathcal{C}_j,\ j = 1, 2, \ldots, n$$

(9-15*b*) $$(\mathcal{C}_j, d\psi) = 0 \qquad j = 1, 2, \ldots, n - 1$$

[Indeed, (9-15*b*) also holds for $j = n$, by the conservation law in Lemma 8-4, Sec. 8-4.] The singularity (9-14) occurs, or

(9-15*c*) $$d\psi = \text{Re}\left[\frac{-\alpha\, dz}{(z - z_0)^2}\right] + \text{reg differential}$$

A sketch of the flow is given in Fig. 9-3. It is like the flow in Fig. 9-2*c* (rather than Fig. 9-2*a*) in that the flow lines have *no interior stagnation points*. Once more the absence of stagnation points in the dipole flow is an intuitive assertion which will follow from Theorem 9-1 (Sec. 9-3).

For further reference we define the dipole function

(9-15*d*) $$D(z;\alpha,z_0) = \psi + i\psi^*$$

DIFFERENTIALS OF THE FIRST KIND

As a final illustration, consider abelian differentials of the first kind on $\overline{\mathcal{R}}_n$, consisting (again) of an application of the first existence theorem (hydrodynamic case) where singularities simply are absent. Hence the prescribed data reduce to

(9-16*a*) $$d\psi^* = 0 \qquad \text{on } \mathcal{C}_j,\ j = 1, 2, \ldots, n$$

(9-16*b*) $$(\mathcal{C}_j, d\psi) = k_j = \text{real const} \qquad j = 1, 2, \ldots, n - 1$$

There would seem to be infinitely many "different" functions thus defined, depending on the values of k_j; but from these, we select as a basis $(n - 1)$ so-called "normal differentials" $d\psi^{(i)}$ $(1 \le i \le n - 1)$.

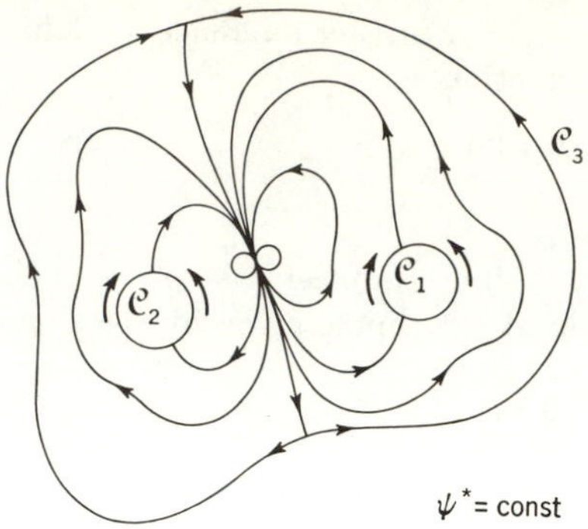

Fig. 9-3 Dipole flow in the region $\mathcal{R}_3$.

Each $d\psi^{(i)}$ is defined by

(9-17a) $$d\psi^{(i)*} = 0 \qquad \text{on } \mathcal{C}_j,\ j = 1, 2, \ldots, n$$

(9-17b) $$(\mathcal{C}_j, d\psi^{(i)}) = 2\pi\delta_{ij} \qquad 1 \le i, j \le n - 1$$

(where δ_{ij} is the Kronecker symbol), and

(9-17c) $$(\mathcal{C}_n, d\psi^{(i)}) = -2\pi$$

Actually, (9-17c) follows from (9-17b) by the conservation law (8-48a). For future reference, we write

(9-17d) $$G^{(i)}(z) = \psi^{(i)} + i\psi^{(i)*}$$

Thus by the uniqueness part of the first existence theorem, the combination

(9-18) $$d\psi = \sum_{j=1}^{n-1} \frac{k_j\, d\psi^{(i)}}{2\pi}$$

defines the solution to (9-16b).

Again we shall see in Theorem 9-1 (Sec. 9-3) that the "normal differentials" $d\psi^{(i)}$ defined here have no stagnation point (see Fig. 9-4).

We normalize $\psi^{(i)}$ further by setting $\psi^{(i)*}(z) = 0$ for z on $\mathcal{C}_n$. Thus we obtain a $(n - 1) \times (n - 1)$ matrix of values (or *modules*)

(9-19) $$a_{ij} = \psi^{(i)*}(z) \qquad \text{for } z \text{ on } \mathcal{C}_j,\ 1 \le i, j \le n - 1$$

This matrix is called a "modular matrix" and the a_{ij} are called "modules." These modules are invariant under conformal mapping of $\mathcal{R}$ onto another (n-tuply-connected) plane domain because of the invariance of harmonic functions (see Theorem 8-4 in Sec. 8-3).

LEMMA 9-1 *The abelian differentials of the first kind on* $\overline{\mathcal{R}}_n$ *are given by real linear combinations* (9-18) *of the* $(n - 1)$ *normal abelian differentials.*

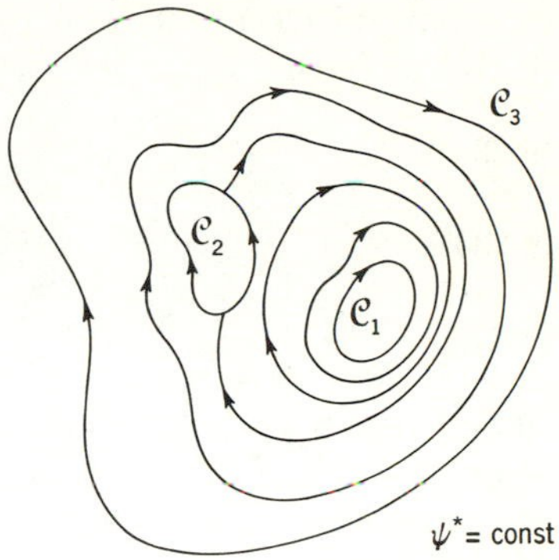

Fig. 9-4 Differential of the first kind for $n = 3$.

The proof follows from the first existence theorem since (9-18) is an actual expression for $d\psi$; but it is repeated as Exercise 7 because of its special later interest (in Sec. 13-1).

METHOD OF IMAGES

Other types of intuitive arguments also are applicable. For example, for $\mathcal{R}$ a circular disk (simply-connected, so $n = 1$), Green's function can be found by the method of images applied to the *first* (*hydrodynamic*) *existence theorem*.

We shall develop this method first for the upper half plane,

$$\mathcal{R}: \quad \text{Im } z > 0 \qquad (\overline{\mathcal{R}}: \quad \text{Im } z \geq 0 \text{ and } z = \infty) \tag{9-20}$$

For this region we verify the conservation law (8-44*d*) is satisfied by singularities $f_t(z)\, dz = dF(z)$ for the following types:

$$F(z) = \frac{\sigma}{(z - \alpha)^m} \qquad \sigma \text{ complex, Im } \alpha > 0,\ m \geq 1 \tag{9-21a}$$

$$F(z) = i\rho \log (z - \alpha) \qquad \rho \text{ real, Im } \alpha > 0 \tag{9-21b}$$

$$F(z) = \rho \log \frac{z - \alpha}{z - \beta} \qquad \rho \text{ real, Im } \alpha > 0, \text{ Im } \beta > 0 \tag{9-21c}$$

It is left as Exercise 12 to show that any set of singular parts satisfying this conservation law can be expressed as the sum of such terms.

The conditions of the first existence theorem require only that $\psi^* =$ constant on the real axis. Hence it is quite clear that

$$G(z) = F(z) + \overline{F(\bar{z})} \tag{9-22}$$

is a function for which $dG = d\psi + i\, d\psi^*$ provides a solution for the first existence theorem. Indeed, Re $G(z) = 2$ Re $F(z)$ for $z = x$ on the real axis, but Im $G(x) = 0$ from (9-22). Actually, for each of (9-21*a*) to (9-21*c*) we have

(9-23*a*) $$G(z) = \frac{\sigma}{(z-\alpha)^m} + \frac{\bar{\sigma}}{(z-\bar{\alpha})^m}$$

(9-23*b*) $$G(z) = i\rho \log \frac{z-\alpha}{z-\bar{\alpha}}$$

(9-23*c*) $$G(z) = \rho \log \frac{(z-\alpha)(z-\bar{\alpha})}{(z-\beta)(z-\bar{\beta})}$$

(Recall that $\log \bar{z} = \overline{\log z}$ by virtue of the choice of the real branch of logarithm for real z.)

For the unit circular disk we can show that (9-22) can be replaced by

(9-24) $$G_0(z) = F(z) + \overline{F(z^*)}$$

where z^* is the image of z with respect to the boundary of such a disk (see Sec. 2-6). If $G_0(z)$ does not introduce new singularities back in the unit circle, then $G_0(z)$ can be taken as $= \psi + i\psi^*$ for Green's function, as before. Otherwise, we make modifications (as in Exercise 13). In any case, we call $\psi + i\psi^* = G(z)$ once again.

(9-25*a*), (9-25*b*), (9-25*c*)
$$F(z) = \begin{cases} \dfrac{\sigma}{(z-\alpha)^m} \\ i\rho \log (z-\alpha) \\ \rho \log \dfrac{z-\alpha}{z-\beta} \end{cases} \qquad G(z) = \begin{cases} \dfrac{\sigma}{(z-\alpha)^m} + \dfrac{\bar{\sigma} z^m}{(1-\bar{\alpha} z)^m} \\ i\rho \log \dfrac{z-\alpha}{1-z\bar{\alpha}} \\ \rho \log \dfrac{(z-\alpha)(1-z\bar{\alpha})}{(z-\beta)(1-z\beta)} \end{cases}$$

with σ complex, ρ real, and $|\alpha| < 1$.

We can recognize the Green's function as the *imaginary* part of (9-23*b*) and (9-25*b*) ($\rho = 1$), thus

(9-23*d*) $$\psi^* = \log \left| \frac{z-\alpha}{z-\bar{\alpha}} \right| \qquad \text{for the upper half plane, Im } z \geq 0$$

(9-23*e*) $$\psi^* = \log \left| \frac{z-\alpha}{1-z\bar{\alpha}} \right| \qquad \text{for the unit circle, } |z| \leq 1$$

EXERCISES

1 Show that the dipole flow (9-14) $dF = f\,dz = -\alpha\,dz/(z-z_0)^2$ transforms into $dF = f_1\,dz' =$ constant dz' if we transform $z = z_0$ to $z' = \infty$ by a linear transformation, say $z' = 1/(z - z_0)$. (Note that this makes the "uniform" flow, $W = U + iV$ constant, a dipole at ∞.)

2 Consider the flow about a unit circle $|z| = 1$ given by $\psi + i\psi^* = F(z)$ where $F(z) = z + 1/z$. Interpret this flow as a dipole flow from ∞ about a unit circle with a dipole at the origin (see Fig. 9-5). Verify that no stagnation point $[F'(z) = 0]$ occurs except (on the circle) at $z = \pm 1$, and verify that the equipotential lines come in at 45° angles there (by reference to Exercise 1 of Sec. 9-1).

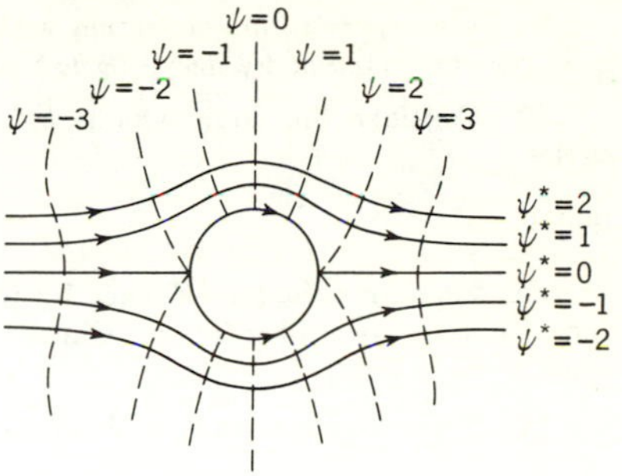

Fig. 9-5 Flow about a unit circle.

3 Show that the abelian differential of the first kind on $\bar{\mathfrak{R}}$ determined by (9-16*a*) and (9-16*b*) satisfies the Riemann period relation (of Sec. 8-5) for $\mathfrak{R}^*$, the result of crosscut dissection, namely

$$\int_{\partial\mathfrak{R}^*} \psi\, d\psi^* = -\sum_{j=1}^{n} (\mathcal{C}_j, d\psi)(\mathcal{K}_j, d\psi^*) = -\sum_{j=1}^{n-1} (\mathcal{C}_j, d\psi)\psi^*(\mathcal{C}_j) \tag{9-26}$$

where $\psi^*(\mathcal{C}_j)$ is the constant value of ψ^* on $\mathcal{C}_j$. [Take $\psi^*(\mathcal{C}_n) = 0$ always, and use the conservation law $(\mathcal{C}_n, d\psi) = k_n = -k_1 - \cdots - k_{n-1}$]. [*Hint:* By referring to the canonical models observe $(\mathcal{K}_j, d\psi^*) = \psi^*(\mathcal{C}_j) - \psi^*(P_0)$ where $\mathcal{K}_j$ joins P_0 to $\mathcal{C}_j$.]

4 For an abelian integral ψ of the first kind, the set of values $\psi^*(\mathcal{C}_j)$ will all vanish ($j = 1, \ldots, n$) if and only if $\psi^* \equiv 0$. [Interpret (9-26) as a Dirichlet integral.]

5 From Exercise 4, show that the set of $(n-1)$ real values $g_j = \psi^*(\mathcal{C}_j)$ ($g_n = 0$ for standardization) can be defined *arbitrarily* by choice of the abelian integral ψ of the first kind. [*Hint:* If we write $\psi = \sum_1^{n-1} \xi_i \psi^{(i)}$ (with normal integrals), then we must solve [see (9-19)] $\Sigma \alpha_{ij}\xi_i = g_j$. Then consider the alternative based on the vanishing of the determinant α_{ij}.]

6 Show the determinant of modules α_{ij} in (9-19) is nonsingular.

7 Show that if $\psi = \sum_1^{n-1} \xi_i \psi^{(i)}$ is substituted into (9-26), then from (9-17*b*) and (9-17*c*)

$$\int \psi\, d\psi^* = -2\pi Q(\xi_1, \ldots, \xi_{n-1}) \tag{9-27}$$

where $Q(\xi_1, \ldots, \xi_{n-1}) = -\Sigma \xi_i \xi_j \alpha_{ij} > 0$ unless all $\xi_i = 0$. Deduce another proof of Lemma 9-1.

8 Show from the Riemann period relation that if ψ and ϕ are two different abelian integrals of the first kind (subject to $\phi^* = \psi^* = 0$ on $\mathcal{C}_n$, as usual),

$$\int_{\partial\mathfrak{R}^*} \psi\, d\phi^* = -\sum_{j=1}^{n-1} (\mathcal{C}_j, d\psi)\phi^*(\mathcal{C}_j) \tag{9-28}$$

Show $\int_{\partial\mathfrak{R}^*} \psi^{(i)}\, d\psi^{(j)*} = -2\pi\psi^{(j)*}(\mathcal{C}_i) = -2\pi\alpha_{ji}$.

9 Prove $a_{ij} = a_{ji}$ in connection with Exercise 8. [Remember, always, that $\mathcal{R}^*$ is simply-connected and $\psi\, d\phi^* - \phi\, d\psi^*$ is closed, by Exercise 9(*b*) of Sec. 8-3.]

10 Consider the doubly-connected region (annulus) contained between two circles

$$\mathcal{R}:\quad (0 <)r_1 < |z| < r_2 \qquad \mathcal{C}_j:\quad |z| = r_j \tag{9-29}$$

There is one normal integral defined by $(\mathcal{C}_1, d\psi) = -(\mathcal{C}_2, d\psi) = 2\pi$ and $\psi^* = 0$ on $\mathcal{C}_2$. Show $\psi + i\psi^* = +i \log (z/r_2)$. Check the signs carefully. Show a_{11} (the only module) $= -\log (r_2/r_1) < 0$.

11 From the invariance of abelian integrals show that the annulus

$$\mathcal{R}':\quad (0 <)r_1' < |z'| < r_2' \tag{9-30}$$

cannot be mapped conformally and biuniquely onto $\mathcal{R}$ when $r_2/r_1 \neq r_2'/r_1'$.

12 Show that any set of singularities for the upper half plane in (9-20) satisfying the conservation laws (8-44*d*) can be expressed in terms of (9-21*a*) to (9-21*c*). [The difficult terms are (9-21*c*); show that a grouping of canceling pairs is always possible.]

13 Verify the formulas (9-25*a*) to (9-25*c*). Note that for (9-25*b*), $G_0(z)$, as described, has the extra singularity term $i\rho \log z$ which must be artificially omitted to produce $G(z)$.

14 Consider the twin-pole flow whose singularities are given by $\int f_t(z)\, dz = \lambda \log [(z - z_0)/(z - z_1)]$ for z_0, z_1 in $\mathcal{R}$. Here λ can be arbitrary. Assuming, as usual, $(\mathcal{C}_j, d\psi) = 0$ $(j = 1, 2, \cdots, n)$, describe the flow given, say, by

$$G(z;\lambda, z_0, z_1) = \psi + i\psi^* \tag{9-31}$$

for λ real (canceling sources) and λ imaginary (canceling vortices).

9-3 Canonical Mappings

If we accept the first existence theorem (on the basis of physical intuition for now), we are almost prepared to prove a large number of mapping theorems of the multiply-connected region† $\mathcal{R}_n$ of the z sphere onto certain "canonical" domains in, say, the H sphere. Certain difficulties now arise which are not due to physical intuition but "topological intuition," and they must be justified carefully.

The simplest case occurs when $\mathcal{R}_n$ is simply-connected ($\mathcal{R}_1 = \mathcal{R}$). Then either hydrodynamic or electrostatic Green's function g ($= \psi_H^*$ or ψ_E, see Sec. 9-2) is defined uniquely by

$$\begin{aligned} g &= \log \frac{1}{|z - z_0|} + \text{reg harm funct} \qquad \text{for } z \in \mathcal{R} \\ g &= 0 \qquad \text{on } \partial\mathcal{R} \end{aligned} \tag{9-32}$$

† The theory also applies to a Riemann manifold topologically equivalent to a plane region $\mathcal{R}_n$. The idea of proving mapping theorems in this fashion is a masterpiece of intuition due to Riemann.

where z_0 of course is the pole (interior to $\mathcal{R}$). Next we form the conjugate h and the analytic function

$$G = -\log(z - z_0) + \text{reg funct} = g + ih \tag{9-33}$$

and we apply the usual conservation law in Lemma 8-4 (8-48*b*). Then $d[-\log(z - z_0)]$ is of residue -1, and

$$(\partial\mathcal{R}, dh) = -2\pi \tag{9-34}$$

Hence G has a period of $-2\pi i$ and from (9-33)

$$H = \exp(-G) = \phi(z)(z - z_0) \tag{9-35}$$

where $\phi(z)$ is analytic and nonvanishing in $\mathcal{R}$. Thus $H(z)$ maps $\mathcal{R}$ onto a neighborhood of $H = 0$ with boundary given as a *subset* of

$$|H| = \exp -g = 1 \qquad \text{on } \partial\mathcal{R}$$

In other words, $H(\mathcal{R})$ is a subset of the interior of the unit circle. We still have a topological problem of proving that $H(\mathcal{R})$ is *precisely* the interior of the unit circle and that the mapping is bianalytic. (This will be handled in the next chapter.)

In the meantime, we shall list various canonical models and outline what is necessary to establish the proofs along similar lines.

CANONICAL MAPPINGS

Let us refer to certain abelian integrals just defined in Sec. 9-2 (above) over an n-tuply-connected region $\mathcal{R}_n$ of the z plane† with boundary curves $\mathcal{C}_1, \mathcal{C}_2, \ldots, \mathcal{C}_n$. In each case, of course, $\psi^* = g_j$ (constant) on each $\mathcal{C}_j$ (with $g_n = 0$ for standardization). We also prescribe $(\mathcal{C}_j, d\psi) = 0$ for $j = 1, 2, \ldots, n - 1$, except for the case (*c*) below where the contrary is specifically stated. *Assuming the first existence theorem*, we have the following single-valued functions $H_t(z)$:

(*a*) *Hydrodynamical Green's function* (9-13*b*)
$G(z,z_0) = \psi + i\psi^* = -i\log(z - z_0) + \cdots$
$H_1(z) = \exp iG(z;z_0)$

(*b*) *Dipole flow* (9-15*d*)
$$D(z;a,z_0) = \psi + i\psi^* = \frac{a}{z - z_0} + \cdots$$
$H_2(z) = D(z;a,z_0)$

(*c*) *Normal differentials of first kind for* $n \geq 2$, (9-17*a*) and (9-17*b*)
$G^{(1)}(z) = \psi^{(1)} + i\psi^{(1)*}$
and $(\mathcal{C}_1, d\psi^{(1)}) = 2\pi$
$H_3(z) = \exp(-iG^{(1)}(z))$

† The theory also applies to a Riemann manifold topologically equivalent to a plane region $\mathcal{R}_n$.

(*d*) *Self-canceling pole function* (9-31)

$$G(z;\lambda,z_0,z_1) = \lambda \log \frac{z - z_0}{z - z_1} + \cdots = \psi + i\psi^*$$

$$H_4(z) = \exp G(z;1,z_0,z_1) \qquad \text{source and sink}$$

$$H_5(z) = \exp \frac{G(z;i,z_0,z_1)}{i} \qquad \text{canceling vortices}$$

Incidentally, in cases (*a*) and (*c*), $(\mathfrak{C}_n,d\psi) = -2\pi$, but all other such contour integrals vanish. (Here ψ also means $\psi^{(1)}$.)

The geometrical characteristics of the mapping from the z plane into the H plane are easily determined.

$H = H_1(z)$ — $z = z_0$ becomes $H_1 = 0$; $\mathfrak{C}_j$ becomes $|H_1| = \text{const}$ (circular arcs) but $\mathfrak{C}_n$ becomes a complete circle surrounding $H(\mathfrak{R})$

$H = H_2(z)$ — $z = z_0$ becomes $H_2 = \infty$; $\mathfrak{C}_j$ becomes $\operatorname{Im} H_2 = \text{const}$ (straight horizontal slits)

$H = H_3(z)$ — $\mathfrak{C}_j$ becomes $|H_3| = \text{const}$ (circular arcs) but $\mathfrak{C}_1$ and $\mathfrak{C}_n$ become complete circles

$H = H_4(z)$ — $z = z_0, z_1$ becomes $H_4 = 0, \infty$; $\mathfrak{C}_j$ becomes $\arg H_4 = \text{const}$ (radial slits)

$H = H_5(z)$ — $z = z_0, z_1$ becomes $H_5 = 0, \infty$; $\mathfrak{C}_j$ becomes $|H_5| = \text{const}$ (circular arcs)

In all cases the circular arcs have centers at the origin. The circular arcs which are asserted to be complete circles must cover complete circles by virtue of the property that, say $(\mathfrak{C}_n,d\psi) = \int_{\mathfrak{C}_n} d\psi = -2\pi$, etc.

THEOREM 9-1 *The five canonical mapping functions* $H_1(z), \ldots, H_5(z)$ *each map* $\mathfrak{R}_n$ *biuniquely onto a subregion of the H sphere bounded by disjoint continua consisting of segments, arcs, etc., as shown in Fig.* 9-6.

A formal proof requires such a stage of accomplishment that in many texts this theorem is the objective of conformal mapping (rather than the Riemann theorem on compact manifolds). Moreover, the best proofs from the modern viewpoint employ minimal principles which are outside the scope of this text. What we shall do is to give a rough (indeed an incorrect) proof and ask what must be done to make it valid.†

Consider $H_2(z)$. We want to show a biunique mapping of $\mathfrak{R}_n$ onto the exterior of n slits by a dipole function (see Fig. 9-6a_2). Even if $H_2(\mathfrak{R}_n)$ does not produce a one-to-one image of $\mathfrak{R}_n$, we can consider (general) points h_0 which are covered by $H_2(\mathfrak{R}_n)$, but not by $H_2(|\partial\mathfrak{R}_n|)$, the image of

† Contrast Appendix A and Lemma 10-2 (Sec. 10-2).

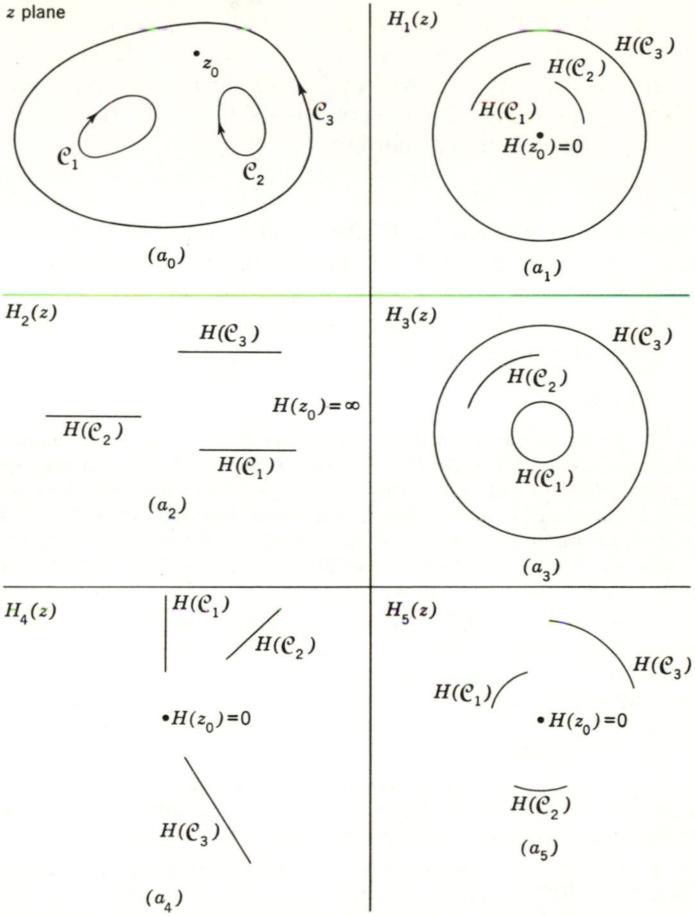

Fig. 9-6 Canonical mappings. The region $\mathfrak{R}_3$ is shown under the five typical mappings of the z plane in Fig. 9-6a_0 onto H planes in Fig. 9-6 (a_1-a_5).

the boundary. We first show such points are covered in one-to-one fashion. Indeed, by the argument principle [since $H_2(z) - h_0$ has one pole in $\mathfrak{R}_n$], the number N of roots of $H_2(z) = h_0$ is given by

$$N - 1 = \frac{1}{2\pi i} \int_{\partial\mathfrak{R}_n} d \log [H_2(z) - h_0] = \text{say } \{\partial\mathfrak{R}_n, h_0\} \tag{9-36}$$

Now this formula is not quite correct since $H_2(|\partial\mathfrak{R}_n|)$ will not constitute an analytic mapping (but we might still replace $\partial\mathfrak{R}_n$ by arbitrarily close curves). But clearly if $H_2(z)$ maps $\partial\mathfrak{R}_n$ on some slits *external* to h_0, the contribution of $\int d \log (\cdot \cdot \cdot)$ is 0 (the log is single-valued on each slit) and $N - 1 = 0$. Thus we have $N = 1$, or biuniqueness.

The next step is to show that the images of the slits $H_2(|\partial \mathcal{R}_n|)$ are unique; i.e., each $\mathcal{C}_j$ is "sewed up" into a slit by $H = H_2(z)$ and each slit is simple and disjoint from every other slit. To do this, we must find a suitable way of counting solutions to

$$H_2(z) = h_0 \tag{9-37}$$

where h_0 is preassigned in $H_2(\partial \mathcal{R}_n)$ and z varies on $\overline{\mathcal{R}}_n$. Obviously, we must be able to extend $H_2(z)$ analytically to a larger region, say $\mathcal{R}_n{}^+$, which includes $\mathcal{R}_n$ as a subregion, and we seek to solve (9-37) for $z \in \mathcal{R}_n{}^+$ $(\supset \overline{\mathcal{R}}_n)$. The argument again will revolve around the formula (9-36), but we shall not pursue it here.†

EXERCISES

1 Let $\mathcal{R}_1$ be the unit disk $|z| < 1$ so that $H_2(z) = \frac{1}{2}(z + 1/z)$ maps $\mathcal{R}_1$ onto the exterior of the slit $-1 \leq \operatorname{Re} H \leq 1$ in the H sphere. Prove that exactly two points z_1, z_2 go into each real h_1 such that $-1 < h_1 < +1$ by letting $\mathcal{R}^+$ be the disk $|z| < a$ $(a > 1)$ and counting the number of solutions of $H(z) = h_1$ for $z \in \mathcal{R}^+$. Explain the peculiar nature of $h_1 = \pm 1$ $(z_1 = z_2)$. [*Hint:* Note that $H_2(\mathcal{R}^+)$ covers the entire H sphere in such a way that the neighborhood of the slit $[-1,1]$ is covered twice.] Show that in (9-36), $\{\partial \mathcal{R}^+, h_1\} = 1$, thus show $N = 2$ by drawing a diagram.

2 On the basis of the canonical mapping $H_3(z)$, show that a region of connectivity n is determined by *no more than* $3(n - 2)$ real parameters if $n \geq 3$ (by one real parameter if $n = 2$, by no real parameters if $n = 1$). Identify the parameters explicitly with the radii of the circles $H_3(\mathcal{C}_1)$, $H_3(\mathcal{C}_n)$ and with the positions of the arcs $H_3(\mathcal{C}_2), \ldots, H_3(\mathcal{C}_{n-1})$. Note that rotations and similarity transformations affect the number of parameters.‡

3 Obtain the number of parameters by counting degrees of freedom for the other mapping functions $H_t(z)$ $(t = 1, 2, 4, 5)$. (Remember that such arguments are boldly unrigorous at the level required here.)

4 Discuss the canonical domain obtained [analogously with $H_4(z)$, $H_5(z)$] by using $H = \exp [G(z;\lambda,z_0,z_1)/\lambda]$ (λ complex).

5 Show that in a certain sense, all canonical-type mappings of $\mathcal{R}_n$ onto a plane region have been accounted for. (Consider the abelian integrals with no more than one pole!)

† See Exercise 1 for a significant special case. In Chap. 11 we show every (noncompact) Riemann manifold $\mathfrak{M}$ is equivalent to a manifold over, say, the z sphere, which is analytically parametrized $z = f(u)$ *across* any part of the boundary $\partial \mathfrak{M}$. Thus (9-36) could be referred to parameter planes and be literally correct.

‡ It is not evident, but *exactly* the indicated number of parameters determine $\mathcal{R}_n$ conformally (see the Bibliography).

PART IV
REAL EXISTENCE PROOFS

EVOLUTION OF SOME INTUITIVE THEOREMS

The next three chapters are devoted to the proof of the two existence theorems which constitute the Dirichlet problem. We have attempted to justify the theorems beforehand (in Chap. 9) by physical intuition, mostly to stress the position of historical urgency which these problems occupied from 1850 to 1910. (If the results are physically obvious, why can't we *prove* them?) The sophisticated reader who would like to reject the physical world cannot wholly reject the intuitive arguments since they state with chastity: "What is consistent from the point of view of differentials (cohomology, if you please, rather than curl and flux) must be mathematically correct!"

The urgency of these problems indeed made it difficult to agree when the proofs had been consummated. Riemann's original argument (1852) was through a faulty minimal principle,† but the essential features of correct proofs were soon given by Schwarz (1870) and Neumann (1877). We shall present a historically tempered version of their proof (coming closer to Picard's presentation in his "Cours d'Analyse"). Actually, the problem was still not considered solved until about 1910 when Hilbert, Weyl, Courant, etc., carried Riemann's original argument to completion!

We shall purposely avoid newer concepts in analysis in order to contrast the adequacy of simple analytic tools with the emergent subtlety of topological concepts. We do not strive for optimal‡ proofs as much as for historical insight.

† Compare Appendix A.

‡ Some students will have encountered the necessary proofs in other contexts (of functional analysis) or will want to omit Part IV on first reading so as to first appreciate the objectives reached in Part V. For this reason, Part IV is made completely independent so it may be omitted once one accepts the physical intuition (or topological intuition!) as the basis of the two existence theorems.

CHAPTER 10
CONFORMAL MAPPING

We now turn our attention to the celebrated Riemann mapping theorem. It states that a finite simply-connected region $\mathcal{R}$ on the z plane can be mapped onto the unit circular disk $|z'| < 1$ by a biunique analytic function $z' = f(z)$, defined for z in $\mathcal{R}$. The original proof process followed a devious historical path, witnessing the interaction of real and complex techniques in analysis. There are now a plethora† of proofs, many quite self-contained. We shall, however, follow a somewhat rambling path of least resistance in order to show the interplay of the fields. In retrospect, we shall finally come to have several shorter, neater proofs by the end of the text.

10-1 Poisson's Integral

Some of the most important properties of harmonic functions arise from the so-called "ordinary Dirichlet problem."‡ This is the problem (involving the second existence theorem) determined by zero integrals (single-valuedness), no prescribed singularities, and continuous prescribed boundary data. Specifically, let ψ be an unknown single-valued (real) harmonic function defined on an n-tuply-connected plane region $\mathcal{R}$ with, say, n simple closed boundary curves. We are asked to solve the boundary value problm for $\psi(x,y)$ or $\psi(z)$ ($z \in \mathcal{R}$),

$$(10\text{-}1) \qquad \frac{\partial^2\psi}{\partial x^2} + \frac{\partial^2\psi}{\partial y^2} = 0 \qquad \text{(harmonic property)}$$

$$\psi(z) \to g(Z_0) \qquad \text{as } z \to Z_0 \text{ on } \partial\mathcal{R} \text{ (boundary data)}$$

where the prescribed boundary data $g(Z)$ are continuous functions on each of the n components of $\partial\mathcal{R}$.

† There are not many other theorems which have so excited the "collectors' interest" in proofs; one might cite, analogously, the Pythagorean theorem (with hundreds of proofs), quadratic reciprocity (with dozens of proofs), etc.

‡ We do not assume the validity of the first or second existence theorems from now on until each one is proved.

The ordinary Dirichlet problem for the *circle* can be solved explicitly by a formula of a general type due to Poisson (1820). We let the circle be a circle of radius R about the origin, and we let

$$\psi = g(R \exp i\Theta) \tag{10-2a}$$

be the given (continuous) boundary values for Θ varying from 0 to 2π. We desire the solution at $\rho = r \exp i\Theta$ $(r < R)$. Thus

$$\psi = \psi(\rho) = \psi(r \exp i\Theta) \qquad r < R \tag{10-2b}$$

is a harmonic function which is to equal $g(r \exp i\Theta)$ when $r = R$.

To solve the problem, we assume a solution to exist and then obtain it *deductively*. After the result is *checked* we shall have proved both existence (because we checked it) and uniqueness (because we deduced it).

First we note that $\psi(0)$ is actually easily obtained from the boundary data $g(R \exp i\Theta)$.

THEOREM 10-1 (*Gauss*) *If* $\psi(z)$ *or* $\psi(x,y)$ *is harmonic in the closed disk in* $\mathcal{C}_r$, *a circle of radius* r *about a point* $z_0 = x_0 + iy_0$, *then* $\psi(z_0)$ *is the average of its values about* $\mathcal{C}_r$ *or*

$$\psi(z_0) = \frac{1}{2\pi}\int_0^{2\pi} \psi(z_0 + r \exp i\theta)\, d\theta \tag{10-3}$$

PROOF We consider the harmonic function ψ as part of the analytic function $w(z) = \psi(x,y) + i\psi^*(x,y)$. We can write Cauchy's representation theorem as

$$w(z_0) = \frac{1}{2\pi i}\int_{\mathcal{C}} \frac{w(z)\, dz}{z - z_0}$$

where w is analytic at $z_0 = x_0 + iy_0$ and $\mathcal{C}$ is a circle described counterclockwise and of radius r. Thus with $z = z_0 + r \exp i\theta$ and $dz = ri \exp i\theta\, d\theta$, (10-3) follows for w, ψ (and ψ^* also). Q.E.D.

We can find $\psi(0)$ as the average over Θ of $g(R \exp i\Theta)$. To find $\psi(\rho)$, we map the circle onto itself so that $z = \rho$ goes to the *origin* $(z' = 0)$. This is done by using

$$\frac{z'}{R} = \frac{z/R - \rho/R}{1 - \bar{\rho}z/R^2} \tag{10-4}$$

which has the required mapping property (Lemma 2-10, Sec. 2-7). Furthermore, the special choice of mapping provides

$$\frac{dz'}{dz} = R^2 \frac{R^2 - |\rho|^2}{(\bar{\rho}z - R^2)^2} > 0 \qquad \text{at } z = \rho$$

thus the slopes are preserved *exactly* (zero rotation) at $z = \rho$, $z' = 0$.

Turning our attention to the z' plane, the boundary values (10-2a) obviously are rearranged by the map (10-4). Let $Z' = R \exp i\lambda$ be the new boundary point of the z' plane corresponding to $Z = R \exp i\Theta$ of the z plane. Then by (10-4) in the new variables

$$R \exp i\lambda = \frac{(R \exp i\Theta - \rho)R^2}{-\bar{\rho}R \exp i\Theta + R^2} \tag{10-5a}$$

The critical step is that by Theorem 8-4 of Sec. 8-3, the solution is invariant under analytic mappings. Thus Gauss' mean value theorem for the z' plane gives us the result that at $z' = 0$, $\psi = \psi(\rho)$ is given by

$$\psi(\rho) = \frac{1}{2\pi}\int_0^{2\pi} g(R \exp i\Theta)\, d\lambda \tag{10-6}$$

Here (10-6) is actually the desired solution, with (10-5a) relating λ and Θ. We must next change the variables to Θ. From (10-5a)

$$i\lambda = \log(R \exp i\Theta - \rho) - \log(-\bar{\rho}R \exp i\Theta + R^2) + \log R$$

Differentiating we find

$$i\, d\lambda = \left\{\frac{R \exp i\Theta}{R \exp i\Theta - \rho} - \frac{R\bar{\rho} \exp i\Theta}{R\bar{\rho} \exp i\Theta - R^2}\right\} i\, d\Theta = P(\rho;\Theta) i\, d\Theta \tag{10-5b}$$

We see [Exercise 1(a)] that $P(\rho;\Theta)$ has the form

$$P(\rho;\Theta) = \left[\frac{R^2 - r^2}{R^2 + r^2 - 2rR\cos(\Theta - \theta)}\right] \qquad \rho = r \exp i\theta \tag{10-7a}$$

The quantity $P(\rho;\Theta)/2\pi$ is called the "Poisson kernel." Thus we obtain the *Poisson integral* solution to the Dirichlet problem (10-1):

$$\psi(\rho) = \frac{1}{2\pi}\int_0^{2\pi} P(\rho;\Theta) g(R \exp i\Theta)\, d\Theta \qquad |\rho| < R \tag{10-7b}$$

Actually, $P(\rho;\Theta)$ is harmonic in $\rho = x + iy$ since (10-5b) consists of analytic functions of ρ and of $\bar{\rho}$. Also, $P(\rho;\Theta)$ is nonsingular if $r = |\rho| < R$. Thus the function (10-7b) is harmonic† in ρ.

We next verify that the solution (10-6) satisfies the boundary values by transferring the picture to the z plane. Here the radial lines of the z' plane (or λ = constant) become transformed into the *normal* circles from $z = \rho$, making an angle λ with the positive direction $z = \rho$ (see Fig. 10-1) in accordance with the function determined by (10-5a). The angles are preserved (by conformality).

† We note that the operations $\partial^2/\partial x^2$, $\partial^2/\partial y^2$, etc., can be carried out under the integral (10-7b) for $|\rho| < R$ since the denominator does not vanish. Indeed, $P(\rho;\Theta)$ is uniformly bounded (and so are its derivatives, etc.) if $|\rho| <$ constant $< R$. See Exercise 1(a).

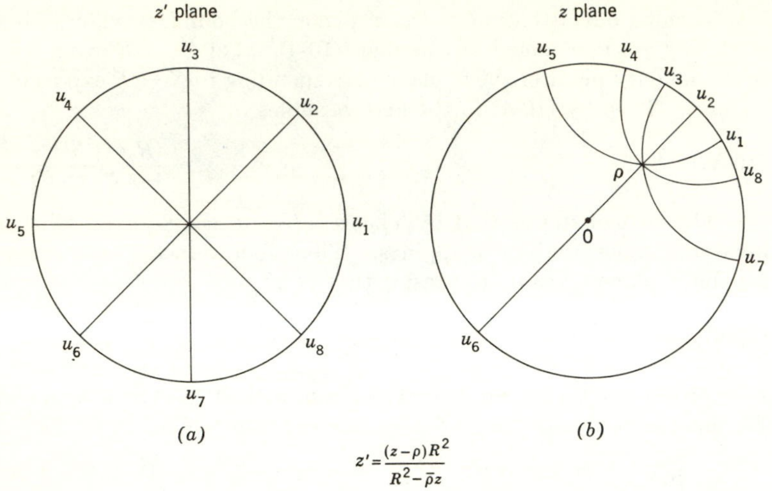

Fig. 10-1 Derivation of Poisson kernel. Note redistribution of boundary values $u_1, \ldots, u_8$ from z plane to z' plane with greater weight given to the boundary values nearest ρ.

Now for the Dirichlet boundary-value problem, the given data $g(R \exp i\Theta)$ are assumed to be continuous in Θ. We shall show $\psi(\rho) \to g(R \exp i\Theta_0)$ as $\rho \to R \exp i\Theta_0 = Z_0$ (subject to $|\rho| < R$). There will be no supposition that the limit $\rho \to Z_0$ must occur only along a radius or in any other special way (only that $|\rho - Z_0| \to 0$ and $|\rho| < R$).

By continuity, for any positive ε, a $\delta(\varepsilon)$ exists such that

$$|g(R \exp i\Theta_0) - g(R \exp i\Theta)| < \varepsilon \qquad \text{for } |\Theta_0 - \Theta| < \delta(<\pi)$$

Also, $|g(R \exp i\Theta)| \leq M$ by continuity.

Thus consider [from (10-6)],

$$\psi(\rho) - g(R \exp i\Theta_0) = \frac{1}{2\pi} \int_0^{2\pi} [g(R \exp i\Theta) - g(R \exp i\Theta_0)]\, d\lambda \tag{10-8}$$

since $g(R \exp i\Theta_0)$ is a constant and $\int_0^{2\pi} d\lambda = 2\pi$.

Thus for the arc

$$(\Theta_0 - \delta) \leq \Theta \leq (\Theta_0 + \delta) \tag{10-9a}$$

we have

$$|g(R \exp i\Theta) - g(R \exp i\Theta_0)| \leq \varepsilon \tag{10-9b}$$

We can let $\Lambda(\rho)$ represent the angle subtended by that arc (10-9a) from a variable point ρ, using orthogonal circles from ρ to the circle

$|z| = R$ (see Fig. 10-1*b*). Easily, $\Lambda(\rho) = \int d\lambda$ where λ is measured equivalently from ρ in the z plane or from the origin in the z' plane. Then, breaking up the interval of integration into the "small" part (10-9*a*) and its complement, we obtain from (10-8):

$$|\psi(\rho) - G(R \exp i\Theta_0)| \leq \frac{1}{2\pi}\left|\int_{\Theta_0-\delta}^{\Theta_0+\delta} \cdots d\lambda\right| + \frac{1}{2\pi}\left|\int_{\Theta_0+\delta}^{\Theta_0-\delta+2\pi} \cdots d\lambda\right|$$

Thus, since $|g(R \exp i\Theta) - g(R \exp i\Theta_0)| \leq 2M$ and

$$\left|\int_{\Theta_0-\delta}^{\Theta_0+\delta} [g(R \exp i\Theta) - g(R \exp i\Theta_0)]\, d\lambda\right| \leq \varepsilon \int_{\Theta_0-\delta}^{\Theta_0+\delta} d\lambda$$

we see

$$|g(\rho) - g(R \exp i\Theta_0)| \leq \frac{2M(2\pi - \Lambda(\rho))}{2\pi} + \frac{\varepsilon\Lambda(\rho)}{2\pi}$$

Finally, let us note that (by elementary geometry) as $\rho \to R \exp i\Theta_0$, $\Lambda(\rho) \to 2\pi$. (See the sequence ρ_1,ρ_2,ρ_3 in Fig. 10-2.) Thus we can surround Θ_0 by a neighborhood $\mathfrak{N}$ of radius δ^* (see shaded zone $\mathfrak{N}$ in Fig. 10-2) such that

(10-10*a*) $$2\pi - \varepsilon < \Lambda(\rho) < 2\pi$$

for $\delta^*(\varepsilon,\delta)$ suitably determined by ε and $\delta(\varepsilon)$ and ρ now subject to

(10-10*b*) $$|\rho - Z_0| < \delta^*(\varepsilon,\delta) \qquad |\rho| < R$$

Thus, (10-10*b*) leads to

(10-11) $$|\psi(\rho) - g(R \exp i\Theta_0)| \leq \varepsilon\frac{2\pi}{2\pi} + \frac{2M\varepsilon}{2\pi} = \varepsilon\left(1 + \frac{M}{\pi}\right)$$

which asserts the continuous approach of $\psi(\rho)$ to the boundary value $g(R \exp i\Theta_0)$.

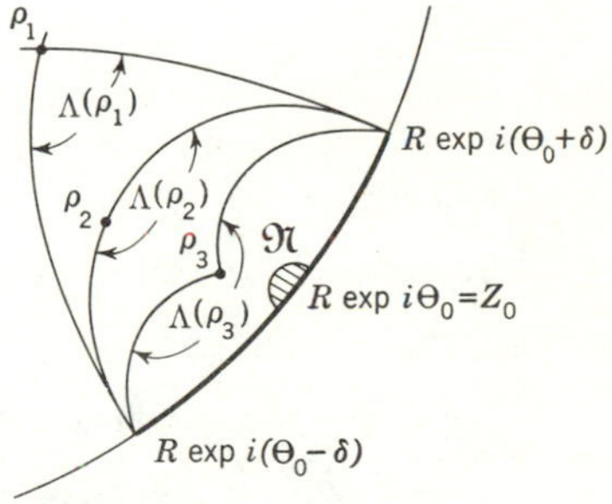

Fig. 10-2 Limiting behavior as $\rho \to Z_0 = R \exp i\Theta_0$.

THEOREM 10-2 (*Schwarz*, 1872) *The equations* (10-7*a*) *and* (10-7*b*) *solve the ordinary Dirichlet boundary-value problem for the circular disk.*

By applying the invariance property of harmonic functions again, we obtain the next result:

COROLLARY 1 *For any region* $\mathcal{R}_1$ *which is the biunique conformal image of the unit circle with mapping continuously extendable to the boundary, the ordinary Dirichlet problem can be solved.*

THEOREM 10-3 *A sequence of harmonic functions which converges uniformly in a region* $\mathcal{R}$ *has a harmonic limit function.*

PROOF Let $\psi_n(z) \to \psi(z)$ uniformly in $\mathcal{R}$ as $n \to \infty$ where $\psi_n(z)$ are harmonic functions. Now any point of $\mathcal{R}$ is contained in some circle of radius R and, say, center β. Then for $z(= \rho + \beta)$ within this circle (10-7*b*) yields

$$\psi_n(\rho + \beta) = \frac{1}{2\pi} \int_0^{2\pi} P(\rho;\Theta)\psi_n(R \exp i\Theta + \beta)\, d\Theta$$

This relation, however, makes $\psi(z)$ harmonic in the limit. Q.E.D.

COROLLARY 1 *A sequence of harmonic functions which is uniformly convergent in a region has uniformly convergent derivatives* $\partial/\partial x$, $\partial/\partial y$, *etc. The limit of the sequence of derivatives is the same as the derivative of the limit of the sequence.*

COROLLARY 2 *A convergent sequence of harmonic functions has a convergent sequence of harmonic conjugate functions, and the limit of the sequence of conjugates is the conjugate of the original sequence, provided the conjugates are adjusted by an additive constant to take a fixed value at some fixed point.*

The proofs are left as Exercise 4. These proofs illustrate the fact that the Poisson integral representation implies "restraints" on harmonic functions just as the Cauchy integral theorem implies restraints on analytic functions. We need the following remarkable property:

THEOREM 10-4 (*Harnack*, 1887) *If a sequence of harmonic functions is monotonic in a region* $\mathcal{R}$ *and if it is bounded at some point a of* $\mathcal{R}$, *then the sequence converges uniformly in any subregion* $\mathcal{R}'$ *of* $\mathcal{R}$ *whose boundary is interior to* $\mathcal{R}$.

The proof is left as Exercises 6 and 7.

EXERCISES

1 (*a*) Derive the Poisson kernel (10-7*a*) from (10-5*b*). (*b*) From the fact that $|\cos(\Theta - \theta)| \leq 1$, show

$$(R - r)^2 < R^2 - 2Rr\cos(\Theta - \theta) + r^2 < (R + r)^2$$

(10-12) $$\frac{R - r}{R + r} \leq p(\rho;\Theta) \leq \frac{R + r}{R - r}$$

(*c*) Show that if $\psi(Z) \geq 0$ on the boundary $|Z| = R$, then at $\rho = r \exp i\theta$ ($|r| \leq R$)

$$\frac{R-r}{R+r} \cdot \psi(0) \leq \psi(\rho) \leq \frac{R+r}{R-r} \cdot \psi(0) \qquad \text{(Harnack's inequality)} \tag{10-13}$$

[*Hint:* Multiply (10-12) by $g(R \exp i\Theta)\, d\Theta$ and integrate using Gauss' mean value theorem.] (*d*) Find a bound for $\partial P/\partial x$ and $\partial P/\partial y$ analogous with (10-12), i.e., in terms of R and r. [*Hint:* We must not differentiate (10-12), but we can set $\rho = x + iy$ and find $\partial P/\partial x = \partial P/\partial \rho \cdot \partial \rho/\partial x + \partial P/\partial \bar{\rho} \cdot \partial \bar{\rho}/\partial x$, using the formula for P [as given in (10-5*b*)] in terms of ρ and $\bar{\rho}$ instead of r and θ.]

2 In the upper half z plane the Dirichlet problem is posed for an unknown harmonic function $\psi(z)$ with boundary values $\psi(X)$ continuous in X (on the real axis) with $\lim \psi(X) = \psi(\infty)$ existing as $|X| \to \infty$. Show the solution is given by

$$\psi(z) = \frac{1}{2\pi} \int_{-\infty}^{\infty} K(z;X)\psi(X)\, dX$$

where $K(z;X) = \partial/\partial X \arg [(z - X)/(\bar{z} - X)] = 2/[(x - X)^2 + y^2]$. [*Hint:* Show where $K(z;X)\, dX = d\theta$, the element of angle subtended from z by the element dX, using arcs of circles through z normal to the real axis. Sometimes $K(z,X)/2\pi$ is called the "Hilbert kernel."]

3 (*a*) Show that we can write

$$P(\rho;\Theta) = 2 \operatorname{Re} \frac{R \exp i\Theta}{R \exp i\Theta - \rho} - 1$$

Hence, separating into real and imaginary parts:

$$\frac{2R \exp i\Theta}{R \exp i\Theta - \rho} = 1 + P(\rho;\Theta) + iQ(\rho;\Theta)$$

We have two complex conjugate functions $P(\rho;\Theta)$, $Q(\rho;\Theta)$ of $\rho = x + iy$, where (using $|\rho| = r$ always)

$$Q(\rho;\Theta) = 2rR \frac{\sin (\theta - \Theta)}{R^2 - 2Rr \cos (\theta - \Theta) + r^2}$$

Show that $\psi^*(\rho) = 1/2\pi \int_0^{2\pi} \psi(R \exp i\Theta)Q(\rho;\Theta)\, d\Theta$ constitutes a conjugate to $\psi(\rho)$ with the property that $\psi^*(0) = 0$. (*b*) Find a bound, like (10-12), to $Q(\rho;\Theta)$. (Is Q necessarily > 0?)

4 Prove Corollary 1 and Corollary 2 of Theorem 10-3 by using Exercises 1(*c*) and 3(*b*).

5 Show, in the usual terminology, ($|\rho| = r$)

$$\psi(\rho) - \psi(0) = \frac{1}{2\pi} \int_0^{2\pi} \psi(R \exp i\Theta) \frac{2rR \cos (\Theta - \theta) - 2r^2}{R^2 + r^2 - 2Rr \cos (\Theta - \theta)}\, d\Theta$$

and from this show

$$|\psi(\rho) - \psi(0)| \leq \left[\frac{2r(R + r)}{(R - r)^2}\right] M$$

where $M = \max |\psi(R \exp i\Theta)|$ as Θ varies from 0 to 2π.

6 Prove Harnack's theorem first in the case where $\mathcal{R}$ is a circular disk of center $a = 0$ and radius R, and a concentric disk $\mathcal{R}'$ of radius R' is covered by $\mathcal{R}(R' < R)$.

Consider the monotonic sequence

$$\psi_1(z) \leq \psi_2(z) \leq \cdots \leq \psi_n(z) \leq \cdots \qquad (z \text{ in } \mathcal{R})$$

bounded at $z = 0$. [Thus $\psi_n(0)$ approaches a limit as $n \to \infty$.] Apply (10-13) to the harmonic function $0 \leq \psi_n(z) - \psi_m(z)$ $(n > m)$ and obtain

$$0 \leq \psi_n(z) - \psi_m(z) \leq \frac{R + R'}{R - R'}[\psi_n(0) - \psi_m(0)] \tag{10-14}$$

Show the convergence at $z = 0$ implies uniform convergence throughout $\mathcal{R}'$.

7 Extend Harnack's theorem to a general region $\mathcal{R}$ $(\supset \mathcal{R}'$, the subregion requiring uniform convergence). The object is to cover $\mathcal{R}'$ by a finite chain of circular disks $\mathfrak{D}$ $(\subset \mathcal{R})$ so that starting with one disk (at $z = a$), each disk has its *center* in some other disk designated as a neighbor, thus convergence in one disk provides convergence in the neighbors. Verify that we can perform this construction by letting δ be the (minimal) separation between the boundaries $\partial\mathcal{R}'$, $\partial\mathcal{R}$ and surrounding each point of the closure of $\mathcal{R}'$ by a disk of radius $\delta/3$. By the Heine-Borel theorem there are at most a finite number, say M, of disks covering $\mathcal{R}'$. Select such a set together with the disk of radius $\delta/3$ centered at a. Taking these $M + 1$ disks we *triple* the radius of each, keeping the centers fixed. This enables us to set up a chain of $M + 1$ disks in $\mathcal{R}$, each one overlapping the center of a neighbor and all covering $\mathcal{R}'$. Show that if we take $R = 3\delta$, $R' = 2\delta$, we obtain the inequality [analogous to (10-14)]

$$0 \leq \psi_n(z) - \psi_m(z) \leq 5^{M+1}[\psi_n(a) - \psi_m(a)]$$

for z in $\mathcal{R}$. (Since M is fixed by $\mathcal{R}$ and $\mathcal{R}'$, this completes the proof.)

8 Prove the maximum-minimum theorem for harmonic functions (see Theorem 3-5, Corollary 2) by Gauss' mean value theorem. [Assume $\psi(z_0 + r \exp i\theta) \geq \psi(z_0)$ in some neighborhood of z_0 and show (10-3) is contradicted unless we have the = sign.]

10-2 Riemann's Theorem for the Disk

We consider a simply-connected region $\mathcal{R}$ in the z plane bounded by a piecewise smooth simple closed curve $\mathcal{C}$, denoted by the variable Z. Let us consider the (real) Green's function $g(z;z_0)$ with pole at z_0 $(\in \mathcal{R})$ defined as follows:

$$\begin{aligned} &\frac{\partial^2 g}{\partial x^2} + \frac{\partial^2 g}{\partial y^2} = 0 \qquad z \neq z_0 \\ &g(z;z_0) - \log\frac{1}{|z - z_0|} = G(z) = \text{harm funct in } \mathcal{R} \\ &g(z;z_0) \to 0 \qquad \text{as } z \to Z(\in \partial\mathcal{R}) \end{aligned} \tag{10-15}$$

LEMMA 10-1 *If the ordinary Dirichlet problem is solvable in $\mathcal{R}$, then a Green's function can be constructed in $\mathcal{R}$ (with any desired pole z_0).*

From the *ordinary* Dirichlet problem the function $\log(1/|Z - z_0|)$ can be defined for Z on $\partial\mathcal{R}$ as a *continuous* function $g(Z)$ (regardless of the logarithmic singularity since $Z \neq z_0$). Then solving the ordinary Dirichlet problem, we obtain a regular harmonic function $G(z)$ with boundary values $g(Z_0)$ as $z \to Z_0$. Then we set

$$g(z;z_0) = \log\left(\frac{1}{|z - z_0|}\right) + G(z) \tag{10-16a}$$

Interestingly enough, the converse of Lemma 10-1 will eventually emerge (after Lemmas 10-2 and 10-3). Thus, although the finding of a Green's function is not so special as it may seem, it leads to the solution of the whole (ordinary) Dirichlet problem.

LEMMA 10-2 *If a Green's function can be found for* $\mathcal{R}$, *then* $\mathcal{R}$ *can be mapped* (*analytically*) *in biunique fashion onto the* (*open*) *unit disk.*

PROOF† Using the Green's function as in (10-16*a*) with the pole $z_0 = a$, we define $h(x,y)$, the harmonic conjugate of $g(x,y)$ by first noting that $-\arg(z - a)$ is the conjugate of $\log(1/|z - a|)$. Therefore,

$$h(x,y) = -\arg(z - a) + H(x,y) \tag{10-16b}$$

where $H(x,y)$ is necessarily single-valued and harmonic as the conjugate of $G(x,y)$ in the simply-connected region $\mathcal{R}$. Then $g + ih$ is an analytic function of z with a pole at $z = a$, and we assert that the required mapping function of $\mathcal{R}$ onto the circle $|w| = 1$ is actually

$$w(z) = \exp[-g(x,y) - ih(x,y)] \tag{10-16c}$$

[The effect of the singularities in (10-16*a*) and (10-16*b*) is easily seen to be that $w(z) = (z - a)F(z)$ where $F(z) \neq 0$ in $\mathcal{R}$.]

This mapping function (10-16*c*) is clearly analytic and the image $w(\mathcal{R})$ lies in the disk $|w| < 1$, since $g > 0$ in $\mathcal{R}$ by the maximum-minimum principle. We must prove $w(z) = w_0$ exactly once for each w_0 (for $|w_0| < 1$) as z varies on the region $\mathcal{R}$.

Since we cannot integrate on $\mathcal{C}_z = \partial\mathcal{R}$, we consider instead the number (N^*) of values of z for which $w(z) = w_0$ for z interior to $\mathcal{R}^*$, any subregion chosen in $\mathcal{R}$, where $\partial\mathcal{R}^* = \mathcal{C}_z^*$, a simple positively oriented closed curve in $\mathcal{R}$. If we take $\mathcal{C}_w'$ in the w plane as the image of $\mathcal{C}_z^*$ under $w(z)$, we do not yet know that $\mathcal{C}_w'$ is a simple closed curve. But we see

$$N^* = \frac{1}{2\pi i}\int_{\mathcal{C}_z^*} \frac{w'(z)\,dz}{w(z) - w_0} = \frac{1}{2\pi i}\int_{\mathcal{C}_w'} \frac{dw}{w - w_0} \tag{10-17}$$

We can have both integrals equal to 1 (as desired for the proof) provided:

(*a*) We can make $\mathcal{C}_w'$ lie arbitrarily close to $|w| = 1$ (for example, closer than $|w_0|$) by selecting $\mathcal{C}_z^*$ sufficiently close to $\partial\mathcal{R}$.

(*b*) We can show by knowledge of $\mathcal{C}_w'$ that the right-hand integral in (10-17) *really counts* the number of values of w (namely 1) in the circular disk where $w = w_0$.

† This is the more careful kind of proof deliberately omitted in Theorem 9-1.

Now (a) is a matter of continuity since $g(x,y) \to 0$ on $\partial\mathcal{R}$ and $g(x,y) > 0$ otherwise. Hence, $g(x,y)$ is small, or $|w(z)|$ is close to 1, at points z close to $\partial\mathcal{R}$ and only at points z close to $\partial\mathcal{R}$ (otherwise g would have a local minimum elsewhere).

But (b) is more difficult since $\mathcal{C}_w'$ is not immediately seen to be a *simple* curve (not to mention orientation), although it lies outside $|w| = |w_0|$. (We do not yet know that the mapping is unique.) What we do know is that since $w = (z - a)F(z)$, we can count the solutions of $w(z) = 0$ by writing (10-17) with w_0 set equal to 0,

$$\frac{1}{2\pi i}\int_{\mathcal{C}_w'} d\log w = \frac{1}{2\pi i}\int_{\mathcal{C}_z^*} \frac{dz}{z-a} + \frac{1}{2\pi i}\int_{\mathcal{C}_z^*} \frac{F'(z)\,dz}{F(z)} = 1 + 0$$

(since $F(z) \neq 0$). But on the other hand, from (10-17),

$$(N^* =)\ \frac{1}{2\pi i}\int_{\mathcal{C}_w'} d\log(w - w_0) = \frac{1}{2\pi i}\int_{\mathcal{C}_w'} d\log w = 1$$

since 0 can be connected to w_0 without intersecting $\mathcal{C}_w'$.

This observation completes the proof of uniqueness and, incidentally, it reassures us (in hindsight only) that $\mathcal{C}_w'$ really was a simple closed curve with positive orientation in the bargain! Q.E.D.

We shall now perform an explicit construction due to Caratheodory (1912) which proves the Riemann mapping theorem.

THEOREM 10-5 *We can construct approximate mappings of the bounded simple-connected region $\mathcal{R}$ into the unit circle in the sense that we can find an infinitude of functions $f_n(z)$ such that:*

(a) *$f_n(a) = 0$ for a fixed point a in $\mathcal{R}$.*

(b) *$f_n(z)$ maps $\mathcal{R}$ onto a region $\mathcal{R}^{(n)}$ (in biunique fashion) so that $\mathcal{R}^{(n)}$ lies inside the unit circle and completely covers the disk $|z| < r_n$ (<1).*

(c) *$r_n \to 1$ as $n \to \infty$.*

PROOF We shall give $f_n(z)$ by an explicit construction. First of all, by reducing the (finite) region $\mathcal{R}$ in size and translating it, the point $z = a$ can be moved to the origin, yielding the region $\mathcal{R}^{(0)}$. Thus, take $f_1(\mathcal{R}^{(0)}) = \mathcal{R}^{(1)}$ as

$$f_1(z) = k(z - a) \qquad k \text{ real} \tag{10-18}$$

We let r_1 be the radius of the maximum disk $\mathcal{D}(r_1)$ which can be contained in $\mathcal{R}^{(1)}$. But, since the boundary points of $\mathcal{R}^{(1)}$ form a closed set, for some such boundary point z_1, $|z_1|$ achieves its maximum value r_1.

Inductively, if r_n is the radius of the maximum disk centered at the origin and contained in $\mathcal{R}^{(n)}$, then $r_n = |z_n|$ for z_n a boundary point of

$\mathcal{R}^{(n)}$ and we define $f_{n+1}(\mathcal{R}^{(n)}) = \mathcal{R}^{(n+1)}$ by the relation,

$$\left(\frac{z_n - f_n(z)}{1 - \bar{z}_n f_n(z)}\right)^{1/2} = \frac{z_n^{1/2} - f_{n+1}(z)}{1 - \bar{z}_n^{1/2} f_{n+1}(z)} \tag{10-19}$$

Obviously (10-19) is based on the general mapping (see Lemma 2-10, Sec. 2-7)

$$\zeta = \frac{a - z}{1 - \bar{a}z} \qquad |a| < 1 \tag{10-20}$$

of the disk $|\zeta| \leq 1$ onto the disk $|z| < 1$, so that $z = a$ becomes $\zeta = 0$ and $d\zeta/dz > 0$. The relation (10-19) furthermore preserves

$$f_n(a) = f_{n+1}(a) = 0$$

To understand the role of (10-19) more completely, we note that z_n is on the boundary of $\mathcal{R}^{(n)}$, hence z_n can be connected with the exterior of $\mathcal{R}^{(n)}$, in particular with $1/\bar{z}_n$ by means of a cut, thus defining

$$\phi_n(z) = \left(\frac{z_n - f_n(z)}{1 - \bar{z}_n f_n(z)}\right)^{1/2} \tag{10-21}$$

a single-valued function for z in $\mathcal{R}$ [or for $f_n(z)$ on $\mathcal{R}^{(n)}$]. Also, we see $|\phi_n(z)| < 1$ for z in $\mathcal{R}$ by using (10-20). Finally, from $\phi_n(z)$, the following relation is used to define the function $f_{n+1}(z)$ whose range (for z in $\mathcal{R}$) still lies in the unit circle:

$$\phi_n(z) = \frac{z_n^{1/2} - f_{n+1}(z)}{1 - \bar{z}_n^{1/2} f_{n+1}(z)} \tag{10-22}$$

Here the choice of radicals is made to satisfy $z_n^{1/2}\bar{z}_n^{1/2} = |z_n|$ (>0).

In Exercise 1, we shall show that if $\mathcal{R}^{(n)}$ contains $\mathfrak{D}(r_n)$, the disk of radius $r_n = |z_n|$ about the origin, then $\mathcal{R}^{(n+1)}$ contains a disk of radius $r_{n+1} \geq G(r_n)$ for a certain continuous function $G(r_n)$ so defined that $G(r) \geq r$ (for $0 < r \leq 1$) with equality at $r = 1$ only. Then

$$r_1 < r_2 < \cdots < r_n < \cdots < 1 \tag{10-23}$$

and it follows that $\lim r_n = r_0$ exists and is ≤ 1. Indeed,

$$r_0 = \lim r_{n+1} \geq \lim G(r_n) = G(r_0)$$

but (earlier) $G(r_0) \geq r_0$. Hence it follows that $r_0 = G(r_0) = 1$.

Q.E.D.

Actually, z_{n+1}, the closest boundary point to the origin in $\mathcal{R}^{(n+1)}$, is not generally $f_n(z_n)$ but a point rather unrelated. Hence this construction generally makes the boundary "rougher" as it comes closer to the unit circle. In principle, the same construction will hold even if the boundary of $\mathcal{R}$ is not smooth.

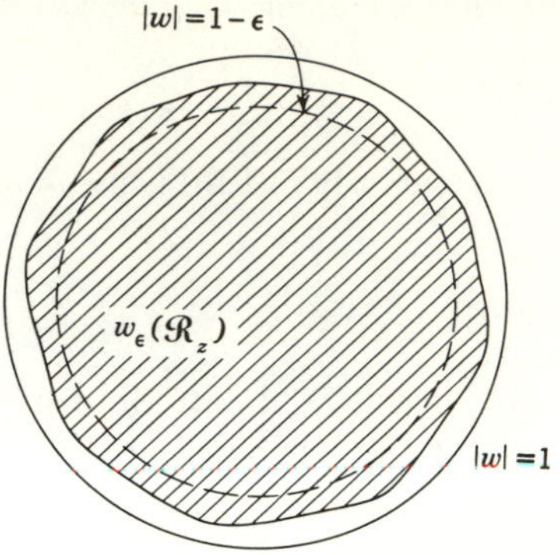

Fig. 10-3 The images w_ϵ of $\mathcal{R}_z$ in Theorem 10-6.

At this point the reader may be tempted to jump to the conclusion that

$$\lim f_n(z) = f(z)$$

exists in $\mathcal{R}$, and that $w = f(z)$ constitutes the mapping function for $\mathcal{R}$ onto the unit disk in the w plane. This indeed is correct† but requires proof.

THEOREM 10-6 (*Riemann mapping theorem*) *A simply-connected region $\mathcal{R}_z$ whose boundary consists of a piecewise smooth, simple closed curve can be mapped onto the interior of a unit circle.*

PROOF We show $\mathcal{R}_z$ has a Green's function.

Take any arbitrarily small positive ε (<1). Let $w_\varepsilon = f(z;\varepsilon)$ be an analytic function which maps $\mathcal{R}_z$ onto $\mathcal{R}_w(\varepsilon)$, some simply-connected region of the w plane whose boundary lies in the annulus given by

$$1 - \varepsilon \leq |w| \leq 1 \tag{10-24}$$

and for which $f(a;\varepsilon) = 0$. (See Fig. 10-3.) Such functions exist (by Theorem 10-5) for each ε. Here an approximation to the Green's function for $\mathcal{R}_w(\varepsilon)$ is given [as in (10-16a)] by the Green's function for the unit disk

$$g_\varepsilon(x,y) = \log \frac{1}{|z - a|} + G_\varepsilon(x,y) = -\log |w|$$

Thus $G_\varepsilon(x,y)$, "the regular part," has the property that on the boundary of $\mathcal{R}_z$,

$$0 < G_\varepsilon(x,y) + \log \frac{1}{|z - a|} < -\log (1 - \varepsilon) \tag{10-25}$$

because $0 \leq -\log |w| \leq -\log (1 - \varepsilon)$ by (10-24).

† For a direct proof see Appendix A, in particular "normal families."

Actually, if ε_1 and ε_2 are two positive quantities, then we can assert from (10-25) that for *z on the boundary of* $\mathcal{R}_z$,

$$|G_{\varepsilon_1}(x,y) - G_{\varepsilon_2}(x,y)| < \max\left(\log\frac{1}{1-\varepsilon_1}, \log\frac{1}{1-\varepsilon_2}\right) \tag{10-26}$$

Yet the maximum principle assures us that (10-26) holds for $z = (x,y)$ in the whole interior of $\mathcal{R}$. Thus $G_\varepsilon(x,y)$, by Cauchy's criterion, is uniformly convergent to a limit function $G(x,y)$. This limit is harmonic (by Theorem 10-3 of Sec. 10-1) and [from the limit of (10-25) as $\varepsilon \to 0$] $\log(1/|z-a|) + G(x,y)$ is the desired Green's function. Q.E.D.

COROLLARY 1 (*Schwarz*) *The only biunique analytic mapping of the interior of the unit disk onto itself is a linear (fractional) transformation.*

PROOF Let us consider such mappings $w = w(z)$ of the disk $|z| < 1$ onto the disk $|w| < 1$. We can assume $w(0) = 0$ [since otherwise a linear transformation of the w disk onto itself will produce this result (see Lemma 2-10 in Sec. 2-7)].

Then consider $q(z) = w(z)/z$ as a function on the z disk. The maximum of $|q(z)|$ for $|z| < 1$ must be attained as $|z| \to 1$ by the maximum-minimum principle [unless $q(z)$ is constant]. Thus, letting $|z| \to 1$, we find $\max |q(z)| \leq \max |w(z)| \leq 1$, or $|w(z)| \leq z$, for all $|z| < 1$.

The role of z and w is an interchangeable one, however; hence $|z| \leq |w(z)|$ and therefore $|w(z)/z| = 1$ for all $|z| < 1$. It is clear, however, that $w(z)/z = \exp i\theta = \text{constant}$ or $w = z \exp i\theta$, since (see Exercise 3) only a constant function can have a constant absolute value. Q.E.D.

COROLLARY 2 *Any two simply-connected regions (with piecewise smooth boundary) can be mapped onto one another conformally and biuniquely.*

PROOF Let $\mathcal{R}$ be such a region in the z plane and $\mathcal{R}_1$ be such a region in the z_1 plane. Then the functions $w = f(z)$ and $w = f_1(z_1)$ perform the biunique mappings of $\mathcal{R}$ and $\mathcal{R}_1$ onto the (open) unit disk. We observe that $f_1'(z) \neq 0$ by biuniqueness (Theorem 2-5). Hence we can invert $z_1 = g_1(w)$ as an analytic function, and thus $z_1 = g_1(f(z))$ performs the mapping. Q.E.D.

In Sec. 11-1 we shall prove a refinement of the Riemann mapping theorem known as the *Riemann mapping theorem with boundary,* to the effect that the mappings $w = f(z)$ referred to in Theorem 10-6 and its corollaries can be extended continuously to a mapping of the boundary of $\mathcal{R}$ (in the z plane) biuniquely and continuously onto the boundary of the unit disk (in the w plane).

LEMMA 10-3 *Riemann's mapping theorem with boundary can be used to prove the solvability of the ordinary Dirichlet problem on $\mathcal{R}$, when $\mathcal{R}$ is simply-connected.*

PROOF We observe that we can solve the ordinary Dirichlet problem on a unit circular disk by Poisson's solution. The mapping theorem *with boundary* can be used to transfer the Dirichlet problem to the disk and to transfer its solution back to $\mathcal{R}$. Q.E.D.

Thus we have four propositions for simply-connected regions:

Riemann's mapping theorem (RMT)
Riemann's mapping theorem with boundary (RMTwb)
Ordinary Dirichlet problem (solvability) (ODP)
Green's function (existence) (GF)

We can summarize our results as follows:

RMTwb	implies	ODP	(Lemma 10-3)
ODP	implies	GF	(Lemma 10-1)
GF	implies	RMT	(Lemma 10-2)
RMT	implies	RMTwb	(Theorem 11-1)

We *entered* this chain of equivalent propositions via the Green's function (GF), but we actually could enter at any stage in many ways (as is amply shown in the literature).

EXERCISES

1 (*a*) Show that the mapping (10-19) can be transformed by (ordinary) rotation to the mapping

$$\left(\frac{a - z}{1 - az}\right)^{1/2} = \frac{a^{1/2} - w}{1 - a^{1/2}w} \qquad 0 < a < a^{1/2} < 1 \tag{10-27}$$

(*b*) Show that (10-27) can be rewritten as

$$z = w\,\frac{A - w}{1 - Aw} \qquad (a^{1/2} <)\ A = \frac{2a^{1/2}}{1 + a}\ (<1) \tag{10-28}$$

(*c*) Show that the point $w = -b$ is transformed into $z = -a$ where

$$b = \frac{-a^{1/2}(1 - a) + [2a(1 + a^2)]^{1/2}}{1 + a} \tag{10-29}$$

(*d*) Show that $a < b < a^{1/2}$ (by manipulating inequalities). (*e*) Show that on the circle $|w| = b$ we can maximize the function $|(A - w)/(1 - Aw)|$ at $w = -b$. [Show as in Fig. 10-4*b* that the circle $\mathcal{Z}$: $|(A - w)/(1 - Aw)| =$ constant, which passes through $w = -b$, completely encloses the locus $|w| = b$.] (*f*) Verify that the disk $|w| \leq b$ is contained in the image of the disk $\mathfrak{D}(a)$: $|z| < a$ under (10-28). (*g*) Finally, verify the fact that (10-29) defines a function $b = G(a)$ satisfying the requirements that the disk $|w| < b$ is completely covered by the image of $\mathfrak{D}(a)$: $|z| < a$ under (10-27). [Here the image of $|z| = a$ is seen to be *one loop* of the lemniscate crosshatched in Fig. 10-4*b*.]

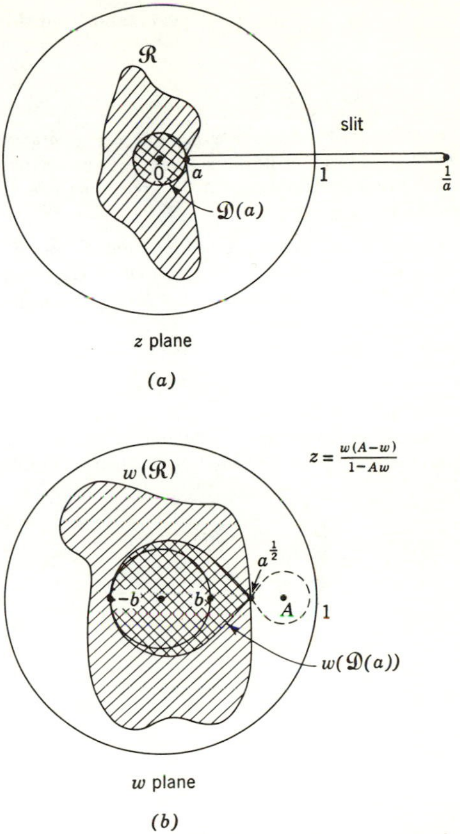

Fig. 10-4 Illustration for Theorem 10-5 and Exercise 1. Note that the disk $|z| < a$ becomes enlarged to the crosshatched disk surrounding $|w| = b$ ($> a$). The "other branch" of the w image of $|z| = a$ is dotted.

2 Prove Corollary 1 to Theorem 10-6 by considering the Green's function with pole at $z = 0$ for the unit disk $|z| < 1$. Check the interrelation of Green's function to the Riemann mapping theorem, and show that this result establishes a *weaker form* of Corollary 1, where we prove uniqueness only for those mappings *which are extendable continuously to the boundary.*

3 Show that if $f(z)$ is an analytic function, then $|f(z)|$ [or $\mathrm{Re}\, f(z)$] can be constant only if $f(z)$ is constant.

4 Let $\psi(Z)$ be the given boundary data for Dirichlet's problem on $\mathcal{R}_z$. Let $w(z)$ be a function mapping $\mathcal{R}_z$ (with boundary) onto the closed unit disk, so that $z = a$ goes into $w = 0$. Show (by Gauss' mean value theorem) that if $\mathcal{C}_z = \partial\mathcal{R}_z$, then

$\psi(a) = 1/2\pi \int_{\mathcal{C}_z} \psi(Z) \text{ Im } [d \log w]$; and from this, show

$$\psi(a) = \frac{1}{2\pi} \int_{\mathcal{C}_z} \frac{\psi(Z)\, \partial g(a;Z)}{\partial n}$$

(*Hint:* $-\partial g/\partial n = \partial h/\partial s$ where $\log w = -g(a;z) - ih$.) Note that this enables us to solve the Dirichlet problem when a Green's function exists, without first mapping $\mathcal{R}_z$ onto the unit disk explicitly! (We must assume the differentiability of the Green's function on $\partial\mathcal{R}_z$.)

5 Show that any biunique analytic mapping $w = w(z)$ of the unit disk $|z| < 1$ onto the unit disk $|w| < 1$, *which is extendable biuniquely and continuously to the boundary,* is linear fractional. As in Theorem 10-6, Corollary 1, assume $w(0) = 0$ and show $g = \log |w(z)|^{-1}$ is the Green's function for the unit disk with pole at $z = 0$, hence g is *unique* $[= \log (1/|z|)]$.

6 Show that the only biunique analytic mapping of the annulus A_z: $r_1 < |z| < r_2$ onto the annulus $r_1 < |w| < r_2$ is a rotation $w = r \exp i\theta$ (assuming such mappings are differentiably extendable to the boundary. [*Hint:* Show Re $\log w$ is an abelian integral of the first kind on A_z, hence is uniquely determined.]

CHAPTER 11
BOUNDARY BEHAVIOR

Having established the Riemann mapping theorem for the interior of the disk, we consider the problem of extending such analytic mappings to the boundary. (Which in our context can be limited to piecewise smooth curves.)

The surprising result is that these extensions are essentially possible in a very strong sense. The analytic mapping function can be extended to the boundary as a continuous function (see Sec. 11-1). In addition, this extension mapping is analytic when the boundary curves are also assumed analytic (see Sec. 11-2). We can then go still further and say that by reference to a suitable choice of parameter planes (the u plane triangles in Sec. 5-3 above) any finite Riemann manifold (with piecewise smooth boundary curves) can be considered to be analytically parametrized on the boundary (see Sec. 11-3). Thus, generally, an analytic mapping becomes extensible across the boundary in this context.

This idea of extending a mapping function (actually an abelian integral) across the boundary of its domain or manifold of definition (or extending a Riemann manifold across its boundary) is not merely an afterthought to conformal mapping. It is seen to become an essential part of the theory of existence of functions on a Riemann manifold.

11-1 Continuity

We saw in Sec. 1-8 that singularities of an analytic function are not as arbitrary as in the case of real functions. In particular, every singularity is infinite or else "removable." This principle holds as well for boundary behavior of analytic mappings (in a region $\mathcal{R}$) as compared with the wild possibilities of boundary behavior of a real mapping. In principle, Theorem 11-1 will show that an analytic mapping is either continuous at a boundary arc or it "spreads the boundary arc out" to cover an infinite image (which is excluded by finite size of the triangles).

LEMMA 11-1 (*Schwarz' inequality*) *If $f(x)$ and $g(x)$ are real continuous functions in an interval $[a,b]$,*

$$\left[\int_a^b f(x)g(x)\,dx\right]^2 \le \int_a^b f^2(x)\,dx \int_a^b g^2(x)\,dx \tag{11-1}$$

To see the proof, integrate

$$\begin{aligned} 0 &\le \iint [f(x)g(y) - f(y)g(x)]^2\,dx\,dy \\ &= \iint f^2(x)g^2(y)\,dx\,dy - 2\iint f(x)g(x)f(y)g(y)\,dx\,dy + \iint f^2(y)g^2(x)\,dx\,dy \\ 0 &\le 2\int f^2(x)\,dx \int g^2(x)\,dx - 2\left[\int f(x)g(x)\,dx\right]^2 \end{aligned}$$

where we note that x and y are merely "dummy" variables of integration. Hence, (11-1) follows. (Compare Exercise 1.)

THEOREM 11-1 (*Caratheodory*, 1913) *Consider two finite regions $\mathcal{R}_z$, $\mathcal{R}_w$ with simple (piecewise smooth) boundaries. Let $w = f(z)$ map the region $\mathcal{R}_z$ onto the region $\mathcal{R}_w$ in biunique conformal fashion. Then the map can be extended to the boundaries in biunique continuous fashion.*

PROOF Let Z_0 be a point on the boundary $|\partial\mathcal{R}_z|$ of $\mathcal{R}_z$. We wish to show that for some W_0 on the boundary of $\mathcal{R}_w$ we have

$$W_0 = f(Z_0)$$

in the sense that the following relationship holds: Given any $\varepsilon > 0$, a $\delta = \delta(\varepsilon) > 0$ exists such that if

$$|z - Z_0| < \delta \qquad (z \text{ in } \mathcal{R}_z)$$

then

$$|f(z) - W_0| < \varepsilon$$

We then say $f(|\partial\mathcal{R}_z|)$ is a continuous mapping into $|\partial\mathcal{R}_w|$, or $f(\mathcal{R}_z) = \mathcal{R}_w$ is a mapping which can be extended continuously to the boundary.

Consider the circular arcs surrounding Z_0, namely $z = Z_0 + r \exp i\theta$. Here the whole circle does not lie in $\mathcal{R}_z$ but an approximate "semicircle,"

$$\mathcal{C}_r\colon\quad \theta_1(r) < \theta < \theta_2(r) \tag{11-2}$$

Under $w = f(z)$, the image $\mathcal{C}_r'$ of $\mathcal{C}_r$ would presumably be some almost semicircular arc in the region $\mathcal{R}_w$, but we do not know this at present. It could look like $\mathcal{C}_r'$ in Fig. 11-1; but it still has the property that as $\theta \to \theta_1(r)$ or $\theta \to \theta_2(r)$, then $f(Z_0 + r \exp i\theta)$ approaches the boundary of $\mathcal{R}_w$ even if it has no unique limit.

The portion $\mathfrak{D}_r$ of $\mathcal{R}_z$ contained in $\mathcal{C}_r$ has an image the area $\mathfrak{D}_r'$. As $r \to 0$, $\mathfrak{D}_r'$ represents a diminishing set of areas. In fact, if $r_1 > r_2$, then $\mathfrak{D}_{r_1}'$ contains $\mathfrak{D}_{r_2}'$ in its interior. Hence the closures of $\mathfrak{D}_r'$ have at least one point in common for all r. One such point will be called W_0. Then for a given $\varepsilon > 0$, one of the following holds true:

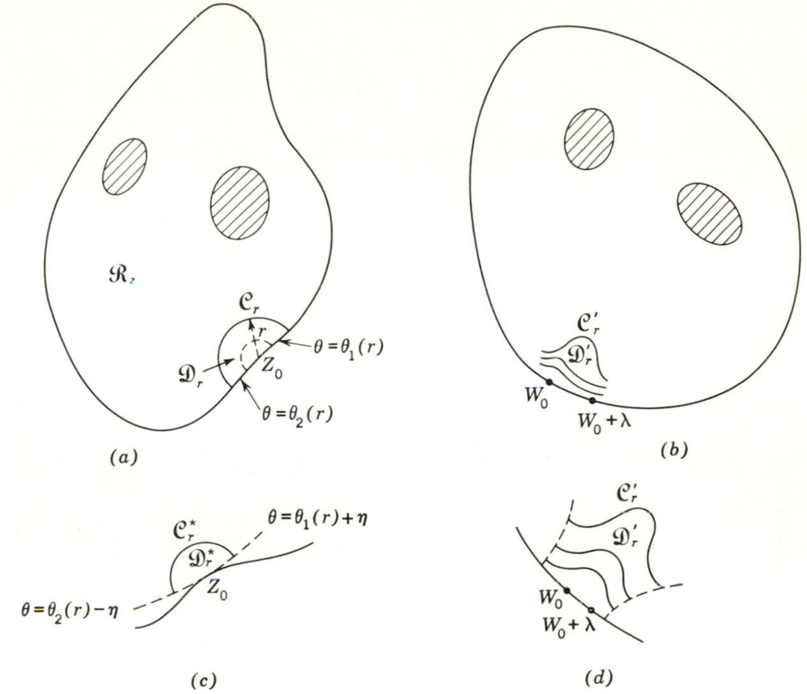

Fig. 11-1 (*a, b*) For Theorem 11-1; (*c, d*) for Exercise 3.

(*a*) All $\mathfrak{D}_r'$ lie in a circle of radius ε about W_0 for $r <$ some $\delta(\varepsilon)$.

(*b*) All $\mathfrak{D}_r'$ contain in their closures a second point in common, W_0'. (This point W_0' must lie on the boundary of $\mathfrak{R}_w$ since $\mathfrak{D}_r'$ approaches the boundary.)

Thus if a second point W_0' exists, then the images of $\mathfrak{C}_r$ all exceed in length some *fixed* λ $(<|W_0 - W_0'|)$, and†

(11-3)
$$\int_{\mathfrak{C}_r} |dw| = \int_{\mathfrak{C}_r} |f'(z)|\,|dz| \geq \lambda \qquad \text{for all } r < \text{(say) } r_0$$

If we use $|dz| = r\,d\theta$,

(11-4)
$$\int_{\theta_1(r)}^{\theta_2(r)} |f'(z_0 + r \exp i\theta)| r\,d\theta \geq \lambda$$

† Theoretically, these integrals may be infinite but the inequalities still may be interpreted as correct, in that $\infty \geq$ any finite or infinite real quantity. (Compare Exercise 3.)

By using the Schwarz inequality on (11-4), we obtain

$$\int_{\theta_1(r)}^{\theta_2(r)} [f'(z_0 + r \exp i\theta)]^2 \, d\theta \int_{\theta_1(r)}^{\theta_2(r)} r^2 \, d\theta \geq \left[\int_{\theta_1(r)}^{\theta_2(r)} |f'(z_0 + r \exp i\theta)| \cdot r \, d\theta \right]^2 \geq \lambda^2$$

Hence, ignoring the middle term of the inequality, we obtain (with λ = constant):

$$2\pi \int_{\theta_1}^{\theta_2} |f'(z_0 + r \exp i\theta)|^2 \, d\theta \geq \frac{\lambda^2}{r^2} \tag{11-5}$$

since $|\theta_2 - \theta_1| < 2\pi$. Finally, taking $\int_r^{r_0} \cdots r \, dr$ on both sides of (11-5), we obtain

$$2\pi \int_{\mathfrak{D}_{r_0}} |f'(z)|^2 r \, dr \, d\theta \geq \int_{\mathfrak{D}_{r_0}} \frac{\lambda^2}{r} \, dr \tag{11-6a}$$

But $|f'(z)|^2 r \, dr \, d\theta = |f'(z)|^2 \, dx \, dy = du \, dv$ (in the $w = u + iv$ plane), hence

$$2\pi \int_{\mathfrak{R}_w} \int du \, dv \geq 2\pi \mathfrak{D}'_{r_0} \geq \lambda^2 \int_0^r \frac{dr}{r} = \infty \tag{11-6b}$$

Hence $\mathfrak{R}_w$ has an infinite area.

We see we obtain a contradiction if the other limit point W'_0 exists. This reduces the alternative to (a), and $\mathfrak{D}'_r$ converges to W_0. Thus $w = f(\partial \mathfrak{R}_z)$ is a continuous mapping. But, since the inverse map exists in $\mathfrak{R}_w$, $z = f^{-1}(\partial \mathfrak{R}_w)$ is likewise continuous. Q.E.D.

Theorem 11-1 essentially discusses the boundary behavior of an *analytic* function in a region. We can ask a corresponding question concerning a *harmonic* function in a region. To start with, we consider isolated singularities and immediately dispose of removable singularities (as in Sec. 1-8).

THEOREM 11-2 *If a harmonic function $\psi(z)$ is defined (single-valued) in a region surrounding a point z_0, then one of the following holds true:*

(a) *$\psi(z)$ becomes unbounded as $z \to z_0$.*

(b) *$\lim \psi(z)$ exists as $z \to z_0$ and, assigning the limit as $\psi(z_0)$, we have a function harmonic at z_0.*

PROOF Consider $\mathfrak{N}_0$ the "deleted" neighborhood

$$0 < |z - z_0| < k$$

Within a neighborhood of any point of $\mathfrak{N}_0$ we can define ψ^*, the conjugate of $\psi(z)$. It need not happen that ψ^* is single-valued in $\mathfrak{N}_0$ (as is ψ). So let the period be $p = \int d\psi^*$, taken around the circle $z - z_0 = r$

$\exp i\theta$ $(0 < r < k)$. The analytic function $\psi + i\psi^*$ likewise is not necessarily single-valued (because of the period), but this can be adjusted by subtracting $i\,(p/2\pi)\arg(z - z_0) = i\,\mathrm{Im}\,[(p/2\pi)\log(z - z_0)]$. Thus the following *analytic* function is single-valued in $\mathfrak{N}_0$:

$$\phi(z) = \psi + i\psi^* - \frac{p}{2\pi}\log(z - z_0) \tag{11-7}$$

and $\phi(z)$ has at most an isolated singularity at $z = z_0$.

We then consider as our *most general* harmonic function in $\mathfrak{N}_0$,

$$\psi = \mathrm{Re}\,\phi(z) + \frac{p}{2\pi}\log|z - z_0| \qquad p \text{ real} \tag{11-8}$$

Then these possibilities can occur:

(*a*) $\phi(z)$ is analytic at $z = z_0$ and $p = 0$.

(*b*) $\phi(z)$ is analytic at $z = z_0$ and $p \neq 0$.

(*c*) $\phi(z)$ has a pole at $z = z_0$.

(*d*) $\phi(z)$ has an essential singularity at $z = z_0$ and $p = 0$.

(*e*) $\phi(z)$ has an essential singularity at $z = z_0$ and $p \neq 0$.

In case (*a*), evidently $\psi(z)$ is definable and harmonic at z_0. We shall show in every other case that the hypothesis is contradicted, since $\psi(z)$ as defined in (11-8) becomes unbounded in the limit as $z \to z_0$ along some sequence of values of z:

(*b*) $\lim \psi(z) = \mathrm{Re}\,\phi(z_0) + (p/2\pi)(-\infty) = \infty$.

(*c*) Let the leading term be $\phi(z) = a/(z - z_0)^n + \cdots$, $(n > 0)$ where $a = A \exp i\alpha \neq 0$, and let $z - z_0 = r \exp i\theta$; then as $r \to 0$,

$$r^n\psi(z) = \mathrm{Re}\,[A \exp i(\alpha - n\theta)] + \frac{p}{2\pi} r^n \log r + \text{vanishing terms}$$

Thus if $\theta = \alpha/n$ and $r \to 0$, then

$$r^n\psi(z) = A\cos(\alpha - n\theta) \neq 0$$

so $\lim \psi(z) = \infty$.

(*d*) We can choose $\phi(z)$ arbitrarily close to any m (>0), thus $\mathrm{Re}\,\phi(z) > m - \varepsilon$ for z arbitrarily close to z_0. If m is made arbitrarily large, $\psi(z)$ becomes arbitrarily large.

(*e*) We do as before but keep m fixed; then $\psi(z)$ behaves like $(p/2\pi)\log r$ as $r \to 0$. Q.E.D.

THEOREM 11-3 *If $\psi(z)$ is harmonic in some region $\mathfrak{R}$ in the upper half plane, part of whose boundary is a segment $\mathfrak{S}$ of the real axis, and if*

$\psi(z)$ can be defined continuously as 0 on $\mathcal{S}$, then the following is a harmonic function in the neighborhood of $\mathcal{S}$:

(11-9)
$$\Psi_0(z) = \begin{cases} \psi(z) & \textit{if } \operatorname{Im} z > 0 \\ 0 & \textit{if } \operatorname{Im} z = 0 \\ -\psi(\bar{z}) & \textit{if } \operatorname{Im} z < 0 \end{cases}$$

PROOF The function $\Psi_0(z)$ described in (11-9) is not yet known to be *harmonic* on the real axis. Let x_0 be a real number, and let r (>0) be chosen so that the upper half of the circle $\mathcal{C}_r$: $|z - x_0| = r$ lies wholly in $\mathcal{R}$. On the whole circumference of $\mathcal{C}_r$, the values $\Psi_0(z)$ describe the data for a Dirichlet boundary-value problem in $\mathcal{C}_r$. Call $\Psi(z)$ the solution, say, by Poisson's integral. If we can show $\Psi(z) = \Psi_0(z)$, then $\Psi_0(z)$ [like $\Psi(z)$] is a harmonic function in a *neighborhood* of x_0.

First consider the fact that $\Psi(z) = 0$ when $\operatorname{Im} z = 0$, which is seen from the symmetry in the Poisson kernel (see Exercise 4) or graphically (Fig. 11-2*a*) from the distribution of the angles subtended by the circles perpendicular to the circle $|z - z_0| = r$. Once we know $\Psi(z) = 0$ when $\operatorname{Im} z = 0$ we can note (as in Fig. 11-2*b*), that $\Psi_0(z)$ and $\Psi(z)$ are harmonic

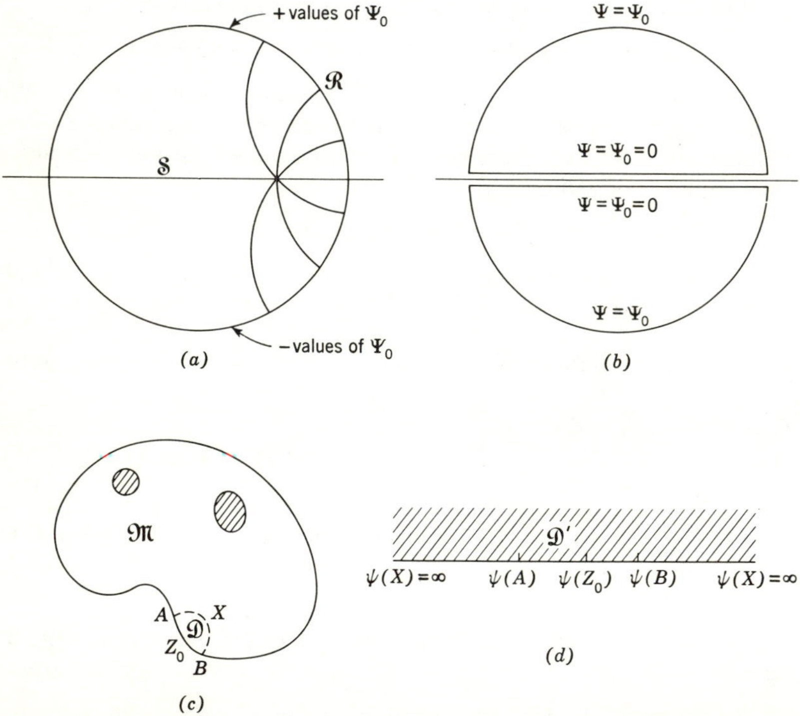

Fig. 11-2 (*a*, *b*) For proof of Theorem 11-3; (*c*, *d*) for proof of Corollary 1.

functions with identical boundary values on semicircle and diameter in the upper half disk (and similarly, in the lower half disk); hence, they coincide by Theorem 8-7, Sec. 8-5. Q.E.D.

COROLLARY 1 *Let $\psi(z)$ be harmonic on a Riemann manifold $\mathfrak{M}$ which has a segment of a smooth boundary curve $\mathcal{C}$, and let $\psi(z) \to 0$ as $z \to Z$ on $\mathcal{C}$ except possibly for Z_0 a point in $\mathcal{C}$. Then one of the following is true:*

(a) $\psi(z)$ *becomes unbounded as* $z \to Z_0$.

(b) $\psi(z) \to 0$ *as* $z \to Z_0$.

To first see that the alternative is realistic, consider $\mathfrak{M}$ to be the unit disk $|z| < 1$ and take $Z_0 = 1$. Then for z on the unit circle, $z + 1$ and $z - 1$ have a purely imaginary ratio (by the elementary theorem on an angle inscribed in a semicircle). Hence set

$$\psi(z) = \operatorname{Re} \frac{z+1}{z-1} = \frac{x^2 - 1 + y^2}{(x-1)^2 + y^2} \qquad \left(= \frac{x+1}{x-1} \text{ when } y = 0 \right) \tag{11-10}$$

Here $\psi(z) = 0$ on $|z| = 1$ except for $z = 1$; but note that $\psi(x,0) \to \infty$ as $x \to 1 - 0$. Thus we are not surprised that $\psi(z) \to 0$ is false as $z \to 1$.

PROOF Since a harmonic function is preserved by conformal mapping, we imagine that $\mathfrak{M}$ is in the upper half plane, $\mathcal{C}$ is a segment of the real axis, and $Z_0 = x_0$. If $\mathfrak{M}$ is not simply-connected, we can cut out a smaller portion $\mathfrak{D}$ around Z_0 which is simply-connected and map the closure of that portion instead, as in Fig. 11-2*c* and *d*. Hence Theorem 11-2 applies to the function $\Psi_0(z)$ defined as in (11-9). Thus unless $\psi(z) \to 0$ as $z \to Z_0$, it will follow that $\Psi_0(z)$ becomes unbounded [and with it, so does $\psi(z)$]. Q.E.D.

With this background we are prepared to treat jump discontinuities. We consider a function $g(s)$ defined for $a \le s \le b$. We say g has (at most) *jump discontinuities* if we can subdivide the interval $[a,b]$ into n (≥ 1) subintervals

$$[s_0,s_1], [s_1,s_2], \ldots, [s_{n-1},s_n]$$

with $s_0 = a < b = s_n$, such that g is continuous on each of the open intervals (s_t,s_{t+1}) $(0 \le t \le n-1)$, and [as $\varepsilon \to 0$ $(\varepsilon > 0)$],

$$\begin{aligned} &g(s_t + 0) \; [= \lim g(s_t + \varepsilon)] \text{ exists} && \text{for } 0 \le t \le n-1 \\ &g(s_t - 0) \; [= \lim g(s_t - \varepsilon)] \text{ exists} && \text{for } 1 \le t \le n \end{aligned}$$

We say the boundary data are *properly prescribed* if

$$g(s_t) = \tfrac{1}{2}[g(s_t - 0) + g(s_t + 0)] \qquad \text{for } 1 \le t \le n-1 \tag{11-11}$$

Of course, if $g(s)$ is defined for s a parameter on a closed curve $\mathcal{C}$ of length L, the definitions are all modified so that $0 \le s < L$ and $s = 0$ is identified with $s = L$.

We now return to the second Dirichlet problem for a (noncompact) manifold $\mathfrak{M}$ in Sec. 8-4, and we consider it modified so that (8-46*e*) (the boundary prescription) is replaced by

(11-12*a*) $\lim_{P \to P_0} \psi(P) = g(s_0)$ for P_0 not a point of discontinuity

(11-12*b*) $|\psi(P)|$ bounded as $P \to P_0$ a point of discontinuity

[Here, we recall, $g(s_0)$ represents the boundary data at point P_0 corresponding to $s = s_0$ on the boundary curves of the manifold $\mathfrak{M}$.] We call this the "second Dirichlet problem with jump data," and we call the corresponding problem for the plane region without (internal) prescribed singularities the "ordinary Dirichlet problem with jump data."

THEOREM 11-4 *The second Dirichlet problem with jump data is uniquely solvable.*

We have not yet established that *one* solution exists, but *two* cannot. This is an extension of Theorem 8-7, Sec. 8-5. (See Exercise 7.)

THEOREM 11-5 *The ordinary Dirichlet problem with jump data is solvable for a circle by Poisson's integral.*

(See Exercise 8.)

COROLLARY 1 *The ordinary Dirichlet problem is solvable for any simply-connected region bounded by a smooth curve.*

THEOREM 11-6 *Assume the second Dirichlet problem with jump data to be solvable on a (noncompact) manifold $\mathfrak{M}$, for a function $\psi(P)$. Then at a point of discontinuity P_0 on $\partial\mathfrak{M}$, the limit*

(11-13) $$\lim_{P \to P_0} \psi(P) = F(\theta)$$

will exist if P is constrained to follow a path to P_0, making a definite angle θ at P_0 (with, say, the real axis) in the parameter plane of the triangulation of $\mathfrak{M}$ at P_0. The function $F(\theta)$ varies linearly with θ between the limits $g(s_0 - 0)$ on the left-hand tangent and $g(s_0 + 0)$ on the right-hand tangent to $\partial\mathfrak{M}$ [where $g(s)$ represents the boundary data of ψ, as usual].

PROOF First consider the case where $\mathfrak{M}$ is a disk $\mathfrak{R}$ in, say, the finite z plane. Let P_0 then denote Z_0 a point on $\mathfrak{C}_1 = \partial\mathfrak{M}$ where $g(s)$ has a jump discontinuity. Then we assert (see Exercise 9) that for some constant k properly chosen,

(11-14) $$\psi_1(P) = \psi(P) + k \arg (z - Z_0)$$

is a harmonic function (single-valued) in $\mathfrak{R}$ near Z_0 [as is $\psi(P)$] but experiencing no jump discontinuity in its boundary data near Z_0. Then Theorem 11-6 is immediate when $\mathfrak{M}$ is a disk.

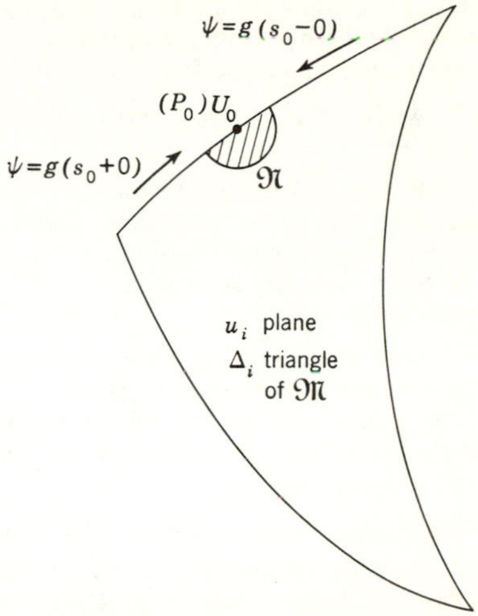

Fig. 11-3 Jump discontinuities near a point of $\partial\mathfrak{N}$.

If $\mathfrak{M}$ is a (noncompact) Riemann manifold, then we can consider P_0 as corresponding to U_0 on, say, the triangle Δ_i in the u_i parameter plane. Let U_0 lie on the side of Δ_i containing a boundary arc $\mathfrak{L}_{ij}$ of $\mathfrak{M}$. (See Sec. 5-3.) We can take U_0 as an interior point of the arc $\mathfrak{L}_{ij}$ by suitable parametrization. Then some neighborhood of U_0 intersects Δ_i in $\mathfrak{N}$ a simply-connected semicircular neighborhood of U_0 (see Fig. 11-3). Thus $\partial\mathfrak{N}$ consists of part of $\mathfrak{L}_{ij}$, together with an arc lying in $\mathfrak{M}$. Hence ψ can be regarded as a solution to a second Dirichlet problem relating to $\partial\mathfrak{N}$ (if we use the values of ψ on $\partial\mathfrak{N}$ as boundary data!). This reduces the problem to the (earlier) simply-connected case. Instead of (11-14) we write

$$\psi_1(u) = \psi(u) + k \arg(u - U_0) \tag{11-15}$$

(where u parametrizes P, of course). Q.E.D.

EXERCISES

1 Consider real A, B, C and the locus in the $\lambda\mu$ plane of $A\lambda^2 + 2B\lambda\mu + C\mu^2 \geq 0$. From elementary analytic geometry, this locus is the whole plane *only* if $B^2 \geq AC$. Using this result, deduce Schwarz' inequality from $\int_a^b (\lambda f + \mu g)^2\, dx \geq 0$.

2 Show that the equality holds in Schwarz' inequality exactly when f and g (assumed continuous) are linearly dependent. (Do this two ways, according to the proof of Lemma 11-1 or according to Exercise 1, but be careful about dividing by zero!)

3 In proving Theorem 11-1, we have not considered the possibility that $\mathcal{C}_r'$ might be infinitely long, so that the left-hand side of (11-3) might be infinite. Show this can be avoided by taking (not $\mathcal{C}_r$ but) $\mathcal{C}_r^*$, which goes from $\theta_2(r) - \eta$ to $\theta_1(r) + \eta$ for a sufficiently small η, and letting $\eta \to 0$ afterward. Repeat the proof with suitable modifications (see Fig. 11-1*c* and *d*).

4 In Theorem 11-3 verify that $\Psi_0(z) = 0$ (z real) directly from the Poisson integral.

5 Let $\psi(z)$ be called *harmonic at* at $z = \infty$ if it is harmonic for $|z|$ large enough and bounded at ∞. (*a*) Show that $\psi(z)$ is harmonic at ∞ if and only if $\psi(1/z)$ is harmonic in a neighborhood of $z = 0$ [using a suitable definition of the *value* of $\psi(\infty)$]. (*b*) Show that $\psi(z)$ is harmonic at ∞ if and only if it is the real part of a function analytic at ∞.

6 Note that $\psi(x,y) = y$ vanishes on the boundary of the upper half plane ($y = 0$) except at the boundary point $z = \infty$, hence it provides an illustration of an infinite discontinuity as effective as (11-10). Show how this illustration is really the same as that in (11-10), by mapping the upper half plane onto the unit circle!

7 Prove Theorem 11-4 by showing how the boundedness condition for jump data (11-12*b*) is used together with Theorem 11-3, Corollary 1.

8 Prove Theorem 11-5 by reviewing the derivation of the Poisson integral (Theorem 10-2). Show in particular that if boundary data $g(R \exp i\theta)$ are substituted in Poisson's integral, then the harmonic function $\psi(z)$ thus defined satisfies (*a*) $|\psi(\rho)| \leq \max g(R \exp i\theta)$ and (*b*) $\psi(\rho) \to g(R \exp i\theta_0)$ as $\rho \to R \exp i\theta_0$ *under the new hypothesis* that g has jump discontinuities, as long as $R \exp i\theta_0$ is not a point of discontinuity.

9 Derive the exact value of k in (11-14) if the boundary point Z_0 corresponds to a boundary point with given jump discontinuity and given (corner) angle in the (piecewise smooth) boundary.

10 Show how Theorem 11-6 would apply to a manifold $\mathfrak{M}$ for which $\partial\mathfrak{M}$ has a *cusp* (a corner of 0° angle), by supposing that at P_0 there are two segments of $\partial\mathfrak{M}$ tangent to one another but having (different) continuous *curvatures* on each segment of the corner. (*Hint:* Consider $\mathfrak{M}$ to be the strip $0 \leq y \leq 1$, $-\infty \leq x \leq +\infty$, so that P_0 is the point at ∞.) Show $\psi_0 = y$ plays the same role as $\psi_0 = \arg(z - Z_0)$ would play in (11-15) for a finite Z_0.

11 In Theorem 11-1, show that if $\mathcal{R}_z$ and $\mathcal{R}_w$ are disks, any triple of boundary points of one can be made to correspond by conformal mapping to any triple of boundary points of the other, provided both triples be in the same *order* (clockwise or counterclockwise). [*Hint:* Prove this for circular disks by Lemma 2-5; for example, $\{z,z_1;z_2,z_3\} = \{w,w_1;w_2,w_3\}$.] (How did the order enter?)

11-2 Analyticity

We define an arc as *analytic* if it is locally the image of a segment of the real axis under a conformal map. Thus each point of an analytic arc in the uv plane lies in a neighborhood of the arc which can be given a parametrization $u = u(x)$, $v = v(x)$, derived from the analytic function $w = f(z)$ when z lies on a segment $\mathcal{S}$ of the real axis:

$$u(x) + iv(x) = f(x) \qquad \text{on } \mathcal{S}\text{: } a \leq x \leq b \tag{11-16}$$

and (for conformality) $f'(x) \neq 0$ on $\mathcal{S}$. We speak, as before, of a curve as an oriented arc. Clearly an arc or curve may be given in one parametrization, say its arc length, but would have to be reparametrized according to (11-16) so as to be analytic in its new parameter. No exception is made for end points; they must be interior to some interval of definition $\mathcal{S}$. In this fashion a closed curve is defined as analytic in a natural way, by having intervals $\mathcal{S}$ such that for some $\mathcal{S}$, $f(\mathcal{S})$ covers any given point of $\mathcal{C}$.

It is clear that an analytic curve is the conformal image (not necessarily biunique) for a single interval such as (11-16). For example, let two intervals of $\mathcal{C} = \mathcal{C}_1 \cup \mathcal{C}_2$ overlap so that analogously with (11-16),

(11-17)
$$\begin{aligned} \mathcal{C}_1 &= f_1(\mathcal{S}_1) \qquad \mathcal{S}_1\colon\ a_1 \le x \le b_1 \\ \mathcal{C}_2 &= f_2(\mathcal{S}_2) \qquad \mathcal{S}_2\colon\ a_2 \le x \le b_2 \end{aligned}$$

and let $\mathcal{C}_2$ overlap the end point $f_1(b_1)$ of $\mathcal{C}_1$. Then the function

$$x' = f_2^{-1}(f_1(x))$$

is analytic near $x = b_1$, and the inverse

(11-18)
$$x = f_1^{-1}f_2(x')$$

is real on the real axis and provides a mapping of $\mathcal{S}_2$ onto an interval $\mathcal{S}_2'$, which continues (and overlaps) $\mathcal{S}_1$ at b_1. Then $\mathcal{C}$ is an analytic continuation of $f_1(x) = f_2(x')$ from $\mathcal{S}_1$ to $\mathcal{S}_1 \cup \mathcal{S}_2'$. This process is continued for any (finite) covering of $\mathcal{C}$.

By using the transformation $z = \exp ix$ we see that a (locally) conformal image of the unit circle is a closed analytic arc, and conversely. (Compare Exercise 14.)

By the chain rule for derivatives, the conformal image of an analytic (closed) curve under a conformal map is an analytic (closed) curve. By Exercise 1, an n-tuply-connected region with piecewise smooth simple boundary can be mapped conformally (with continuous biunique boundary behavior) onto one with n closed analytic curves† as its boundary.

We now build up a system of image points which generalizes the concept of images with respect to Möbius circles. In Sec. 3-5, we proved the *weak reflection principle* for an analytic function $w = f(z)$ which is analytic on the real axis and above it. That result was called "weak" because we then presupposed knowledge of $f(z)$ for *z on the reflecting curve* (the real axis). The following result assumes less:

LEMMA 11-2 (*Strong reflection principle*) *Let $\mathcal{S}_z$ be a segment of the real z axis which serves as part of the boundary of a region $\mathcal{R}_z$ lying, say, above it. Let $w = f(z)$ be bounded and analytic in $\mathcal{R}_z$, and let it take on real continuous limiting values $f(x)$ for z $(= x)$ on $\mathcal{S}_z$. Then $w = f(z)$ can be extended analytically across $\mathcal{S}_z$ so as to take values satisfying*

(11-19)
$$f(\bar{z}) = \overline{f(z)}$$

† We continue to exclude a region with isolated points as part of its boundary (at least until Appendix B).

PROOF From Theorem 11-3 we can extend $\operatorname{Im} f(z) = v(z)$ across $\mathcal{S}_z$ by the imaginary component of (11-19); i.e., by

$$v(\bar{z}) = -v(z) \qquad \operatorname{Im} z > 0 \tag{11-20}$$

Now that v is harmonic across $\mathcal{S}_z$, so is its conjugate, and indeed, so is the analytic function $f(z)$ extendable across $\mathcal{S}_z$ (see Theorem 8-3, Sec. 8.3). We are now in possession of the stronger hypothesis for the weak reflection principle (see Theorem 3-10 of Sec. 3-5). Q.E.D.

If we put the proof in terms of real variables, we should find that we have for $z \in \mathcal{R}_z \cup \mathcal{S}_z$ ($y \geq 0$), analytic continuation by equating the following functions:

$$\begin{aligned} f(z) &= u(x,y) + iv(x,y) \\ \overline{f(\bar{z})} &= u(x,-y) - iv(x,-y) \end{aligned} \tag{11-21}$$

Thus $u(x,y)$ continues into $u(x,-y)$, and $v(x,y)$ into $-v(x,-y)$ [quite naturally since $v(x,0) = 0$].

IMAGES

We also have a weak image principle. Let $\mathcal{C}$ be a simple analytic arc which is defined by (11-16) as an image of $f(z)$ for z on $\mathcal{S} = [a,b]$, a segment of the real axis. Then $w = f(z)$ is biunique near $\mathcal{S}$ [since $f'(x) \neq 0$ on $[a,b]$] in some small circle with center on the real axis. We can define the image relationship

$$I_f\colon \quad [w = f(z)] \leftrightarrow [w^* = f(\bar{z})] \tag{11-22}$$

as a unique correspondence, say $w \leftrightarrow w^*$, of points on opposite sides of $\mathcal{C}$ and sufficiently close to $\mathcal{C}$. It would seem that the relations between w^* and w might depend on f, but this is not so!

THEOREM 11-7 (*Weak image principle*) *Consider a curve $\mathcal{C}$ defined by $\mathcal{C} = f(\mathcal{S})$ where $\mathcal{S}$ is a segment of the real axis on which $f' \neq 0$. Then the image relationship* (11-22) *depends* (*for points sufficiently close to* $\mathcal{C}$) *only on $\mathcal{C}$ and not on f the parametrizing function.*

PROOF Let $\mathcal{C}_w$ have two parametrizations $f_1(z_1)$, $f_2(z_2)$, namely

$$\begin{aligned} \mathcal{C}_w\colon \quad w &= f_1(z_1) \qquad z_1 \in \mathcal{S}_1 = [a_1,b_1] \qquad \text{(real interval)} \\ \mathcal{C}_w\colon \quad w &= f_2(z_2) \qquad z_2 \in \mathcal{S}_2 = [a_2,b_2] \qquad \text{(real interval)} \end{aligned} \tag{11-23}$$

By local uniqueness, these mappings can be inverted. Thus [as in (11-21)],

$$z_2 = f_2^{-1}(w) = f_2^{-1}f_1(z_1) \tag{11-24}$$

and $\phi = f_2^{-1}f_1$ is a biunique mapping of $\mathcal{S}_1$ onto $\mathcal{S}_2$.

Now suppose the theorem were false. Then w_1, as defined by (11-23), would have two images, w_1^* (under f_1) and w_2^* (under f_2). On the other hand $w_1^* = f_1(\bar{z}_1)$ and $w_2^* = f_2(\bar{z}_2)$ by (11-22).

If we put (11-24) in the form $z_2 = \phi(z_1)$, then ϕ satisfies the conditions for the weak reflection principle (Theorem 3-10). Therefore, as in (11-17),

$$\bar{z}_2 = \phi(\bar{z}_1) = f_2^{-1}f_1(\bar{z}_1)$$

Thus

$$(w_2^* =) \quad f_2(\bar{z}_2) = f_1(\bar{z}_1) \quad (= w_1^*)$$

Q.E.D.

COROLLARY 1 *Any conformal biunique mapping of the neighborhood of one analytic curve $\mathcal{C}_1$ onto another analytic curve $\mathcal{C}_2$ preserves the relation of images with regard to either curve.*

(For proof, see Exercise 7.)

Now we cannot have a strong image principle in the sense of an image principle for nonanalytic curves. We can still show that an analytic mapping "naturally extends" over an analytic boundary. This is really the same as a strong image principle.

THEOREM 11-8 *Let us suppose that two finite Riemann manifolds, $\mathfrak{M}$ over the z sphere and $\mathfrak{M}'$ over the w sphere, are conformally equivalent under an analytic mapping $\mathfrak{M}' = f(\mathfrak{M})$ (defined on $\mathfrak{M}$) and that $\partial\mathfrak{M}$ and $\partial\mathfrak{M}'$ correspond biuniquely on analytic curves $\mathcal{C}$ ($\subseteq \partial\mathfrak{M}$) and $\mathcal{C}'$ ($\subseteq \partial\mathfrak{M}'$). Then the mapping $w = f(z)$ can be locally extended analytically across $\partial\mathfrak{M}$ and $\partial\mathfrak{M}'$ (at $\mathcal{C}$ and $\mathcal{C}'$).*

PROOF Use a triangulation of $\mathfrak{M}$ and $\mathfrak{M}'$ such that some portion of $\mathcal{C}$ and $\mathcal{C}'$ is given by segments $\mathcal{S}_1$, $\mathcal{S}_2$ of the real axis (in, say, the triangulations on the ζ_1 and ζ_2 planes). Then, as in (11-23),

$$z = \phi_1(\zeta_1) \qquad w = \phi_2(\zeta_2) \tag{11-25a}$$

for functions ϕ_1, ϕ_2 conformal on (i.e., surrounding) segments $\mathcal{S}_1$, $\mathcal{S}_2$ of the real axis of the ζ_1 and ζ_2 planes. Thus

$$\mathcal{C} = \phi_1(\mathcal{S}_1) \qquad \mathcal{C}' = \phi_2(\mathcal{S}_2) \tag{11-25b}$$

It is clear, however, by the definition of analyticity, that ϕ_1 and ϕ_2 are analytic in *regions surrounding* $\mathcal{S}_1$ and $\mathcal{S}_2$.

Now since $w = f(z)$,

$$\phi_2(\zeta_2) = f(\phi_1(\zeta_1))$$

on, say, the *upper half* ζ_1 and ζ_2 planes near $\mathcal{S}_1$ and $\mathcal{S}_2$ [since $w = f(z)$ is valid only *inside* $\mathfrak{M}$]. Thus, $\zeta_2 = \phi_2^{-1}f\phi_1(\zeta_1)$ is a mapping satisfying Lemma 11-2, and $\phi_2^{-1}f\phi_1 = \psi$ is analytic across the real ζ_1 axis! From this last relation, $f = \phi_2\psi\phi_1^{-1}$ is analytic across $\mathcal{C}$ (that is, $\partial\mathfrak{M}$). Q.E.D.

COROLLARY 1 *A biunique analytic mapping of the interior of a Möbius circle onto itself is analytic on the boundaries and can be determined by assigning the images of any three points of one circle on the other circle (preserving orientation of boundaries).*

(See Exercise 11 of Sec. 11-1 and Exercise 8.)

POLYGONAL MAPPINGS

We apply Theorem 11-8 to the mapping of a nonanalytic boundary, a polygon. First consider the following function

$$w = A \int_{+\infty}^{z} \frac{dz}{(z - x_1)^{\alpha_1} \cdots (z - x_n)^{\alpha_n}} B+ \tag{11-26}$$

in the upper half z plane. Here x_t and α_t are real numbers such that

$$x_1 < x_2 < \cdots < x_n$$

and the α_t satisfy

$$\alpha_1 + \alpha_2 + \cdots + \alpha_n = 2 \qquad \text{and} \qquad -1 < \alpha_t < +1$$

Then it can be verified that if the function $w(z)$ defined in the upper half plane is extended to the real axis, the *image of the real z axis is a broken (polygonal) path in the w plane with exterior angle $\pi\alpha_t$ at w_t, the image of $z = x_t$.*

To see this, write

$$\frac{dw}{dz} = \frac{A}{(z - x_1)^{\alpha_1} \cdots (z - x_n)^{\alpha_n}}$$

Thus the argument of dw/dz is fixed in the (real) intervals $x_t < z < x_{t+1}$ $(1 \le t \le n - 1)$, as well as in the two remaining intervals (where

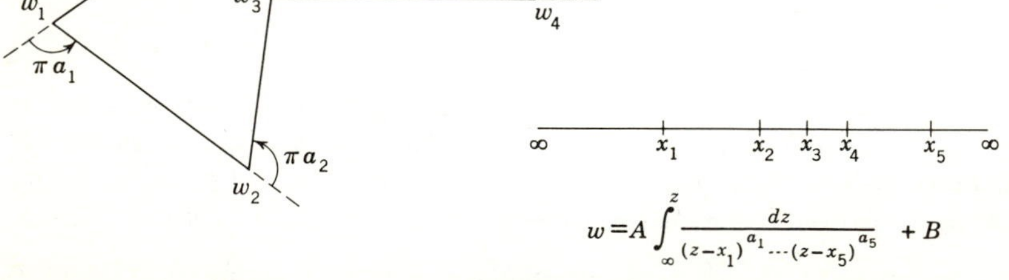

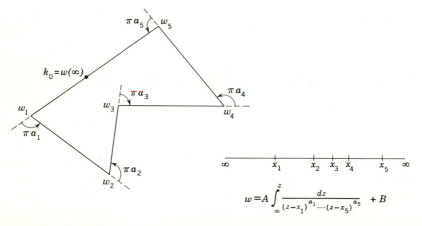

Fig. 11-4 Schwarz-Christoffel theorem. Note here $\alpha_3 < 0$ (the other α_i are $> \theta$).

the argument agrees): $z < x_1$, $z > x_n$. If we form the ratio (see Fig. 11-4)

$$\frac{\left(\dfrac{dw}{dz}\right)_{z=x_t-0}}{\left(\dfrac{dw}{dz}\right)_{z=x_t+0}} = \Lambda$$

we find $\arg \Lambda = \arg [1/(z - x_t)^{\alpha_t}]$ as $z \to x_t - 0 + 0i$. Thus

$$\arg \Lambda = \arg \frac{1}{(\exp \pi i)^{\alpha_t}} = -\pi\alpha_t$$

or

$$\arg (dw)_{x_t-0} - \arg (dw)_{x_t+0} = -\pi\alpha_t \tag{11-27}$$

which is a statement about the angles in the polygon. We note that $z = \infty$ does not correspond to any angle, as w is meromorphic at $z = \infty$:

$$w = A \int_{\mathrm{e}}^{z} \frac{dz}{z^2}\left(k_0 + \frac{k_1}{z} + \frac{k_2}{z^2} + \cdots\right) + B$$

Thus the *closed* z axis (including $z = \infty$) has a closed w image.

Now Theorem 3-7 of Sec. 3-4 tells us that the image of the boundary determines the image of the interiors. [The result holds even despite the nonanalyticity of $w(z)$ at $z = x_t$, since we can take arbitrarily close curves near the real axis.]

It therefore follows that if through a fortuitous choice of the x_t and α_t the w image of the closed z axis is a simple (nonintersecting) polygonal path, then (11-26) *provides a mapping of the interior of the polygon onto the upper half plane.*

This result is rather strong since α_t can be negative. (Consider the effect of a reentrant polygon on Fig. 11-4.) Nevertheless, the true value of the mapping (11-26) is in this stronger statement:

THEOREM 11-9 (*Schwarz-Christoffel,* 1867–1869) *Given any nonintersecting but possibly reentrant simple closed polygonal path in the w plane, there exist real quantities x_t and α_t such that the equation* (11-26) *provides the mapping of the interior of the polygon onto the upper half z plane.*†

PROOF Let us define $w(z)$ as the required mapping function. (It is known to exist by the Riemann mapping theorem!) Choose it for convenience so that $z = \infty$ does not become a vertex of the polygon. Let x_t be the point on the real axis corresponding to the vertex at w_t (with

† The actual choice of the x_t is extremely difficult, as in the module problem of the canonical regions in Sec. 9-3.

exterior angle πa_t). Form

$$\frac{dw}{dz}(z - x_1)^{a_1} \cdots (z - x_n)^{a_n} = \phi(z) \tag{11-28}$$

We need merely show $\phi(z)$ is constant ($\neq 0$)!

Here Theorem 11-8 applies despite our professed interest in the "interior" as stated in Theorem 11-9.

First of all, $w(z)$ can be *extended* to the lower half plane by analytic continuation through any segment $x_t < z < x_{t+1}$ (but such analytic continuations *vary with each segment* because of the lack of analyticity at x_t, the vertices). Nevertheless, at $z = \infty$, w is analytic since

$$a_1 + \cdots + a_t = 2$$

Thus

$$w = k_0 + \frac{k_1}{z} + \cdots \qquad \text{and} \qquad \frac{dw}{dz} = -\frac{k_1}{z^2} + \cdots$$

Thus from (11-28), $\lim_{z\to\infty} \phi(z) = \text{constant}$. This is true for the upper and lower half planes; that is, $\phi(z)$ is analytic in z at the point at ∞.

We now show $\phi(z)$ is analytic *everywhere*. We need only consider the vertices. Thus at vertex z_t there is the interior angle $\pi(1 - a_t)$. Therefore, let

$$\Omega(z) = (w - w_t)^{1/(1-a_t)} = k_1(z - x_t) + k_2(z - x_t)^2 + \cdots \tag{11-29}$$

We now have a straight segment of the real axis in the Ω plane mapped bicontinuously onto the straight segment (x_{t-1}, x_{t+1}) of the real axis containing $z = x_t$.

We apply Theorem 11-8.

Hence, $\Omega(z)$ can be defined analytically in the neighborhood of $z = x_t$. From (11-29), the function $(dw/dz)(z - x_t)^{a_t}$ is analytic in a neighborhood of $z = x_t$ and by (11-28), so is $\phi(z)$. Thus, $\phi(z)$ is definable in the whole z plane and bounded at ∞, hence a constant by Liouville's theorem. (The constant $\neq 0$, since otherwise $dw/dz \equiv 0$.) Q.E.D.

EXERCISES

1 An n-tuply-connected region $\mathcal{R}$ with n simple closed, piecewise smooth boundary curves $\partial\mathcal{R} = \mathcal{C}_1 + \cdots + \mathcal{C}_n$ can be mapped conformally (with continuous biunique boundary behavior) on a like region with n closed analytic boundary curves. When $n = 2$, let $\mathcal{R}$ lie inside $\mathcal{C}_2$ and outside $\mathcal{C}_1$, and map the interior of $\mathcal{C}_2$ onto the interior of a unit w disk by $w = f_2(z)$. Then map the exterior of $f_2(\mathcal{C}_1)$ onto the exterior of the unit disk by $W = f_1(w)$. Generalize this to arbitrary n. Show how to interpret this method when $\partial\mathcal{R}$ is not simple (say a slit).

2 What can be said about a curve $\mathcal{C}$ whose natural parametrization (by arc length s) agrees with its parametrization by x [see (11-16)]?

3 The reflection principle for harmonic functions is sometimes stated as $u(x,y) = u(x,-y)$ if $\partial u/\partial y = 0$ at $y = 0$. Express this more precisely in *weak* form (assume necessary analyticity), and show how it relates to (11-20) and (11-21).

4 Verify that (in Sec. 2-6) the construction of images with regard to the Möbius circle is a specialization of (11-22), where $f(z)$ maps the real axis onto, say, a circle.

5 Show that the relationship (11-22) has the form $w = \bar{\phi}(w^*)$ for ϕ analytic.

6 Show $w = w^*$ in Theorem 11-7 if and only if $w \in \mathcal{C}$ (assuming w sufficiently close to $\mathcal{C}$). Where do we use the assumption that w is close to $\mathcal{C}$?

7 Prove Theorem 11-7, Corollary 1, by the indirect method. Let $\mathcal{C}_2 = \phi(\mathcal{C}_1)$ and let $\mathcal{C}_i = f_i(\mathcal{S}_i)$ serve as parametrization [as in (11-23)]. Then $\mathcal{C}_2 = \phi f_1(\mathcal{S}_1)$ and $\mathcal{C}_2 = f_2(\mathcal{S}_2)$ are two parametrizations of $\mathcal{C}_2$. Show that Theorem 11-7 gives the desired result.

8 Let $w = \phi(z)$ be a mapping of the interior of the unit disk onto itself. Using $1/\bar{w} = \phi(1/\bar{z})$, show we obtain a mapping of the whole z sphere onto the whole w sphere by using Theorems 11-1 and 11-8. From these remarks construct an alternate proof of Theorem 10-6, Corollary 1, of Sec. 10-2.

9 Note that a polygon in the w plane has $2n$ real parameters (two for each point). Show the mapping (11-26) has $2n + 3$ *real* parameters in x_t, a_t, A, B. Explain the indeterminacy of three parameters!

10 Consider the function

$$w = \int_\infty^z \frac{dz}{\sqrt{(z - x_1)(z - x_2)(z - x_3)(z - x_4)}} \qquad x_1 < x_2 < x_3 < x_4$$

as a mapping of the upper half z plane onto a rectangle in the w plane, by considering *dw/dz directly* (without using Theorem 11-9). Show that *four* replicas of the rectangle constitute a period parallelogram for the elliptic integral involved here.

11 Explain what happens in (11-26) if $\Sigma a_t \neq 2$; in particular, consider what happens if Σa_t lies between 1 and 3. Consider the change of variables $z = 1/z'$ (assuming $x_t \neq 0$).

12 Map the upper half plane onto the interior of an equilateral triangle.

13 Assume that a segment of the boundary of a disk is analytic, specifically that it is given by an analytic function $z = x + iy = f(t)$ [with $f'(t) \neq 0$ and t real in the interval $t_0 < t < t_1$ of the real axis]. Assume that the Dirichlet boundary-value problem is given by an analytic function $u = g(t)$ (real for t in the same interval). Show that $u(z)$ can then be continued analytically across this segment. [*Hint:* Show $u(z) - \operatorname{Re} g(f^{-1}(x + iy)) = 0$ on the segment in question; then consider $u(f(t)) - \operatorname{Re} g(t)$ near the t axis.]

14 *Every closed analytic curve is the (locally) conformal image of a unit circle.* Prove it first for a simple curve by conformal mapping. (The more general result follows from the fact that the exterior of the curve can be taken as the boundary of a disklike Riemann manifold at ∞.)

11-3 Schottky Double

We show that the noncompact Riemann manifold $\mathfrak{M}$ can be regarded as being imbedded in a certain special compact manifold $\hat{\mathfrak{M}}$. Thus, among other things, the degree of smoothness of the boundary and bound-

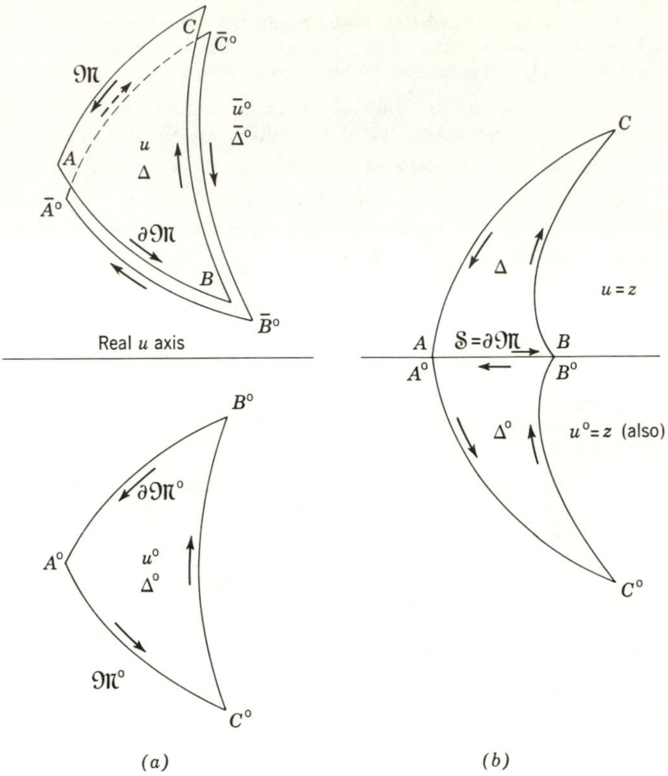

Fig. 11-5 Definition of the Schottky double.

ary behavior of differentials on 𝔐 and $\bar{𝔐}$, etc., were never any serious difficulty.

We start with a finite noncompact manifold 𝔐 with piecewise smooth boundary, and we transform it locally so that any assigned point of the boundary ∂𝔐 can be regarded as lying on an analytic segment of ∂𝔐. This can be done in the fashion of Fig. 11-2*c* and *d* using further triangulations of 𝔐. The triangulations must be considered carefully so as to take care of the annoying case where P is a vertex of two adjacent triangles Δ_1, Δ_2. (Then we construct an equivalent retriangulation which "moves the boundary" of Δ_1 and Δ_2 so that it does not pass through† P.)

We first construct $𝔐^0$, the replica (or image) of 𝔐. We consider

† This was done in Lemma 7-1, to take care of the case where a function f (defined on 𝔐) has a singularity at a point where Δ_1 and Δ_2 join. In effect, this means we have a *class* of triangulations and a *class* of matching functions which are consistent (whenever we define a Riemann manifold). Here, particularly, the concept of a manifold would be useless without parametrizations.

a triangle Δ_i with parameter u_i in $\mathfrak{M}$. For each Δ_i we construct a replica† $\bar{\Delta}_i^0$ with negatively oriented boundary and with parameter $\bar{u}_i^0$. Thus the conjugate of $\bar{\Delta}_i^0$ or Δ_i^0 would be positively oriented and appear as in Fig. 11-5*a* with regard to Δ_i if we were to lay the u_i^0 plane on top of the u_i plane. We, of course, do not have a manifold until we give parametrizations to determine neighboring triangles of Δ_i or Δ_i^0.

The parametrizations which connect, say Δ_i and Δ_j, are retained (compare Sec. 5-3). Such parametrizations might take the form of biunique mappings of Δ_i and Δ_j onto the z sphere:

$$z = f_i(u_i) \qquad z = f_j(u_j) \qquad u_i \in \Delta_i,\ u_j \in \Delta_j \tag{11-30}$$

and these transformations might map a common side $\mathfrak{L}$ between Δ_i and Δ_j on to a common side between $f_i(\Delta_i)$ and $f_j(\Delta_j)$. The corresponding mappings to connect Δ_i^0 and Δ_j^0 will be the identical ones (11-30) applied to the conjugates

$$\bar{z} = f_i(\bar{u}_i^0) \qquad \bar{z} = f_j(\bar{u}_j^0) \qquad \bar{u}_i^0 \in \bar{\Delta}_i^0,\ \bar{u}_j^0 \in \bar{\Delta}_j^0 \tag{11-31}$$

The more important step is to decide that if Δ_i has a boundary side, say AB, then $\bar{\Delta}_i^0$ shall join Δ_i precisely on that common side ($AB = A^0B^0$). The parametrization, however, requires care. It consists of first mapping Δ_i and $\bar{\Delta}_i^0$ so that AB is the segment $\mathfrak{S}$ of the real axis. Then the mapping which connects $\mathfrak{M}$ and $\mathfrak{M}^0$ is merely

$$z = u_i \qquad z = u_i^0 \qquad u_i \in \Delta_i^0,\ u_i \in \Delta_j^0 \tag{11-32}$$

(See Fig. 11-5*b*.) To take care of matching the vertex, say A, of $\partial\mathfrak{M}$ with A^0, the vertex of $\partial\mathfrak{M}^0$, we would have to reparametrize the manifold so that A and A^0 are inner points of some arc (such as AB) of a new system of triangles.

Thus we have defined a manifold $\hat{\mathfrak{M}} = \mathfrak{M} \cup \mathfrak{M}^0 \cup \partial\mathfrak{M}$, the *Schottky double*. It is compact (i.e., it no longer has a boundary).

LEMMA 11-3 *The Schottky double $\hat{\mathfrak{M}}$ of $\mathfrak{M}$ has genus*

$$\hat{p} = 2p + (n - 1) \tag{11-33}$$

where $\mathfrak{M}$ has p handles and n boundary contours.

PROOF We find it easiest to work with the rectangular model for $\mathfrak{M}$ (see Sec. 7-2), which consists of the region exterior to p nonintersecting rectangles (with opposite sides identified) and also exterior to n disks surrounded by $|\mathfrak{C}_1|, \ldots, |\mathfrak{C}_n|$. If we join the models for $\mathfrak{M}$ and for $\mathfrak{M}^0$ on the contours $\mathfrak{C}_1, \ldots, \mathfrak{C}_n$, we have $\hat{\mathfrak{M}}$.

† The symbol $\bar{\Delta}$ means "conjugate of Δ" here and not "closure of Δ." We shall not need a conjugation symbol for $\mathfrak{M}$ or $\mathfrak{M}^0$, however, so $\bar{\mathfrak{M}}$ always refers to closure.

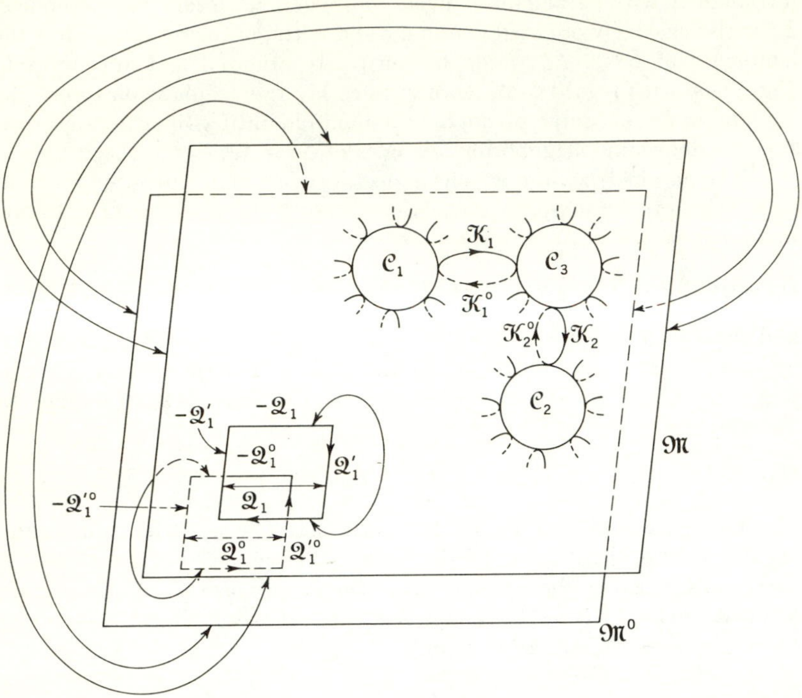

Fig. 11-6 Model of $\hat{\mathfrak{M}}$ (when $p = 2, n = 3$). $\mathcal{K}_1, \mathcal{K}_2$ are in the top sheet $\mathfrak{M}$. $\mathcal{K}_1^0, \mathcal{K}_2^0$ are in the bottom sheet $\mathfrak{M}^0$. $\mathcal{C}_1, \mathcal{C}_2, \mathcal{C}_3$ are where $\mathfrak{M}$ and $\mathfrak{M}^0$ join. (Note the canceling orientations in $\mathfrak{M}$ and $\mathfrak{M}^0$.)

Here we observe Fig. 11-6 which shows how to consummate the proof: The p handles in $\mathfrak{M}$ are matched by (but not connected to) the p handles in $\mathfrak{M}^0$. If we observe $\mathcal{C}_1, \ldots, \mathcal{C}_n$, however, we see that $n - 1$ of them, say $\mathcal{C}_j$ $(1 \le j \le n - 1)$, can each be connected with $\mathcal{C}_n$ by a chain of segments $\mathcal{K}_j$ in $\mathfrak{M}$ and $\mathcal{K}_j^0$ in $\mathfrak{M}^0$. Thus we obtain $n - 1$ additional pairs of crosscuts $\mathcal{C}_j$ and

$$\hat{\mathcal{K}}_j(= \mathcal{K}_j + \mathcal{K}_j^0) \qquad 1 \le j \le n - 1$$

We choose all $\mathcal{K}_j$ so as not to intersect $\mathcal{C}_j$, $\mathcal{Q}_i$, $\mathcal{Q}_i'$, or each other (except on $\partial\mathfrak{M}$). As in Figs. 11-5 and 11-6, the orientation of any 1-chain in $\mathfrak{M}$ is reversed for the corresponding 1-chain in $\mathfrak{M}^0$. This is what "cancels" the $\mathcal{C}_j$ with a $\mathcal{C}_j^0$ (not shown) to produce compactness.

We now obtain a set of $\hat{p} = 2p + (n - 1)$ pairs of crosscuts renamed $\mathcal{G}_k$ $(1 \le k \le 2p + n - 1)$,

$$\text{(11-34)} \qquad \{\mathcal{G}_k\} = \mathcal{Q}_i, \mathcal{Q}_i'; \mathcal{Q}_i^0, \mathcal{Q}_i^{0\prime}; \mathcal{C}_j, \hat{\mathcal{K}}_j \qquad 1 \le i \le p, 1 \le j \le n - 1$$

It remains to show they form a homology basis; this means that any 1-cycle $\mathcal{Z}^1$ is expressible *uniquely* as

(11-35) $$\mathcal{Z}^1 = \sum_1^p m_k \mathcal{G}_k$$

where m_k are integers. (See Exercise 1.) Q.E.D.

The interesting feature about an *abelian* differential $d\psi$ on $\mathfrak{M}$ is that it automatically extends to an abelian differential on $\hat{\mathfrak{M}}$ (and this applies to such things as canonical mappings). To see this, we need only show the possibility of extending $d\psi$ across $\partial\mathfrak{M}$ in such a manner that when we proceed to define $d\psi$ on $\mathfrak{M}^0$, we *encounter only poles* as singularities. Indeed, we shall extend $d\psi$ onto $\hat{\mathfrak{M}}$ in such a manner that the singularities encountered on $\mathfrak{M}^0$ are precisely replicas of those on $\mathfrak{M}$.

We show first how to extend $d\psi$ across $\partial\mathfrak{M}$ (from $\mathfrak{M}$ to $\mathfrak{M}^0$). Since no singularities of $d\psi$ are permitted on $\partial\mathfrak{M}$, then

(11-36a) $$d\psi = \operatorname{Re} df \qquad df = d\psi + i\, d\psi^*$$

(11-36b) $$f = \int df + \text{const}$$

for f an analytic function defined along some portion of $\partial\mathfrak{M}$. Take this portion of $\partial\mathfrak{M}$ to be the segment $\mathcal{S}$ in Fig. 11-5*b*. We can then adjust the constant in (11-36*b*) so that $\operatorname{Im} f = 0$, if we recall

(11-36c) $$d\psi^* = \operatorname{Im} df = 0 \qquad \text{on } \partial\mathfrak{M}$$

(by definition of an abelian differential). Then f can be extended across $\partial\mathfrak{M}$ (that is, across $\mathcal{S}$) by the usual reflection:

(11-37) $$f(\bar{z}) = \overline{f(z)}$$

As $f(z)$ is extended analytically through $\mathfrak{M}^0$ we get precisely the conjugate power series and (in a sense) the conjugate (polar) singularities of df on $\mathfrak{M}^0$. We must be careful about notation, however. If

$$f(z) = \frac{a}{z - z_0} + \text{reg funct} \qquad \operatorname{Im} z > 0, \operatorname{Im} z_0 > 0$$

then

$$\overline{f(z)} = \frac{\bar{a}}{\bar{z} - \bar{z}_0} + \text{reg funct}$$

Therefore, we see by (11-37)

$$f(\bar{z}) = \frac{\bar{a}}{\bar{z} - \bar{z}_0} + \cdots$$

Thus in terms of z ($\operatorname{Im} z < 0$), $f(z)$ has residue $\bar{a}$ at $z = \bar{z}_0$.

LEMMA 11-4 *If an abelian integral $f(z)$ on $\mathfrak{M}$ is extended to $\mathfrak{M}^0$ analytically, then the residues at corresponding poles z_* and $z_*{}^0$ are conjugates*

(11-38) $$\operatorname{res} f'(z_*)\, dz = \overline{\operatorname{res} f'(z_*{}^0)\, dz^0}$$

LEMMA 11-5 *The extension of a real abelian differential $d\psi$ onto $\mathfrak{M}^0$ is precisely the same as $d\psi$ on $\mathfrak{M}$, under the change of variables where u is changed to $\bar{u}^0$ (or the identification of corresponding points of the replica).*

For proof it suffices to take the situation of Fig. 11-5*b* once more (see Exercise 3) where $u = z$ and $\bar{u}^0 = \bar{z}$. Then from (11-37),

$$f(z) = \psi(x,y) + i\psi^*(x,y) \tag{11-39a}$$

$$\overline{f(\bar{z})} = \psi(x,-y) - i\psi^*(x,-y) \tag{11-39b}$$

and thus $\psi(x,y) = \psi(x,-y)$. Hence

$$d\psi(x,y) = d\psi(x,-y) \tag{11-39c}$$

(Here it is easiest to think of the same value of ψ as being "pinned" on u as on $\bar{u}^0$, so that $d\psi$ is invariant as $u \to \bar{u}^0$, etc.)

Thus for an abelian differential $d\psi$ on $\mathfrak{M}$ we can consider the integrals $(\mathcal{C},d\psi)$ over a curve $\mathcal{C}$ (or a 1-chain). Here $\mathcal{C}^0$, the replica of $\mathcal{C}$ on $\mathfrak{M}^0$, is the same as $\mathcal{C}$ (under $u \to \bar{u}^0$) but is negatively oriented as part of the manifold $\hat{\mathfrak{M}}$. Hence

$$(\mathcal{C},d\psi) = -(\mathcal{C}^0,d\psi) \tag{11-40}$$

We therefore define the *Schottky conjugate* of any real abelian differential $d\psi$ on $\mathfrak{M}$ as the formal result $d\psi^0$ of replacing u by $\bar{u}^0$.

LEMMA 11-6 *A meromorphic differential $d\psi = d\psi^0$ is its Schottky conjugate if and only if it is abelian. (We exclude singularities on $\partial\mathfrak{M}$.)*

PROOF The "if" part comes from Lemma 11-5. To prove the "only if" part, construct $df = d\psi + i\,d\psi^*$. If we go back again to (11-39*a*) and (11-39*b*), we note that if $\psi(x,y) = \psi(x,-y)$, then by the uniqueness of the (ordinary) conjugate

$$\psi^*(x,y) = -\psi^*(x,-y) + \text{const}$$

Thus by setting $y = 0$, we see $\psi^*(x,0) =$ constant on the real axis, whence $d\psi^* = 0$ (or $d\psi$ is abelian). Q.E.D.

ABELIAN DIFFERENTIALS

THEOREM 11-10 *The second existence theorem for a compact manifold implies the first existence theorem (for abelian differentials) on a noncompact manifold.*

PROOF Consider the requirements of the first existence theorem for a noncompact manifold $\mathfrak{M}$ [see (8-44*a*) to (8-44*d*)]:

$$d\psi^* = 0 \qquad \text{on } \mathcal{C}_j \tag{11-41a}$$

$$(\mathcal{Q}_i,d\psi) = h_i \qquad (\mathcal{Q}_1',d\psi) = h_1' \qquad (\mathcal{C}_j,d\psi) = k_j \qquad 1 \le j \le n-1 \tag{11-41b}$$

$$d\psi = \operatorname{Re} f_t(\zeta_t)\,d\zeta + d\psi_t \qquad 1 \le t \le s \tag{11-41c}$$

for $d\psi_t$ regular at singularities ($\zeta_t \in \mathfrak{M}$) for which

$$\text{(11-41}d\text{)} \qquad \operatorname{Re} \sum_1^s \operatorname{res} f_t(\zeta_t)\, d\zeta = 0$$

According to (11-40), we extend these conditions to $\hat{\mathfrak{M}}$ as follows:

$$\text{(11-41}e\text{)} \qquad \begin{aligned} (\mathfrak{Q}_i^0, d\psi) &= -h_i \qquad (\mathfrak{Q}_i^{0\prime}, d\psi) = -h_i' \\ (\hat{\mathfrak{K}}_j, d\psi) &= 0 \end{aligned}$$

and (11-41c) is extended by (11-38). (See Fig. 11-6, and recall

$$\hat{\mathfrak{K}}_j = \mathfrak{K}_j + \mathfrak{K}_j^0)$$

Thus we have s additional singularities ζ_t in $\mathfrak{M}^0$ ($s + 1 \leq t \leq 2s$), and summing over *all* of them, we find

$$\text{(11-41}f\text{)} \qquad \sum_1^{2s} \operatorname{res} f_t(\zeta_t)\, d\zeta = 0$$

since the imaginary parts of the residues cancel at the Schottky images by Lemma 11-4.

Now by hypothesis the second existence theorem as applied to (11-41b) to (11-41f) has a solution $d\psi$. It also has another solution $d\psi^0$ (the Schottky conjugate). By symmetry of the conditions (11-41b) to (11-41f), $d\psi = d\psi^0$; hence, we are dealing with an (abelian) integral satisfying (11-41a). We now have a solution to the first existence problem for (11-41a) to (11-41d). Q.E.D.

EXERCISES

1 Complete the proof that (11-34) constitutes a homology basis analogously with the case of Theorem 7-5 of Sec. 7-5 on the homology basis of $\mathfrak{M}$ itself. Note that any cycle on $\hat{\mathfrak{M}}$ can be broken up into arcs which connect a point of $\mathfrak{C}_i$ to a point of $\mathfrak{C}_j$, etc.

2 In Sec. 9-2, some abelian functions are listed for the upper half plane by the use of (conjugate) images. Verify that these illustrations are in accord with Lemma 11-4.

3 Justify the use of the parameters z and ($\bar{z}^0 =$)$\bar{z}$ to prove Lemma 11-5 by invoking analytic continuation.

4 Consider some abelian differential on the Riemann sphere $\hat{\mathfrak{M}}$ which is not abelian with respect to the upper half plane $\mathfrak{M}$, and discuss two differentials which are Schottky conjugates.

5 Define by suitable means the Schottky conjugate of a complex differential, and apply it to Lemma 11-6.

CHAPTER 12

ALTERNATING PROCEDURES

The preceding chapter narrows down the problem sufficiently so that if we want to prove the first existence theorem on abelian integrals for finite manifolds, it would suffice to prove the second existence theorem for compact manifolds (since then there is no distinction between the first and second existence theorems). We shall find that the second existence theorem needs to be proved for *noncompact manifolds first*, before we can prove it for compact manifolds.

The method used here refers to the alternating procedures of Schwarz (1870) and Neumann (1877). It provided the first rigorous proof of Riemann's existence theorem (except for the fact that the earliest versions of the proof had too many topological assumptions, which were overcome by later use of canonical models). This method, in addition, required no techniques of *analysis* above the level of difficulty of uniform convergence. Hence, for the novice it remains preferable to the broader techniques of functional analysis (see Appendix A).

12-1 Ordinary Dirichlet Problem

We first consider a noncompact plane manifold $\mathfrak{M}$ ($p = 0$, $n > 0$) on which a nonsingular (ordinary) second type Dirichlet problem is formulated (see Sec. 8-4).

Specifically, let $\partial\mathfrak{M} = \mathfrak{C}_1 + \cdots + \mathfrak{C}_n$ in the usual fashion, and let $g(Z)$ [or $g_t(Z)$ ($t = 1, 2, \ldots, n$)] denote real piecewise continuous boundary data on $\partial\mathfrak{M}$. We wish to find a single-valued abelian integral $u(z)$ for which $g(Z)$ serves as boundary value.†

Let us now consider the situation where a plane manifold $\mathfrak{M}$ is the union of two overlapping plane manifolds $\mathfrak{M}_1$, $\mathfrak{M}_2$, such that the ordinary Dirichlet problem is solvable in either one separately and such that neither of them contains the boundary of the other completely.

† Specifically, $u(z) \to g(Z)$ as $z \to Z$ for $z \in \mathfrak{M}$ and $Z \in |\partial\mathfrak{M}|$, except that at points z_0 where $g(Z)$ is prescribed as discontinuous, $u(z)$ must merely remain bounded as $z \to Z_0$.

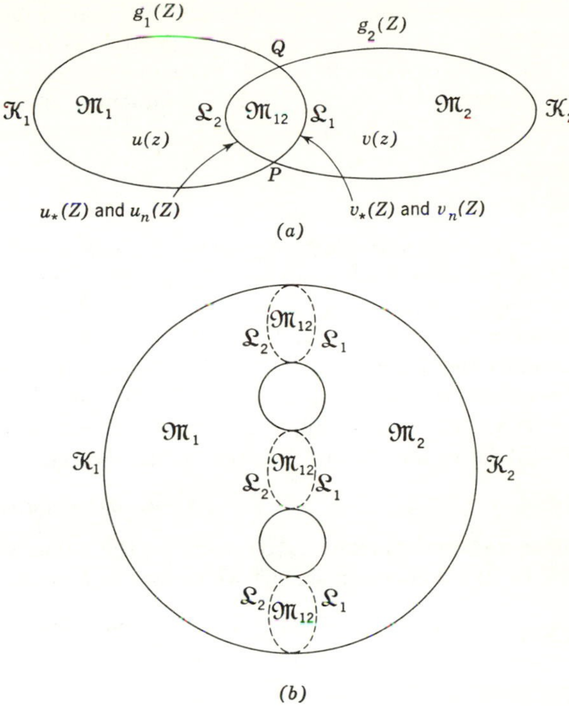

Fig. 12-1 Applications of the alternating procedure.

Thus (as in Fig. 12-1*a* and *b*), the solutions are

$$\begin{array}{lll} u(z) \text{ in } \mathfrak{M}_1 & \partial\mathfrak{M}_1 = \mathfrak{K}_1 + \mathfrak{L}_1 & \mathfrak{L}_1 = |\partial\mathfrak{M}_1| \cap \mathfrak{M}_2 \\ v(z) \text{ in } \mathfrak{M}_2 & \partial\mathfrak{M}_2 = \mathfrak{K}_2 + \mathfrak{L}_2 & \mathfrak{L}_2 = |\partial\mathfrak{M}_2| \cap \mathfrak{M}_1 \end{array}$$

where $\mathfrak{K}_1$, $\mathfrak{K}_2$, $\mathfrak{L}_1$, $\mathfrak{L}_2$ are collections of nonempty boundary arcs. We prescribe $g_1(z)$ on $\mathfrak{K}_1$ and $g_2(z)$ on $\mathfrak{K}_2$.

The situation which we describe is really very broad. In Fig. 12-1*a* we see the prototype situation to which the further symbols will refer. In Fig. 12-1*b* we see how a multiply-connected region $\mathfrak{M}$ can be built from simply-connected regions $\mathfrak{M}_1$, $\mathfrak{M}_2$.

We make a simplifying assumption that the boundaries $\partial\mathfrak{M}_1$, $\partial\mathfrak{M}_2$ are *not tangent* at P, Q, their points of intersection; thus, $\mathfrak{L}_2$ is not tangent to $\partial\mathfrak{M}_1$ (and $\mathfrak{L}_1$ is not tangent† to $\partial\mathfrak{M}_2$).

† In Exercises 1 to 3, we see that this assumption is really unnecessary, but we wish to establish a "ratio of convergence" $q_1q_2 < 1$ for the alternating procedure [see (12-12*a*) and (12-12*b*)]. This enables us to proceed more uniformly in the several additional cases which will arise here in Secs. 12-2 and 12-3.

We presuppose that the ordinary Dirichlet problem is solvable for $\mathfrak{M}_1$ (and $\mathfrak{M}_2$). Thus if we prescribe boundary values $v_*(Z)$ on $\mathfrak{L}_1$ (or $u_*(Z)$ on $\mathfrak{L}_2$) we have a (single-valued) harmonic function $u(z)$ [or $v(z)$] uniquely determined in $\mathfrak{M}_1$ (or $\mathfrak{M}_2$). The uniqueness is represented by certain functionals $u[\cdot\ \cdot\ \cdot](z)$ and $v[\cdot\ \cdot\ \cdot](z)$, namely

$$u(z) = u[g_1,v_*](z) \tag{12-1a}$$

$$v(z) = v[g_2,u_*](z) \tag{12-1b}$$

These unknown functions $u(z)$, $v(z)$ are necessarily determined by the (as yet unknown) functions $u_*(Z)$, $v_*(Z)$. The functionals $u[g_1,v_*](z)$ and $v[g_2,u_*](z)$ are quite complicated. For example, if $\mathfrak{M}_1$ and $\mathfrak{M}_2$ were circles, then $u[\cdot\ \cdot\ \cdot](z)$ and $v[\cdot\ \cdot\ \cdot](z)$ would be expressed by Poisson integrals (see Exercise 5). These functionals, nevertheless, are linear in the argument pairs $[g_1,v_*]$ and $[g_2,u_*]$. In solving the Dirichlet problem for $\mathfrak{M}_1 \cup \mathfrak{M}_2$, we are actually solving the iteration based on substituting (12-1b) for v_* into (12-1a). Thus (12-1a) becomes

$$u_*(Z) = u[g_1(Z),v[g_2(Z),u_*(Z)]](Z) \tag{12-2}$$

for an *unknown function* $u_*(Z)$, namely, the value of u on $\mathfrak{L}_1$.

In time-honored fashion,† we iterate as follows:

$$\begin{aligned} u_0(Z) &\equiv 0 \qquad \text{(a convenient choice)} \\ u_{n+1}(Z) &= u[g_1(Z),v[g_2(Z),u_n(Z)]](Z) \qquad \text{for } n \geq 0 \end{aligned} \tag{12-3}$$

this gives us a Dirichlet problem on $\mathfrak{M}_1$. We expect that $\lim u_n(z)$ (as $n \to \infty$) will provide the correct solution [and also $u_*(Z)$].

To make the notation simpler in (12-3), we define

$$u_0(Z) \equiv 0 \qquad Z \in \mathfrak{L}_1 \tag{12-4a}$$

$$v_1(z) = v[g_2,0](z) \qquad z \in \mathfrak{M}_2 \tag{12-4b}$$

and, in general, for $n \geq 0$

$$u_{n+1}(z) = u[g_1,v_{n+1}(Z)](z) \qquad z \in \mathfrak{M}_1,\ Z \in \mathfrak{L}_1 \tag{12-5a}$$

$$v_{n+1}(z) = v[g_2,u_n(Z)](z) \qquad z \in \mathfrak{M}_2,\ Z \in \mathfrak{L}_2 \tag{12-5b}$$

Together, these constitute (12-3). Since we alternate the Dirichlet problem from one region to another, this is called an *alternating procedure.*

To prove convergence of $u_n(z)$ and $v_n(z)$ in $\mathfrak{M}_1$ and $\mathfrak{M}_2$ (as $n \to \infty$), we consider the difference in two successive terms of (12-5a).

$$\begin{aligned} u_{n+1}(z) - u_n(z) &= u[g_1,v_{n+1}(Z)](z) - u[g_1,v_n(Z)](z) \\ u_{n+1}(z) - u_n(z) &= u[0,v_{n+1}(Z) - v_n(Z)](z) \qquad z \in \mathfrak{M}_1,\ Z \in \mathfrak{L}_1 \end{aligned} \tag{12-6a}$$

by linearity of $u[g_1,v_*]$ in g_1 and v_*. Likewise from (12-5b),

$$v_{n+1}(z) - v_n(z) = v[0,u_n(Z) - u_{n-1}(Z)](z) \qquad z \in \mathfrak{M}_2,\ Z \in \mathfrak{L}_2 \tag{12-6b}$$

† This method goes back to the medieval "rule of the false conjecture."

In either case, by the maximum-minimum principle for $Z \in \mathfrak{M}_1$,

(12-7a) $$\max |u_{n+1}(z) - u_n(z)| \le \max |u_{n+1}(Z) - u_n(Z)| \qquad Z \in \mathfrak{L}_1$$

and for $z \in \mathfrak{M}_2$,

(12-7b) $$\max |v_{n+1}(z) - v_n(z)| \le \max |v_{n+1}(Z) - v_n(Z)| \qquad Z \in \mathfrak{L}_2$$

Next let us consider the functional

$$u(z) = u[0,1](z) \qquad \text{in } \mathfrak{M}_1$$

the solution to the Dirichlet problem formed by prescribing (discontinuous) data for u [in (12-1)],

$$g_1(Z) = 0 \qquad \text{on } \mathfrak{K}_1 \qquad \text{and} \qquad v_*(Z) \equiv 1 \qquad \text{on } \mathfrak{L}_1$$

The solution would be a harmonic function u defined in $\mathfrak{M}_1$ with the property that as z varies on $\mathfrak{L}_2$ (see Fig. 12-1a)

(12-8) $$\begin{aligned} &u[0,1](z) < 1 \qquad \text{for } z \in \mathfrak{L}_2 \text{ (excluding end points } P,Q \text{ of } \mathfrak{L}_2) \\ &\lim u[0,1](z) < 1 \qquad \text{as } z \to P \text{ or } Q \text{ (for } z \in \mathfrak{L}_2) \end{aligned}$$

The second condition follows from Theorem 11-6 on discontinuous boundary-value problems. We see why we exclude any possible tangency of $\mathfrak{L}_2$ and $|\partial\mathfrak{M}_1|$. From (12-8), it follows that for some constant q_1 (<1)

(12-9a) $$u[0,1](z) \le q_1 < 1 \qquad \text{for } z \in \mathfrak{L}_2$$

Likewise, if we define the functional $v(z) = v[0,1](z)$ in $\mathfrak{M}_2$, then this function yields

(12-9b) $$v[0,1](z) \le q_2 < 1 \qquad \text{for } z \in \mathfrak{L}_1$$

Thus it easily follows from (12-6a) that for $z \in \mathfrak{M}_1$,

(12-10) $$|u_{n+1}(z) - u_n(z)| \le \max_{Z \in \mathfrak{L}_1} |v_{n+1}(Z) - v_n(Z)| \cdot u[0,1](z)$$

Hence, if we specialize z by the restriction $z \in \mathfrak{L}_2$, we have from (12-10),

(12-11a) $$\max_{Z \in \mathfrak{L}_2} |u_{n+1}(Z) - u_n(Z)| \le q_1 \max_{Z \in \mathfrak{L}_1} |v_{n+1}(Z) - v_n(Z)|$$

and from (12-6b),

(12-11b) $$\max_{Z \in \mathfrak{L}_1} |v_{n+1}(Z) - v_n(Z)| \le q_2 \max_{Z \in \mathfrak{L}_2} |u_n(Z) - u_{n-1}(Z)|$$

Thus

(12-12a) $$\max_{Z \in \mathfrak{L}_2} |u_{n+1}(Z) - u_n(Z)| \le q_1 q_2 \max_{Z \in \mathfrak{L}_2} |u_n(Z) - u_{n-1}(Z)|$$

By using (12-11b) and rewriting (12-11a) with index $n - 1$ instead of n, we obtain

(12-12b) $$\max_{Z \in \mathfrak{L}_1} |v_{n+1}(Z) - v_n(Z)| \le q_1 q_2 \max_{Z \in \mathfrak{L}_1} |v_n(Z) - v_{n-1}(Z)|$$

From the maximum-minimum principle it is clear that

$$\max_{z \in \mathfrak{M}_1} |u_{n+1}(z) - u_n(z)| \leq \max_{Z \in \mathfrak{L}_1} |u_{n+1}(Z) - u_n(Z)| \tag{12-13a}$$

and likewise

$$\max_{z \in \mathfrak{M}_2} |v_{n+1}(z) - v_n(z)| \leq \max_{Z \in \mathfrak{L}_2} |v_{n+1}(Z) - v_n(Z)| \tag{12-13b}$$

Thus it suffices to estimate the expressions in (12-11a) and (12-11b). Clearly, the series

$$\begin{aligned} \lim_{n \to \infty} u_n(z) &= u_0(z) + [u_1(z) - u_0(z)] + \cdots \\ &\qquad + [u_n(z) - u_{n-1}(z)] + \cdots \\ \lim_{n \to \infty} v_n(z) &= v_1(z) + [v_2(z) - v_1(z)] + \cdots \\ &\qquad + [v_n(z) - v_{n-1}(z)] + \cdots \end{aligned} \tag{12-14}$$

both converge uniformly in the closures $\bar{\mathfrak{M}}_1$, $\bar{\mathfrak{M}}_2$, by comparison with a geometric series of ratio $q_1 q_2 < 1$.

It is now clear that the limits in (12-14) exist by uniform convergence and *harmonic* functions are thus defined by these limits; i.e.,

$$\begin{aligned} \lim u_n(z) &= u_\infty(z) \qquad z \in \mathfrak{M}_1 \\ \lim v_n(z) &= v_\infty(z) \qquad z \in \mathfrak{M}_2 \end{aligned} \tag{12-15}$$

The next step is to verify that

$$u_\infty(z) \equiv v_\infty(z) \qquad \text{for } z \in \mathfrak{M}_1 \cap \mathfrak{M}_2 \tag{12-16}$$

then it will follow that u_∞ and v_∞ are the "same" harmonic function in the sense of analytic continuation. To verify (12-16), we consider $u_{n+1}(z) - v_{n+1}(z)$ within the intersection $\mathfrak{M}_1 \cap \mathfrak{M}_2 = \mathfrak{M}_{12}$. Clearly, $\mathfrak{M}_{12}$ is bounded by $\mathfrak{L}_1$ and $\mathfrak{L}_2$ and from (12-5a)

$$u_{n+1}(Z) - v_{n+1}(Z) = 0 \qquad \text{for } Z \in \mathfrak{L}_1 \tag{12-17a}$$

From (12-15b) $v_{n+1}(Z) = u_n(Z)$ for $Z \in \mathfrak{L}_2$, hence

$$u_{n+1}(Z) - v_{n+1}(Z) = u_{n+1}(Z) - u_n(Z) \qquad \text{for } Z \in \mathfrak{L}_2 \tag{12-17b}$$

Thus from the maximum-minimum principle ($\partial\mathfrak{M}_{12} = \mathfrak{L}_1 \cup \mathfrak{L}_2$),

$$\max_{z \in \mathfrak{M}_{12}} |u_{n+1}(z) - v_{n+1}(z)| \leq \max_{Z \in \mathfrak{L}_2} |u_{n+1}(Z) - u_n(Z)| \tag{12-18}$$

and the right-hand side of (12-18) surely approaches 0 as $n \to \infty$.

The final step is far from trivial: We still must show that the common function given in (12-16) actually has $g_1(Z)$ and $g_2(Z)$ as boundary values at points of continuity (and is bounded near discontinuous† boundary values of g_1 or g_2).

† The condition of boundedness follows from the uniform convergence of the series in (12-14), since g_1 and g_2 are bounded to begin with.

To dispose of this last step, we note that the series in (12-14) converge uniformly in $\bar{\mathfrak{M}}_1$, $\bar{\mathfrak{M}}_2$ (the closures). This justifies the interchange of limits involved here. For example, if $g_1(Z)$ is continuous at $Z = Z_0$, then for $Z_0 \in \mathcal{K}_1$, consider the boundary value of $u_\infty(z)$ at Z_0:

$$\lim_{z\to Z_0} u_\infty(z) = \lim_{z\to Z_0}\lim_{n\to\infty} u_n(z) = \lim_{n\to\infty}\lim_{z\to Z_0} u_n(z) = g_1(Z_0) \tag{12-19}$$

[since it is trivial to see that for each n we have $u_n(z) \to u_0(Z_0) = g_1(Z_0)$ as $z \to Z_0$].

Thus we see, from Fig. 12-1*b*, *that the ordinary Dirichlet problem is solvable for an n-tuply-connected plane region by an application of its solution for a simply-connected plane region.* It remains to consider regions on manifolds.

Historically, this method was put to use to prove the very basic Riemann mapping theorem (for a simply-connected disk $\mathfrak{R}$ onto a circle). As we saw in Sec. 10-2 this mapping theorem is equivalent to the solvability of the ordinary Dirichlet problem for $\mathfrak{R}$. The Poisson integral takes care of the case where $\mathfrak{R}$ is a circle. Hence the alternating principle enables us to take care of the case where $\mathfrak{R}$ is a region which is a conglomeration (union) $\mathcal{S}_n$ of n (open) circular disks. It is clear, however, that for an arbitrary region $\mathfrak{R}$ and ε (>0), we can approximate $\mathfrak{R}$ by a region $\mathcal{S}_n$ [where $n = n(\varepsilon)$], such that

$$\mathfrak{R} \supset \mathcal{S}_n \qquad \text{(distance from } |\partial\mathfrak{R}| \text{ to } |\partial\mathcal{S}_n|) < \varepsilon \tag{12-20}$$

(We must observe the precaution that the boundary circles are not tangent, however.)

It is somewhat simpler to note that a polygon can be built up by the process of intersections of half planes, hence we can solve the Dirichlet problem for $\mathfrak{R}$ by approximating $\mathfrak{R}$ by $\mathcal{S}_n$, which is now an n-sided *polygon* subject to (12-20). The region $\mathfrak{R}$ and the polygons $\mathcal{S}_n$ need not even be convex. The proof of the legitimacy of such methods follows the outlines of our proof of Riemann's mapping theorem.† (Recall that in Sec. 10-2, Theorem 10-6 was proved by approximating the Green's function for $\mathfrak{R}$ by the Green's function for $\mathcal{S}_n$.)

EXERCISES

In Exercises 1 to 3, assume that the boundaries $|\partial\mathfrak{M}_1|$ and $|\partial\mathfrak{M}_2|$ *might* be tangent where they intersect. Assume $g_1(Z) \geq 0$, $g_2(Z) \geq 0$ (by adding a constant to the boundary data if necessary).

1 Justify the iterative procedure in (12-4*a*) and (12-4*b*), and (12-5*a*) and (12-5*b*) by noting that $u_n(z)$ and $v_n(z)$ are monotonic and hence uniformly convergent by Harnack's theorem (see Sec. 10-1) in any subregion of $\mathfrak{M}_1$ or $\mathfrak{M}_2$.

† A procedure for "sweeping out" an area is often called "balayage" (French).

2 Show that the limits $u_n(z) \to u_\infty(z)$ and $v_n(z) \to v_\infty(z)$ have the necessary boundary values at, say Z_0, a point of continuity of $g_1(Z)$ on $\partial\mathfrak{M}$. First show (by continuity at the boundary) that for any $\varepsilon > 0$, a neighborhood $\mathfrak{N}_\varepsilon$ of Z_0 exists such that for *any n*

$$u_n(z) \leq g_1(Z) + \varepsilon \qquad z \in \mathfrak{N}_\varepsilon \cap \mathfrak{M} \tag{12-21}$$

For example, to get rid of n, majorize $u_n(z)$ by solving (12-1a) and (12-1b) where v_* is amply large, say $\geq \max\,[g_1(Z),g_2(Z)]$ $(Z \in \mathcal{K}_1 \cup \mathcal{K}_2)$. From (12-21) prove that as $z \to Z_0$, $\limsup u_\infty(z) \leq g_1(Z_0)$, and from (14$ab$) prove that $\liminf u_\infty(z) \leq g_1(Z_0)$ [since $u_n(z)$ is monotonic].

3 Prove that $u_\infty(z) \equiv v_\infty(z)$ in $\mathfrak{M}_{12}$ by using (12-18).

4 What is wrong with the following "proof" of Exercise 3? $u_\infty(z) \equiv v_\infty(z)$ on $\mathfrak{L}_1$, thus the identity holds for infinitely many (hence all) values?

5 Write out the functionals in (12-1) for the case of two intersecting unit disks. Write out the integral equation (12-2), using as many abbreviations as convenient.

6 Consider a *punctured* region $\mathfrak{A}'_n$ formed by taking a multiply-connected region with simple boundary $\mathfrak{R}_n$ and removing t (isolated) interior points, $a_1, \ldots, a_t$, of $\mathfrak{R}_n$. Show that the Dirichlet boundary-value problem for $\mathfrak{A}'_n$ [$u(z)$ prescribed on $\partial\mathfrak{R}_n$ and on $a_1, \ldots, a_t$] is solvable only if the corresponding solution for $\mathfrak{R}_n$ [$u(z)$ prescribed on $\partial\mathfrak{R}_n$ only] has the prescribed values on $a_1, \ldots, a_t$.

12-2 Nonsingular Noncompact Problem

We are now in a position to prove the second existence theorem for a *noncompact* Riemann manifold of arbitrary genus ($p \geq 0$, $n > 0$), *with 2p periods* but without singularities. As a first step let us observe that the alternating procedure as described in Sec. 12-1 takes care of Jordan manifolds ($p = 0$, $n > 0$).

We begin by noticing that the arbitrary Riemann manifold is defined for an aggregate of F parametric planes u_i $(1 \leq i \leq F)$ with one (finite) triangle Δ_i on each plane. The Dirichlet problem is intrinsically solvable for the (simply-connected) region Δ_i (no periods or singularities) by the conformal mapping procedure in Sec. 10-2. To see that it is solvable for an arbitrary Jordan manifold, we proceed by induction.

LEMMA 12-1 *The triangles $\Delta_1, \ldots, \Delta_F$ of a (noncompact) Jordan manifold $\mathfrak{M}$ can be so numbered that the succession of closed domains*

$$\begin{aligned} \overline{\mathfrak{M}}_1 &= \bar{\Delta}_1 \\ \overline{\mathfrak{M}}_2 &= \bar{\Delta}_1 \cup \bar{\Delta}_2 \\ &\cdots\cdots\cdots \\ \overline{\mathfrak{M}}_k &= \bar{\Delta}_1 \cup \cdots \cup \bar{\Delta}_{k-1} \cup \bar{\Delta}_k \;(=\overline{\mathfrak{M}}_{k-1} \cup \bar{\Delta}_k) \\ \mathfrak{M} = \overline{\mathfrak{M}}_F &= \bar{\Delta}_1 \cup \cdots \cup \bar{\Delta}_F \end{aligned}$$

has the property that $\overline{\mathfrak{M}}_k$ does not surround $\bar{\Delta}_{k+1}$ $(1 < k < F)$ [*that is, some portion of $|\partial\bar{\Delta}_{k+1}|$ is free from $|\partial\overline{\mathfrak{M}}_k|$, and some portion of $|\partial\overline{\mathfrak{M}}_k|$ is free from $|\partial\bar{\Delta}_{k+1}|$*].

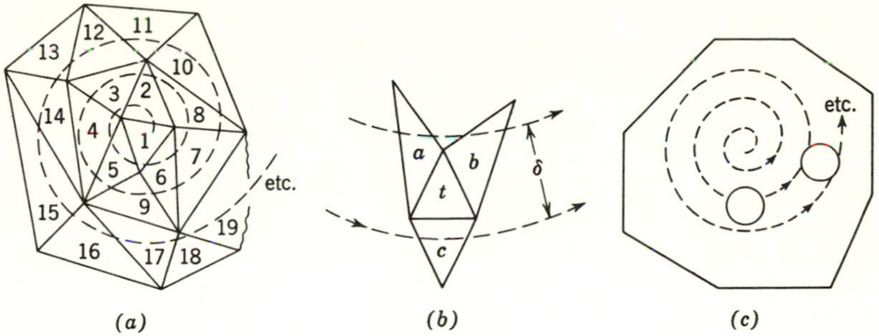

Fig. 12-2 (*a*) Spiral formed in the polygonal model. (*b*) The triangles Δ_a, Δ_b, Δ_c cannot all be enumerated before Δ_t (i.e., t cannot exceed a,b,c). (*c*) The whole Jordan manifold cut by a spiral.

PROOF The result would seem clear since $\mathfrak{M}$ can be represented simplicially as the interior of a *plane* region $\mathfrak{R}_n$ with n (≥ 1) boundary curves. The triangulation of the region $\mathfrak{R}_n$, however, can be very complicated, and unless we number the triangles carefully we can "trap" a $\bar{\Delta}_{k+1}$ inside $\mathfrak{M}_k$. This would happen if we mistakenly worked our way inside from the perimeter of $\mathfrak{M}$ and enclosed the center from outside.

Therefore we choose a systematic method to work from the *inside out* in the manner of peeling the top half of an apple in one strip by starting at the pole and working our way around to the equator. This is illustrated in Fig. 12-2*a* or *c*.

What we shall do is to consider some distance δ less than the shortest altitude of any triangle in the simplicial subdivision. Next, we consider a spiral progressing with (parallel distance) displacement δ from the origin of the polygon. Then we enumerate the triangles in the order in which the spiral penetrates (touches an inner point of) each triangle.

Under this procedure we assert no triangle can be enclosed, as shown in Fig. 12-2*b*. Indeed, the three adjoining neighbors of a triangle Δ_t cannot be all penetrated by the spiral before Δ_t is penetrated. Q.E.D.

PURE BOUNDARY-VALUE PROBLEM FOR JORDAN REGION $\mathfrak{M}$

We next note than an ordinary Dirichlet problem solvable in $\mathfrak{M}_k$ is solvable in $\mathfrak{M}_{k+1}$. The adjunction process is shown in Fig. 12-3*a* to *c*. We note (in Fig. 12-3*a*, for example) that the definition of a Riemann manifold requires that if $\mathfrak{M}_k$ adjoins Δ_{k+1}, then $\mathfrak{M}_k$ and Δ_{k+1} can be imbedded in slightly larger regions ($\mathfrak{M}'_k \supset \mathfrak{M}_k$ and $\Delta'_{k+1} \supset \Delta_{k+1}$ shown in dotted outline in Fig. 12-3). These larger regions overlap, and there is a biunique analytic correspondence of $\mathfrak{M}'_k$ and Δ'_{k+1} where they overlap in some parameter plane [see (5-5)]. If Δ_{k+1} completes a corner (as shown at P in Fig. 12-3*b*), we must refer to Fig. 5-2, which shows how the tri-

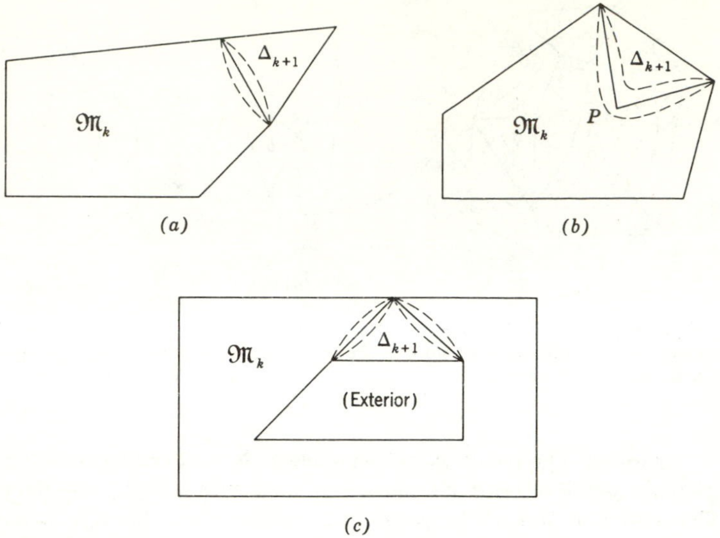

Fig. 12-3 (a) Simple adjunction $\overline{\mathfrak{M}}_k \cup \bar{\Delta}_{k+1}$; (b) enclosing a corner; (c) changing connectivity (for $\mathfrak{M}_k$, $n = 1$, but for $\mathfrak{M}_{k+1}$, $n = 2$).

angles fit together on a cyclic neighborhood. In Fig. 12-3c we see how the adjunction of Δ_{k+1} to $\mathfrak{M}_k$ produces a region $\mathfrak{M}_{k+1}$ of higher connectivity.

PERIODS ON CROSSCUTS

We now consider the case of a noncompact manifold $\mathfrak{M}$ of genus $p \geq 1$ $(n > 0)$. Let $\mathfrak{M}^p$ denote the canonically cut Jordan manifold corresponding to the canonical manifold depicted in Fig. 7-1. The boundary of $\mathfrak{M}^p$ consists of p curvilinear "rectangles" and n closed curves. Actually, the rectangles are effectively closed curves of $\partial\mathfrak{M}^p$, but if the opposite sides were identified, $\mathfrak{M}^p$ would reduce to $\mathfrak{M}$. The nonsingular Dirichlet problem of the second kind is now to find a harmonic integral ψ on $\mathfrak{M}^p$, such that $d\psi$ is a harmonic differential on $\mathfrak{M}$ and the periods on the $2p$ crosscuts are prescribed

$$(12\text{-}22a) \qquad (\mathcal{Q}_i, d\psi) = h_i \qquad (\mathcal{Q}_i', d\psi) = h_i' \qquad i = 1, 2, \ldots, p$$

and the piecewise continuous boundary data are also prescribed for $\partial\mathfrak{M}$, such that some harmonic integral $\int d\psi = \psi$ on $\mathfrak{M}^p$ satisfies

$$(12\text{-}22b) \qquad \psi(Z) = g(Z) \qquad \text{for } Z \text{ on } \partial\mathfrak{M}$$

The situation is amply illustrated if we take a manifold $\mathfrak{M}$ for which $p = 1$, $n = 1$ (indeed, we may even take the canonically cut manifold $\mathfrak{M}^1 = \mathfrak{R}$ to be an actual rectangle with a hole cut out, as shown in Fig. 12-4a). We prescribe $\psi = g(Z)$ on the boundary $\mathcal{C}_1$ inside $\mathfrak{R}$. The other boundary values are to have periods h_1 and h_1' as shown. We obviously do not know the values $\gamma_0(Z)$ and $\gamma_1(Z)$ which ψ shall take on the rectangular boundary. We do know that $\gamma_0(Z)$ and $\gamma_1(Z)$ are to be *chosen* so that the boundary data corresponding to

$$\psi = g(Z) \qquad \text{on } \mathcal{C}_1 \tag{12-23a}$$

$$\begin{array}{llll} \psi = \gamma_1 & \text{on } \mathfrak{Q}_1 & \psi = \gamma_0 + h_1 & \text{on } \mathfrak{Q}_1' \\ \psi = \gamma_1 + h_1' & \text{on } -\mathfrak{Q}_1 & \psi = \gamma_0 & \text{on } -\mathfrak{Q}_1' \end{array} \tag{12-23b}$$

all determine a differential $d\psi$ on $\mathfrak{M}$ (not just on $\mathcal{R} = \mathfrak{M}^1$).

Otherwise expressed, we are to choose $\gamma_1(Z)$, $\gamma_0(Z)$ so that the harmonic integral ψ determined by (12-23*a*) and (12-23*b*) has the periods (12-22*a*) by *analytic continuation* across the opposing pairs of sides of $\partial\mathcal{R}$ (or across the canonical crosscuts of $\mathfrak{M}^p$).

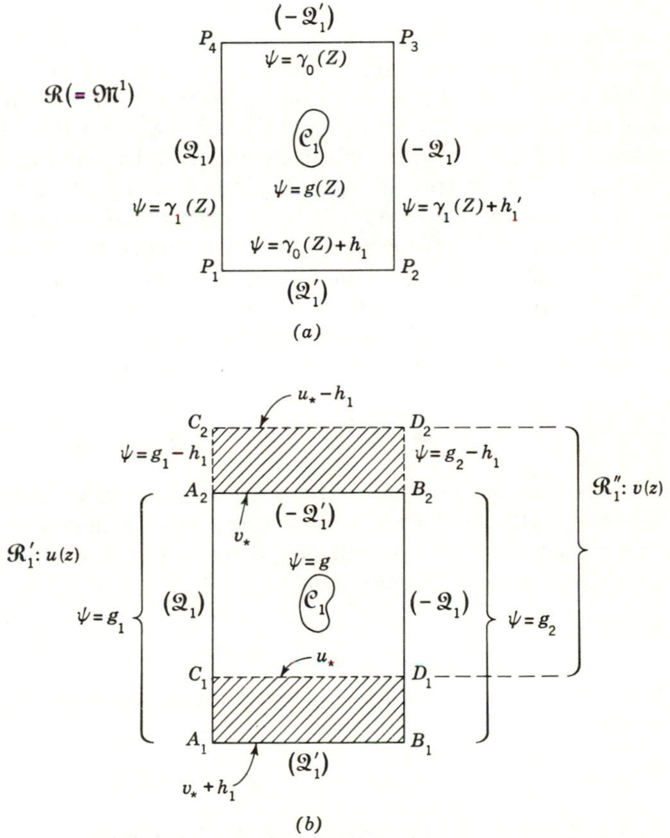

Fig. 12-4 (*a*) The double period problem for $p = 1$, $n = 1$; (*b*) the single period problem for $\mathcal{R}_1'$ and $\mathcal{R}_1''$.

INTERMEDIATE PROBLEM

We solve the original problem in two steps. We first do the intermediate problem for $\mathcal{R}_1$, the manifold formed by dissociating $\mathcal{Q}_1$ and $-\mathcal{Q}_1$ while $\mathcal{Q}_1'$ and $-\mathcal{Q}_1'$ are still identified (i.e., in Fig. 12-4*b* $A_1B_1 \equiv A_2B_2$). Thus

(12-24*a*) $$(\mathcal{Q}_1,d\psi) = h_1$$

(12-24*b*) $$\psi = g_1(Z) \qquad \text{on } A_1A_2(\mathcal{Q}_1) \qquad \psi = g_2(Z) \qquad \text{on } B_1B_2(-\mathcal{Q}_1)$$

and, as usual,

(12-24*c*) $$\psi = g(Z) \qquad \text{on } \mathcal{C}_1$$

Of course, $\mathcal{R}_1$ is a Jordan region topologically, namely, an annulus† (or cylinder) with a hole $\mathcal{C}_1$ cut out (compare Fig. 12-5*a*).

We next consider $\mathcal{R}_1'$, obtained from $\mathcal{R}_1$, by now dissociating the common side ($A_1B_1 = A_2B_2$) formed by $\mathcal{Q}_1'$ and $-\mathcal{Q}_1'$ in $\mathcal{R}_1$. The choice of AB (for $\mathcal{Q}_1'$ and $-\mathcal{Q}_1'$) was, of course, arbitrary. We could have used a neighboring curve $C_1D_1 = C_2D_2$ for $\mathcal{Q}_1'$ and $-\mathcal{Q}_1'$, and if we had done so, we should have the same effect as obtained by cutting off rectangle $A_1B_1C_1D_1$ at the bottom of $\mathcal{R}_1'$ (in Fig. 12-4*b*) and pasting it at the top of $\mathcal{R}_1'$. This alternate choice (of C_1D_1 instead of A_1B_1) would result in the Jordan region $\mathcal{R}_1''$. We associate with the shaded region of $\mathcal{R}_1''$ on A_2C_2 and B_2D_2 the boundary data $\psi = g_1 - h_1$ and $\psi = g_2 - h_1$, where g_1 and g_2 are the values on A_1C_1 and B_1D_1 (as shown in Fig. 12-4*b*). This takes care of $\mathcal{Q}_1$ and $-\mathcal{Q}_1$, the *distinct* sides of $\mathcal{R}_1$.

In other words, on $\mathcal{R}_1''$, (12-24*b*) is modified to

(12-24*b*′) $$\begin{aligned} \psi &= g_1 - h_1 \quad \text{on } A_2C_2 \qquad \psi = g_1 \quad \text{on } C_1A_2 \\ \psi &= g_2 - h_2 \quad \text{on } B_2D_2 \qquad \psi = g_2 \quad \text{on } D_1B_2 \end{aligned}$$

We now set up the appropriate alternating procedure to solve (12-24*a*) to (12-24*c*) for region $\mathcal{R}_1$. If we were to know $\psi = v_*$ on A_2B_2, we could solve (12-24*b*) and (12-24*c*), together with the following equation [as a substitute for (12-24*a*)] in $\mathcal{R}_1'$:

(12-24*d*) $$\begin{aligned} \psi &= v_* \qquad \text{on } A_2B_2 \\ \psi &= v_* + h_1 \qquad \text{on } A_1B_1 \end{aligned}$$

The solution would be a complicated functional analogous to (12-1*a*) and (12-1*b*).

(12-25*a*) $$u = u[g_1,g_2,g,v_*,h_1](z) \qquad z \in \mathcal{R}_1'$$

† In Exercises 1 and 2 we indicate a specific illustration to remind the reader that under the present conditions with prescribed data, we must always deal with discontinuous harmonic functions.

(Here g_1, g_2, g, v_* are *functions* on boundary segments and h_1 is a *constant.*) Likewise, if we were to know $\psi = u_*$ on C_1D_1, we could solve (12-24b') and (12-24c) together with

(12-24d')
$$\begin{aligned} \psi &= u_* && \text{on } C_1D_1 \\ \psi &= u_* - h_1 && \text{on } C_2D_2 \end{aligned}$$

We would obtain

(12-25b)
$$v = v[g_1,g_2,g,u_*,h_1](z) \qquad z \in \mathcal{R}_1''$$

Obviously, we want to solve (12-25a) and (12-25b) simultaneously [analogously with (12-2)] by setting v_* in (12-25a) equal to the expression for (12-25b) on A_2B_2 and by setting u_* in (12-25b) equal to the expression for (12-25a) on C_1D_1.

We therefore set up the alternating procedure:

(12-26a)
$$u_0(z) \equiv 0 \qquad z \in \mathcal{R}_1'$$
(12-26b)
$$v_1(z) = v[g_1,g_2,g,0,h_1](z) \qquad z \in \mathcal{R}_1''$$

More generally, for $n \geq 0$, we have the *ordinary* boundary-value problems:

(12-26c)
$$u_{n+1}(z) = u[g_1,g_2,g,v_{n+1}(Z),h_1](z) \qquad z \in \mathcal{R}_1',\ Z \in A_2B_2$$
(12-26d)
$$v_{n+1}(z) = v[g_1,g_2,g,u_n(Z),h_1](z) \qquad z \in \mathcal{R}_1'',\ Z \in C_1D_1$$

To prove convergence, we follow the usual steps:

(12-27a)
$$u_{n+1}(z) - u_n(z) = u[0,0,0,v_{n+1}(Z) - v_n(Z),0](z) \qquad z \in \mathcal{R}_1'$$
(12-27b)
$$v_{n+1}(z) - v_n(z) = v[0,0,0,u_n(Z) - u_{n-1}(Z),0](z) \qquad z \in \mathcal{R}_1''$$

by subtracting two successive cases of (12-26c) and of (12-26d), respectively.

Let us observe the meaning of $u[0,0,0,f(Z),0](z)$. This is a solution to a problem in $\mathcal{R}_1'$ "almost homogeneous," except that $u = f(Z)$ on A_1B_1 and A_2B_2. Clearly, we can replace $f(Z)$ by the constant max $|f(Z)| = M$ for $Z \in A_2B_2$, and we obtain

(12-28a)
$$|u[0,0,0,f(Z),0](z)| \leqq u[0,0,0,M,0](z) = Mu[0,0,0,1,0](z)$$

Certainly, the maximum-minimum principle can be applied to the harmonic function $u[0,0,0,1,0](z)$ which vanishes on $A_2A_1 \cup B_1B_2 \cup \mathcal{C}_1$ and equals unity on A_1B_1 and A_2B_2. Thus

(12-28b)
$$\max u[0,0,0,1,0](z) = q_1 < 1 \qquad \text{for } z \in C_1D_1$$

since C_1D_1 is *interior* to $\mathcal{R}_1'$.

Hence from (12-27a)

(12-29)
$$|u_{n+1}(z) - u_n(z)| \leq q_1 \max |v_{n+1}(Z) - v_n(Z)| \qquad z \in \mathcal{R}_1',\ Z \in A_2B_2$$

This is a result quite analogous to (12-10). We can then proceed to prove convergence of $u_n(z)$ and $v_n(z)$ in $\mathcal{R}_1'$ and $\mathcal{R}_1''$, respectively. Call $\lim u_n(z) = u_\infty(z)$, $\lim v_n(z) = v_\infty(z)$ (as $n \to \infty$).

We next wish to show that $u_\infty(z) = v_\infty(z)$ in $\mathcal{R}_*' = \mathcal{R}_1' \cap \mathcal{R}_1''$. To this end, compare $u_{n+1}(z)$ and $v_{n+1}(z)$, particularly with reference to (12-26c). Certainly, $u_{n+1}(z) - v_{n+1}(z)$ vanishes on $\mathcal{C}_1$ and on the unmatched sides A_2C_1 and B_2D_1. Actually, $u_{n+1}(z) - v_{n+1}(z)$ also vanishes by (12-26c) on A_2B_2, and by (12-26d) $u_{n+1}(z) - v_{n+1}(z)$ reduces to $u_{n+1}(z) - u_n(z)$ on C_1D_1. This latter quantity converges to zero. The common function $u_\infty(z) \equiv v_\infty(z)$ then has analytic continuations in the shaded regions (see Fig. 12-4) differing by h_1 so as to have the required periodicity (see Exercise 3). Thus we have found $\psi = u_\infty$ (or v_∞).

DOUBLE-PERIOD PROBLEM

Having solved the problem in (12-24a) to (12-24c) we next turn our attention to the original problem as given by

(12-24a') $$(\mathcal{Q}_1, d\psi) = h_1 \qquad (\mathcal{Q}_1', d\psi) = h_1'$$

together with (12-24c). Since we can solve the intermediate problem (12-24a) to (12-24c), we can solve (12-24a') and (12-24c) together by repeating the solution to the intermediate problem with one modification. The given values g_1 and g_2 in (12-25a) and (12-25b) are no longer arbitrary; they are chosen under the condition that ψ is harmonic with period h_1' as z is extended from A_1A_2 to B_1B_2 (or $g_2 = g_1 + h_1'$ at corresponding points). This would lead to a modification of the functionals (12-25a) and (12-25b) on $\mathcal{R}_1'$, so that g_1 and g_2 are no longer prescribed but the period h_1' is now prescribed in the $\mathcal{Q}_1'$ direction, and we are required to choose u_* and v_* so as to produce the period in the $\mathcal{Q}_1$ direction.

(12-25$'$) $$u = u[g, v_*, h_1, h_1'](z)$$
$$v = v[g, u_*, h_1, h_1'](z)$$

We see (12-25) is a problem with two periods, hence a problem on the original manifold $\mathfrak{M}$. We know it is possible to solve the modified problems (12-25) by setting up an iteration pattern analogous to that of (12-26c) and (12-26d). For the new pattern, at each step of the iteration, the solutions are known to exist by problem (12-24a) to (12-24c) on $\mathcal{R}_1$. The only difference is that in (12-24a) to (12-24c) we had produced a period h_1 across $\mathcal{Q}_1$ with the remaining boundary data prescribed in the ordinary sense, whereas now in (12-25$'$) we produce a period across $\mathcal{Q}_1$ with the other data already producing a period because of the earlier problem (12-24a) and (12-24b).

We now have achieved a solution for a non-Jordan region.

The crucial step in the whole convergence proof is the analog to (12-28b). Recall that we knew $q_1 < 1$ by virtue of the fact that C_1D_1 is interior to $\mathcal{R}_1'$ and that $\mathfrak{M}$ is *not compact*. This is still true even if we identify $\mathcal{Q}_1$ and $-\mathcal{Q}_1$. Of course, when we set $v_* = 1$ in (12-25′) (with g, h_1, h_1' vanishing), we must obtain $u[0,1,0,0](z) \leq q_1 < 1$, and this requires the fact that $u[0,1,0,0](z) = 0$ for z on $\partial\mathfrak{M} = \mathcal{C}_1$.

It is clear that we can solve the original problem contained in (12-22*a*) and (12-22*b*) for genus $p = 1$ and arbitrary n by the same process. It is almost obvious how to solve it for arbitrary genus p.

ARBITRARY GENUS

The original problem in (12-22*a*) and (12-22*b*) for arbitrary p and n can now be solved by induction. Let us suppose we can solve (12-22*a*) and (12-22*b*) for genus $p - 1$ (and arbitrary n). Under the rectangular representation we can regard the pth pair of crosscuts as the $(n + 1)$st component of $\partial\mathfrak{M}$. We can then repeat the procedure for solving the intermediate problem for ψ harmonic with period h_p across $\mathcal{Q}_p$ and $-\mathcal{Q}_p$. From this point we can solve the original problem (for h_p and h_p') analogously with the double-period problem just completed. (The reader should again review the facts to recall why $\mathfrak{M}$ must be noncompact!)

Therefore, all that really remains in the noncompact case is to show how to solve the second Dirichlet problem when singularities occur. (The compact region will be a special case.) We complete this in the next section.

EXERCISES

1 Show that the Dirichlet problem with periods can be solved for an annulus by using the same method of proof as for the intermediate problem (12-24*a*) to (12-24*c*) for the manifold $\mathcal{R}_1$. (See Fig. 12-5*a*.) Identify $\mathcal{R}_1'$ and $\mathcal{R}_1''$ by appending the shaded region to the rest of the annulus along CD and AB, respectively. Note that the curve $\mathcal{C}_1$ need not be present for the intermediate problem to be solvable (the region $\mathcal{R}_1$ can be a true annulus).

2 Show that the harmonic function $\lambda(z)$, given by

$$\begin{aligned} \lambda &= 0 && \text{for } |z| = 1,\ |z| = R > 1 \\ \lambda &= \pi && \text{for } \arg z = \pi - 0 \\ \lambda &= -\pi && \text{for } \arg z = -\pi + 0 \end{aligned}$$

(described in Fig. 12-5*b*), yields a harmonic differential $d\lambda$ on the annulus; i.e., the values of $d\lambda$ agree under analytic continuation across $\arg z = \pm\pi$. Would this still be true if we prescribed arbitrary constant values on the (circular and straight) boundary segments? Show why $d\lambda$ does not satisfy a second type Dirichlet problem.

3 Complete the verification that in the intermediate problem $\int d\psi = h_1$. Refer to the shaded regions of Fig. 12-4*b* and compare $v_n(z)$ in $A_2B_2C_2D_2$ with $u_n(z)$ in $A_1B_1C_1D_1$. [*Hint:* The proof can be visualized more easily by comparing the values of $v_n(z)$ and $u_n(z)$ in the (single) shaded region of Fig. 12-5*a*; we can then show that $u_n(z) - v_n(z) \to h$ in the common region by showing $|u_n(z) - v_n(z)| \leq |u_n(z) - u_{n-1}(z)|$, using the boundary values.] (Compare (12-18).)

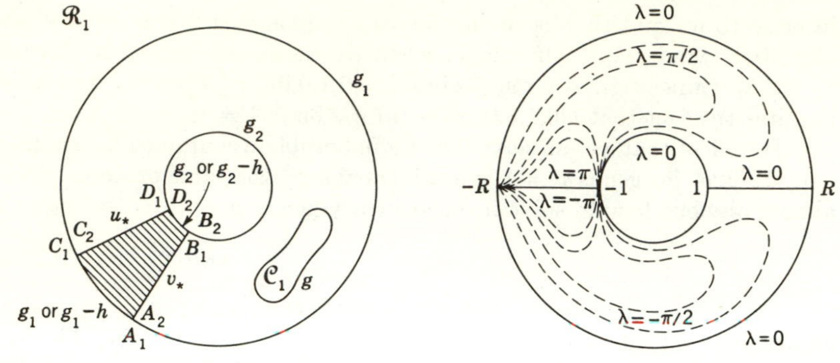

(*a*) (*b*)

Fig. 12-5 (*a*) For Exercise 1; (*b*) for Exercise 2. (Contours for constant λ are shown as dotted lines.)

12-3 Planting of Singularities

We shall now consider two problems which (together with the previous results) finally complete the second existence theorem:

(*a*) *Noncompact problem.* We consider a noncompact manifold $\mathfrak{M}$ ($n > 0$, $p \geq 0$), for which a meromorphic differential $d\psi$ is to be found for the following prescribed data:

(12-30) $$(\mathcal{C}_j, d\psi) = 0 \qquad j = 1, 2, \ldots, n$$

(12-31) $$(\mathcal{Q}_i, d\psi) = (\mathcal{Q}'_i, d\psi) = 0 \qquad i = 1, 2, \ldots, p$$

(12-32) $$\psi = 0 \qquad \text{on } \mathcal{C}_j \text{ (for } \psi \text{ defined on } \mathfrak{M}^p)$$

(12-33) $$d\psi = \operatorname{Re} f_t(\zeta)\, d\zeta + \text{reg diff} \qquad \text{near } \zeta = \zeta_t \ (t = 1, 2, \ldots, s)$$

subject to

(12-34) $$\operatorname{Im} \sum_{t=1}^{s} \operatorname{res} f_t(\zeta_t)\, d\zeta = 0$$

(*b*) *Compact problem.* We consider a compact manifold $\mathfrak{M}$ ($n = 0$, $p \geq 0$), for which a meromorphic differential $d\psi$ is to be found for the following prescribed data:

(12-35) $$(\mathcal{Q}_i, d\psi) = h_i \qquad (\mathcal{Q}'_i, d\psi) = h'_i$$

(12-36) $$d\psi = \operatorname{Re} f_t(\zeta)\, d\zeta + \text{reg diff} \qquad \text{near } \zeta = \zeta_t \ (t = 1, 2, \ldots, s)$$

subject to

(12-37) $$\sum_{t=1}^{s} \operatorname{res} f_t(\zeta_t)\, d\zeta = 0$$

Let us recall that the condition (12-32) means $\psi = 0$ except possibly for a finite number of boundary points near which ψ is bounded. [See

Exercise 2 of Sec. 12-2, where $\psi = 0$ except at discontinuities, but $(\mathcal{C}_j, d\psi) \neq 0$.] Thus (12-32) and (12-30) are not redundant conditions!

For the noncompact problem the case where no singularities occur has already been treated. For the compact problem, however, even the nonsingular case (symbolically $s = 0$) has not yet been treated in the last section. In either case it is seen by superposition that the two problems given above are all that remain to be solved in order that the second existence theorem be established.

NONCOMPACT PROBLEM

Let us first consider singularities occurring in a single triangle of $\mathfrak{M}$, a noncompact manifold ($n > 0$, $p \geq 0$). Indeed, let us assume that the singularities lie within a circle γ relative to the local coordinate of that triangle†, say z. The singularity is given by the (harmonic) meromorphic differential $d\psi_{\text{sing}}(z)$ which is restricted only by the condition that $\psi_{\text{sing}}(z)$ is *single-valued outside of* γ. Thus we can have $\psi_{\text{sing}}(z) = \log |z|$, Re (σ/z^m) (σ complex, $m = 1, 2, \ldots$), or ρ arg $[(z - a_1)/(z - a_2)]$, with ρ real and a_1, a_2 complex numbers *inside* γ. We let r be the radius of γ, and we let γ lie within a larger circle Γ of radius $R > r$ within which $\psi_{\text{sing}}(z)$ is also defined.

The alternating principle is applied to work between two manifolds, as usual. The manifolds are:

$\mathfrak{M}^\theta$: the manifold $\mathfrak{M}$ with the interior of γ removed
$\mathfrak{D}$: the disk formed by the interior of Γ

Thus, $\mathfrak{M}^\theta \cup \mathfrak{D} = \mathfrak{M}$ and $\mathfrak{M}^\theta \cap \mathfrak{D} = \mathfrak{A}$, the annulus lying between radii r (of γ) and R (of Γ). Clearly, $\mathfrak{M}^\theta$ is a manifold for which p is the same as for $\mathfrak{M}$, but for which n is increased by 1. (See Fig. 12-6.)

Our problem once again is to determine values u_* and v_* of a real function on Γ and γ, respectively, such that the solution ψ to the noncompact problem has values $u_*(Z_\Gamma)$ on Γ and $v_*(Z_\gamma)$ on γ. Then the iterative procedure becomes

(12-38*a*) $$\psi = u[v_*(Z_\gamma)](z) \qquad z \in \mathfrak{M}^\theta$$
(12-38*b*) $$\psi - \psi_{\text{sing}}(z) = v[u_*(Z_\Gamma) - \psi_{\text{sing}}(Z_\Gamma)](z) \qquad z \in \mathfrak{D}$$

where v_* and $(u_* - \psi_{\text{sing}})$ are defined on γ and Γ, respectively. The functional $v[\cdot \cdot \cdot](z)$ in (12-38*b*), of course, is the interior Poisson integral, but the functional $u[\cdot \cdot \cdot](z)$ in (12-38*a*) is the exterior one established in Sec. 12-2 for the ordinary Dirichlet problem on $\mathfrak{M}^\theta$.

† The singularities are assumed to lie interior to the given triangle Δ (not on the boundaries) by a convenient "redrawing" of boundaries (if necessary).

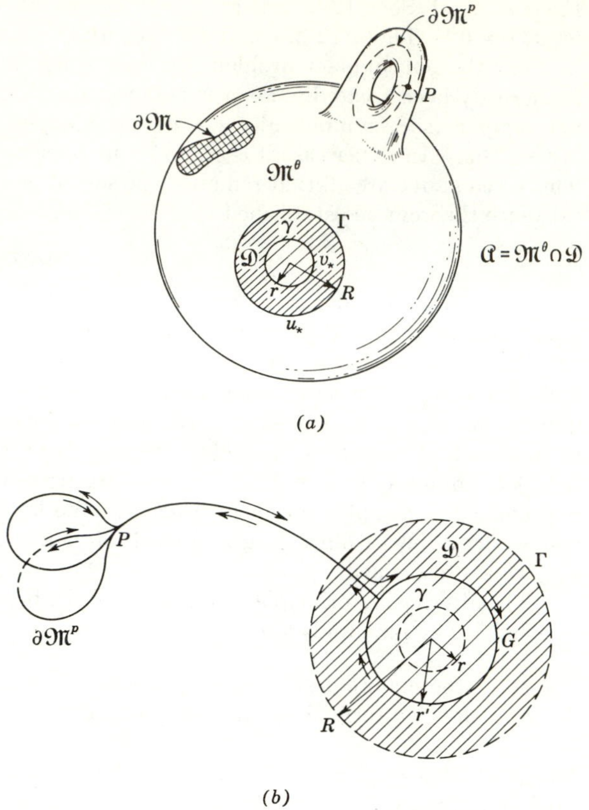

Fig. 12-6 (*a*) Noncompact case; (*b*) compact case (not $\partial\mathfrak{M}_G{}^*$). In either case, $u(z)$ is defined in $\mathfrak{M}^\theta$ (exterior to γ) and $v(z)$ is defined in $\mathfrak{D}$ (interior to Γ).

If we iterate (12-38*a*) and (12-38*b*), we obtain the following functional equation for $v_*(Z_\gamma)$:

(12-39) $$v_*(Z_\gamma) = \psi_{\text{sing}}(Z_\gamma) + v[u[v_*(Z_\gamma)](Z_\Gamma) - \psi_{\text{sing}}(Z_\Gamma)](Z_\gamma)$$

As usual, we solve this by iteration. We start with

(12-40*a*) $$u_0(z) \equiv 0 \qquad z \in \mathfrak{D}$$
(12-40*b*) $$v_1(z) = \psi_{\text{sing}}(z) + v[u_0(Z_\Gamma) - \psi_{\text{sing}}(Z_\Gamma)](z) \qquad z \in \mathfrak{D}$$

and we iterate for $n \geq 0$ by

(12-41*a*) $$u_{n+1}(z) = u[v_{n+1}(Z_\gamma)](z) \qquad z \in \mathfrak{M}^\theta$$
(12-41*b*) $$v_{n+1}(z) = \psi_{\text{sing}}(z) + v[u_n(Z_\Gamma) - \psi_{\text{sing}}(Z_\Gamma)](z) \qquad z \in \mathfrak{D}$$

The convergence question follows the usual line of approach, but we must be careful because the inequalities are quite distinctive in the present noncompact problem (as compared with the later compact problem):

(12-42*a*) $$u_{n+1}(z) - u_n(z) = u[v_{n+1}(Z_\gamma) - v_n(Z_\gamma)](z) \qquad z \in \mathfrak{M}^\theta$$

(12-42*b*) $$v_{n+1}(z) - v_n(z) = v[u_n(Z_\Gamma) - u_{n-1}(Z_\Gamma)](z) \qquad z \in \mathfrak{D}$$

In either equation it is clear that the maximum absolute value occurs on $z = Z_\Gamma$ and $z = Z_\gamma$, respectively. We estimate

(12-43*a*) $$\max |u_{n+1}(Z_\Gamma) - u_n(Z_\Gamma)| \leq q \max |v_{n+1}(Z_\gamma) - v_n(Z_\gamma)|$$

(12-43*b*) $$\max |v_{n+1}(Z_\gamma) - v_n(Z_\gamma)| \leq \max |u_n(Z_\Gamma) - u_{n-1}(Z_\gamma)|$$

for some positive $q < 1$. The presence of this convergence factor q is due to the fact that if a harmonic function $\phi = (u_{n+1} - u_n)$ vanishes on $\partial\mathfrak{M}$ (remember $\mathfrak{M}$ is *not* compact) and if $|\phi|$ has a maximum of 1 on γ, then by the maximum-minimum principle the maximum of $|\phi|$ on an interior curve Γ must be some $q < 1$ (for otherwise, ϕ = constant). There is no way, as yet, to "insert a factor q" in (12-43*b*) since Γ is the *only* boundary curve of $\mathfrak{D}$.

It is left as Exercise 1 to complete the details, but it is now clear that u_n and $v_n - \psi_{\text{sing}}$ converge to harmonic integrals, $u_\infty(z)$ and $v_\infty(z)$, which agree in the annulus $\mathfrak{A}$. Thus the particular single-valued singularity ψ_{sing} is planted. It is clear by superposition that any finite number of such singularities may be planted. The noncompact problem is now completed once we prove the next result.

LEMMA 12-2 *A set of a singularities* (12-33) *subject to* (12-34) *can be obtained by superposition of a finite number of singularities described by functions which may be multivalued in disks of arbitrarily small radius but are single-valued outside such disks.*

PROOF The only singularities which create any difficulty are those of the type $\psi_{\text{sing}} = \rho_s \arg (z - z_s)$ (ρ_s real). The other ones are single-valued. The condition (12-34), however, asserts that $\Sigma\rho_s = 0$. Thus we interpolate a new chain of S singularities connecting $z_1, \ldots, z_s$ so that any two neighboring singularities are arbitrarily close. We can renumber the indices of the singularities so that the singularities occurring at

(12-44*a*) $$z_1, \ldots, z_S$$

are the singularities at which $\rho_1 \ldots, \rho_S$ are the imaginary parts of the residues; e.g., the local singularity is

(12-44*b*) $$\rho_1 \arg (z - z_1), \ldots, \rho_S \arg (z - z_S)$$

but at the interpolated points $\rho_i = 0$. (Thus only s of the S values are nonzero.) Still, $\rho_1 + \rho_2 + \cdots + \rho_S = 0$. If we introduce new symbols,

$$\rho_1 = P_1, \rho_1 + \rho_2 = P_2, \ldots, \rho_1 + \cdots + \rho_S = P_S \ (=0)$$

then we can write our superposition of singularities as

$$(12\text{-}44c) \qquad P_1 \arg \frac{z - z_1}{z - z_2},\ P_2 \arg \frac{z - z_2}{z - z_3},\ \ldots,\ P_{s-1} \arg \frac{z - z_{s-1}}{z - z_s}$$

This is seen to produce the same effect as (12-44*b*). Q.E.D.

COMPACT PROBLEM

Here we consider a compact manifold $\mathfrak{M}$ on which we have an annulus $\mathfrak{A}$ bounded by Γ (of radius R) and γ (of radius r). We assume that $R > r$ are two sufficiently small quantities such that Γ lies entirely in one triangle Δ of $\mathfrak{M}$. For reasons to be explained later on we assume $r \leq R/5$. We next consider a singular function ψ_{sing} defined in γ but single-valued and regular in $\mathfrak{A}$, the annulus between γ and Γ. We require furthermore that

$$(12\text{-}45) \qquad \int_G \psi_{\text{sing}}(z)\, d\theta = 0 \qquad G\colon z = r' \exp i\theta$$

over any circle G of $\mathfrak{A}$ concentric with γ and Γ. This requirement is fulfilled (by symmetry) by any of the following functions:

$$(12\text{-}46a) \qquad \psi_{\text{sing}} = 0$$

$$(12\text{-}46b) \qquad \psi_{\text{sing}} = \operatorname{Re}\left(\frac{a}{z^m}\right) \qquad a \text{ complex, } m = 1, 2, \ldots$$

$$(12\text{-}46c) \qquad \psi_{\text{sing}} = \operatorname{Re}\left(a \log \frac{z - \beta}{z + \beta}\right) \qquad a, \beta \text{ complex, } |\beta| < r$$

We note that (12-46*a*) does not constitute a singularity (strictly speaking), but it will produce a harmonic integral (regular on $\mathfrak{M}$) with given periods. Here (12-46*b*) obviously satisfies (12-45); for example,

$$\psi_{\text{sing}} = \text{const } (r')^{-m} \exp(-m\theta - \text{const})$$

The case for (12-46*c*) is less evident but it follows from the symmetry of

$$\psi_{\text{sing}} = \log \left| \frac{z - \beta}{z + \beta} \right| \qquad \text{or} \qquad \arg \frac{z - \beta}{z + \beta}$$

under the change of z to $-z$. (See Exercise 6.)

We defined $\mathfrak{M}^\theta$ as the manifold $\mathfrak{M}$ with γ cut out (hence $n = 1$, p is unchanged); and we define $\mathfrak{D}$ as the interior of Γ. The iterative procedures are identical in form to the noncompact case as seen in (12-40*a*) and (12-40*b*), (12-41*a*) and (12-41*b*). (Details are referred to Exercise 2.) The basic difference emerges when we consider the *convergence* process, in particular the analogs of equations (12-43*a*) and (12-43*b*).

LEMMA 12-3 *Consider the iterative process formally based on* (12-40*a*) *and* (12-40*b*) *and* (12-41*a*) *and* (12-41*b*) *as applied to* $\mathfrak{M}$, *now a compact manifold. Then each* $u_n(z)$ *and* $v_n(z)$ *has the property that its mean value is zero over any circle* G *in the closure of the annulus* $\mathfrak{A}$.

PROOF Since $u_0(z) \equiv 0$ and $\psi_{\text{sing}}(z)$ has the mean value property, we need only show that "mean zero" property passes

(a) From $u_n - \psi_{\text{sing}}$ on Z_Γ to $v_{n+1} - \psi_{\text{sing}}$ on Z_γ

and

(b) From v_{n+1} on Z_γ to u_{n+1} on Z_Γ

according to equations (12-41*a*) and (12-41*b*). To see (*a*), we simply invoke the Gaussian mean value theorem (Poisson integral) in (12-41*b*).

We confine our attention to (*b*). Let $\mathfrak{M}_G{}^\theta$ be the region $\mathfrak{M}^\theta$ as modified so that G (and not γ) serves as $\partial\mathfrak{M}_G{}^\theta$. Then if we perform a *polygonal* canonical dissection of $\mathfrak{M}^p$ and connect the common point P (of the dissection) to G by a retraced curve, we have a boundary for $\mathfrak{M}_G^*$, the *simply-connected region* formed by a canonical dissection of $\mathfrak{M}_G{}^\theta$. (See Fig. 12-6*b*.) Thus referring to the conjugate u_{n+1}^*, we see

$$\int_{\partial\mathfrak{M}^*_G} du_{n+1}^* = 0$$

by the *single-valued* nature of the differential on a simply-connected region. If we discount the self-canceling nature of the retraced parts of $\partial\mathfrak{M}_G^*$, we see

$$0 = \int_G du_{n+1}^* = \int_G \frac{\partial u_{n+1}^*}{\partial s}\, ds = -\int_G \frac{\partial u_{n+1}}{\partial n}\, ds = -r' \int_0^{2\pi} \left(\frac{\partial u_{n+1}}{\partial r}\right)_{r=r'} d\theta$$

(12-47)

using† the Cauchy-Riemann equations ($\partial\phi^*/\partial s = -\partial\phi/\partial n$). But from (12-47),

(12-48) $$\frac{\partial}{\partial r'} \int_0^{2\pi} u_{n+1}(r' \exp i\theta)\, d\theta = 0$$

for every r' satisfying $r < r' \leq R$. Thus the mean value of $u_{n+1}(r' \exp i\theta)$ is constant; this constant will finally be seen to be zero by letting $r' \to r$ ($G \to \gamma$) where u_{n+1} has the boundary values of v_{n+1}, whose mean is assumed zero.

To complete the proof we must interchange a limit and an integral to show that

$$\lim_{r' \to r} \int_0^{2\pi} u_{n+1}(r' \exp i\theta) = \int_0^{2\pi} u_{n+1}(r \exp i\theta)\, d\theta$$

(where we know that at $r' = r$, or on γ, $u_{n+1} = v_{n+1}$). It is possible to prove that in the problem under consideration $u_{n+1} \to$ "boundary values" in a uniform manner. (See Exercise 3.) Q.E.D.

LEMMA 12-4 *In the ordinary Dirichlet problem for $u(z)$ on a circular disk of radius R, let the boundary data have mean value* 0 *and maximum*

† Note that n is used momentarily as the normal direction (as well as the index of iteration).

absolute value M. Then for $0 \le r < R$

$$|u(r \exp i\theta)| \le \left[\frac{2r(R+r)}{(R-r)^2}\right] M \tag{12-49}$$

For proof see Exercise 5 of Sec. 10-1.

It follows from Lemma 12-3, that if $r = R/5$ (as we assumed), then within γ, $u(z) \le \frac{3}{4}$. Thus we can rewrite analogs of the critical inequalities (12-43*a*) and (12-43*b*), namely

$$\max |u_{n+1}(Z_\Gamma) - u_n(Z_\Gamma)| \le \max |v_{n+1}(Z_\gamma) - v_n(Z_\gamma)| \tag{12-50a}$$
$$\max |v_{n+1}(Z_\gamma) - v_n(Z_\gamma)| \le \tfrac{3}{4} \max |u_n(Z_\Gamma) - u_{n-1}(Z_\Gamma)| \tag{12-50b}$$

Once again, we obtain a combined convergence factor of $\frac{3}{4}$ (<1) for both $u_n(z)$ and $v_n(z)$. [Note that (12-50*a*) is the trivial result of the maximum-minimum principle, for if a harmonic function ϕ $(= u_{n+1} - u_n)$ has a maximum of 1 on $\mathfrak{M}^\theta$, we *cannot* say it has a maximum of q (<1) on any interior curve Γ in $\mathfrak{M}^\theta$ when $\mathfrak{M}^\theta$ is *compact.*]

We have proven convergence. It remains to see that the instances (12-46*a*) to (12-46*c*) satisfy the requirements of all admissible singularities.

LEMMA 12-5 *A set of s singularities* (12-36) *subject to* (12-37) *can be obtained by superposition of a finite number of* (*real*) *singularities, of mean value zero type* [*satisfying* (12-45)], *within disks* Γ *of arbitrary small radius R.*

The proof [that types (12-46*b*) and (12-46*c*) suffice] is left as Exercise 4.

EXERCISES

1 Show from (12-41*a*) and (12-41*b*) that

$$u_{n+1} - v_{n+1} = 0 \qquad \text{on } \gamma$$
$$u_{n+1} - v_{n+1} = u_{n+1} - u_n \qquad \text{on } \Gamma$$

Hence, show that $\lim u_n = \lim v_n$ interior to $\mathfrak{A} \equiv \mathfrak{M}^\theta \cap \mathfrak{D}$.

2 Show that the mechanism of proof in the noncompact case carries over to the compact case (despite the fact that nonzero periods occur in the compact case). In particular, show how the periods h_i, h_i' in (12-35) enter into the functionals (12-38*a*) and (12-38*b*) or (12-41*a*) and (12-41*b*) when extended to the compact case, but such periods "subtract out" in the convergence.

3 Show by reviewing Poisson's integral and the alternating procedures, that if a harmonic function has continuous boundary values on an analytic boundary curve, then the boundary values are approached uniformly with respect to distance to the boundary curve. [First prove this for the Poisson integral by uniform convergence, and then show that the alternating procedure preserves this property, by referring to (12-19).]

4 Complete the proof of Lemma 12-5 by analogy with Lemma 12-2.

5 Show that under the restriction $R/r >$ constant $> 2 + \sqrt{5}$, it follows from (12-49) that (12-43*a*) and (12-43*b*) have a convergence factor $q <$ constant < 1.

6 Show that (12-46*c*) has the property (12-45) by writing out the integrals and establishing symmetry.

PART V
ALGEBRAIC APPLICATIONS

RESURGENCE OF FINITE STRUCTURES

We have amply labored the observation that the use of real variables was a necessary interlude in the development of complex analysis, in particular in the development of existence theorems for algebraic functions.

Equally remarkable is the fact that in the process of developing existence theorems we see the importance of *finite* structures and *finite* processes: A manifold is determined topologically by a finite number of constants (p and n); indeed, the manifold has a finite triangulation. The functions defined on a manifold are algebraic functions defined by polynomial equations (consisting of a finite number of terms with complex† coefficients).

The aspect of finiteness contrasts sharply with the tradition of analysis based on rational mechanics. There, unable to solve all the problems in "closed form," analysts were only too glad to think of a function so broadly that ultimately a function was viewed as a "correspondence between dependent and independent variable." Now we are ready to approach a *finitistic* world with algebraic functions "based on polynomials only," and to find this world still replete with unresolved questions.

† Indeed the highest form of the art is achieved when the coefficients are taken from a *finite* number field (instead of the complex numbers).

CHAPTER 13
RIEMANN'S EXISTENCE THEOREM

We are ultimately concerned with proving the existence of algebraic functions on a compact Riemann manifold $\mathfrak{M}$; indeed, we are concerned with deriving algebraic functions which determine a preassigned $\mathfrak{M}$. We shall do this by specifying singularities and asking for *functions* "with prescribed singularities," but in principle we are asking for the solution to a largely unsolved problem.

What we shall actually do is ask not for functions but for the larger class of *integrals* F, single-valued not on $\mathfrak{M}$ but on $\mathfrak{M}^p$, the dissected polygonal model. (This corresponds to the differentials dF on $\mathfrak{M}$.) We shall contrive to have

$$(\mathcal{Q}_i, dF) = (\mathcal{Q}'_i, dF) = 0 \qquad i = 1, 2, \ldots, p$$

Then such an integral F becomes a (single-valued) meromorphic function.

The first step, however, is to reenter the realm of complex variable theory by considering *complex* (rather than harmonic) integrals and differentials.

13-1 Normal Integrals

We start with *real normal harmonic integrals (or differentials) of the first kind* on a complex manifold of genus p. For convenience we relabel the $2p$ homology basis (of 1-cycles) as $\mathcal{H}_1, \ldots, \mathcal{H}_{2p}$

$$\begin{aligned} \mathcal{H}_1 &= \mathcal{Q}_1, \ldots, \mathcal{H}_p = \mathcal{Q}_p \\ \mathcal{H}_{p+1} &= \mathcal{Q}'_1, \ldots, \mathcal{H}_{2p} = \mathcal{Q}'_p \end{aligned} \tag{13-1}$$

Then we can define $2p$ real *normal* harmonic differentials $d\psi_j$ (or integrals ψ_j)($j = 1, 2, \ldots, 2p$) by the relations

$$(\mathcal{H}_i, d\psi_j) = \delta_{ij} \qquad 1 \leq i, j \leq 2p \tag{13-2}$$

for δ_{ij}, the Kronecker delta function. The first existence theorem assures us that such $d\psi_j$ are uniquely defined.

LEMMA 13-1 *The most general real harmonic differential of the first kind is a unique linear combination of the $2p$ real harmonic normal differentials.*

PROOF By the first existence theorem (Sec. 8-4) a real harmonic differential of the first kind $d\psi$ is uniquely determined by the $2p$ values $(\mathcal{H}_i,d\psi)$. Hence if we set $d\psi = \sum_{j=1}^{2p} (\mathcal{H}_j,d\psi)\, d\psi_j$, then both sides of the equation have the same set of $2p$ values of $(\mathcal{H}_i,d\psi)$. The uniqueness follows from Sec. 8-5. Q.E.D.

We are now in a position to define *p normal (complex abelian) integrals of the first kind*, denoted by

$$F_1^{(1)}, \ldots, F_p^{(1)}$$

[or the corresponding differentials $dF_i^{(1)}$ ($i = 1, 2, \ldots, p$)] by the relations

$$(\mathcal{Q}_i,dF_j^{(1)}) = 2\pi i\delta_{ij} \qquad 1 \le i, j \le p \tag{13-3}$$

(The justification for the $2\pi i$ will appear presently.) If we write

$$dF_j^{(1)} = d\phi_j + i\, d\phi_j^*$$

it is clear that we are looking for p real harmonic differentials $d\phi_j$ with the property that

$$(\mathcal{Q}_i,d\phi_j) = 0 \qquad (\mathcal{Q}_i,d\phi_j^*) = 2\pi\delta_{ij} \qquad 1 \le i, j \le p \tag{13-4}$$

In establishing the existence of these differentials ($dF_j^{(1)}$ or $d\phi$, etc.), it is no more difficult to prove that for any real vector

$$P = (\rho_1, \rho_2, \ldots, \rho_{2p})$$

we can find a unique $d\phi$ (of the first kind) such that

$$(\mathcal{Q}_i,d\phi) = \rho_i \qquad (\mathcal{Q}_i,d\phi^*) = \rho_{i+p} \qquad i = 1, 2, \ldots, p \tag{13-5}$$

Since any such $d\phi$ has the form

$$d\phi = \lambda_1\, d\psi_1 + \cdots + \lambda_{2p}\, d\psi_{2p} = \sum_{j=1}^{2p} \lambda_j\, d\psi_j \tag{13-6}$$

we are confronted with a choice of $2p$ values of λ_j such that the $2p$ values of ρ_i are achieved in (13-5). Thus every real vector

$$\Lambda = (\lambda_1, \ldots, \lambda_{2p})$$

is transformed by (13-6) and (13-5) into a real vector P by a linear transformation (whose exact form is unimportant). This transformation can be written as $2p$ equations:

$$\sum_{j=1}^{2p} \lambda_j(\mathcal{Q}_i, d\psi_j) = \rho_i \qquad \sum_{j=1}^{2p} \lambda_j(\mathcal{Q}_i, d\psi_j^*) = \rho_{i+p} \qquad 1 \leq i \leq p \tag{13-7}$$

The question of whether (13-7) can be solved for Λ (for a given P) is decided on the basis of the abstract version of Cramer's rule.† Specifically, suppose system (13-7) [or (13-5)] has only the *trivial* solution $\Lambda = 0$ (or $d\phi = 0$) when it is made homogeneous by setting $P = 0$. Then Cramer's rule asserts that (13-7) [or (13-5)] has a *unique* solution in general (whether or not it is made homogeneous).

Let us therefore verify that if $P = 0$, any $d\phi$ satisfying (13-5) vanishes identically. We refer to the Riemann period relation (Theorem 8-5)

$$\int_{\partial\mathfrak{M}^p} G\,dF = \sum_{i=1}^{p} \begin{vmatrix} (\mathcal{Q}_i, dG) & (\mathcal{Q}_i, dF) \\ (\mathcal{Q}'_i, dG) & (\mathcal{Q}'_i, dF) \end{vmatrix} \tag{13-8}$$

where F and G are (real or complex) integrals defined on $\mathfrak{M}^p$. If we take $\phi\, d\phi^*$ for $G\, dF$, we see from (13-8) that $\int \phi\, d\phi^* = 0$ if $P = 0$. Thus by the reasoning of Theorem 8-5 again,

$$\iint \left[\left(\frac{\partial \phi}{\partial x}\right)^2 + \left(\frac{\partial \phi}{\partial y}\right)^2 \right] dx\, dy = 0 \qquad \text{and} \qquad d\phi \equiv 0$$

Thus condition (13-3) determines $dF_j^{(1)}$ uniquely.

By the same reasoning, the p *complex* values

$$(\mathcal{Q}_j, dF^{(1)}) \qquad j = 1, 2, \ldots, p \tag{13-9}$$

determine any integral $F^{(1)}$ of the first kind uniquely (i.e., to an additive constant) on $\mathfrak{M}^p$ (or these values determine its differential $dF^{(1)}$ uniquely).

LEMMA 13-2 *An arbitrary (complex abelian) differential of the first kind on $\mathfrak{M}$ is some unique linear combination (with complex coefficients) of the p normal differentials of the first kind.*

In the case of the period parallelogram in the $z = x + iy$ plane, it is not hard to see that the real integrals of the first kind are linear combinations of the harmonic functions x and y, whereas the complex integrals of the first kind are complex multiples of z. (See Exercise 1 for the normal forms.)

In the case of Riemann manifolds of algebraic functions generated by the polynomial $\Phi(w,z) = 0$, the abelian integrals are ready-made combinations of the generating polynomials‡. The advantage to Riemann's

† The concrete version makes reference to the vanishing of the determinant as the criterion of solvability of the homogeneous system in nonzero Λ.

‡ Some trivial illustrations are given in the exercises but the matter is one for refined techniques in the theory of algebraic curves. (See Bibliography.)

approach is that abelian integrals are constructed from the geometry of $\mathfrak{M}$ (by applied mathematics, i.e., potential theory) without any knowledge of the polynomial $\Phi(z,w) = 0$. (Indeed, the z and w are constructed in Sec. 13-2 from such abelian integrals.)

Generally, we shall call abelian *integrals* G of the second kind or third kind on $\mathfrak{M}^p$ *normalized* if they are single-valued on $\mathfrak{M}^p$ (except, of course, for those of the third kind with multivaluedness due to logarithmic singularities) and if

$$(\mathfrak{Q}_j,dG) = 0 \qquad j = 1, 2, \ldots, p \tag{13-10}$$

Recall that in no case (first, second, or third kind) have we tried to prescribe the value of $(\mathfrak{Q}'_j,dG)$ in advance, when G is a *complex* abelian integral. We define normal *differentials* dG in the same fashion.

Lemma 13-3 *Every differential of the second or third kind differs from some normal differential by a differential of the first kind.*

For proof, replace dG by $dG - \sum_{t=1}^{p} (\mathfrak{Q}_t,dG)\, dF_t^{(1)}/2\pi i = dG_0$. Then a direct verification yields $(\mathfrak{Q}_s,dG_0) = 0$ for each s with $s = 1, 2, \ldots, p$.

We now can use the first existence theorem to define integrals of the second and third kind corresponding to the singularities consisting of a pole of order n or a close source-sink combination. Thus a normal integral can be constructed with the singularities shown:

$$F_{P^n}{}^{(2)}(\zeta) = \frac{1}{(\zeta - \zeta_P)^n} + \text{reg funct of } \zeta \tag{13-11a}$$

$$F_{PQ}{}^{(3)}(\zeta) = \log \frac{\zeta - \zeta_P}{\zeta - \zeta_Q} + \text{reg funct of } \zeta \tag{13-11b}$$

and [as in (13-10)] $(\mathfrak{Q}_j,F_{P^n}{}^{(2)}) = 0$ or $(\mathfrak{Q}_j,F_{PQ}{}^{(3)}) = 0$, respectively, for $j = 1, 2, \ldots, p$.

For cases of genus 0, trivially, the functions $F^{(1)}$ are constants, whereas the functions (13-11a) and (13-11b) are necessarily single-valued, as $\mathfrak{Q}_j$ and $\mathfrak{Q}'_j$ are nonexistent. [We allow, of course, for a cut joining P and Q for (13-11b).]

The definition (13-11b) seems to imply that P and Q lie in one triangle (Δ parametrized by the variable ζ). Actually, if P and Q lie in different triangles, they can be imbedded in a chain of points of $\mathfrak{M}^p$,

$$P = P_0, P_1, \ldots, P_s = Q \tag{13-12a}$$

such that P_i and P_{i+1} are either in the same triangle or are sufficiently close that they lie in the same neighborhood in the system used to define the Riemann manifold. Then we can define

$$F_{PQ}{}^{(3)}(\zeta) = \sum_{i=0}^{s-1} F^{(3)}_{P_iP_{i+1}}(\zeta) \tag{13-12b}$$

It can be verified (see Exercise 5) that the definition (13-12*b*) is independent of choice of the chain (13-12*a*) once the manifold is dissected.

LEMMA 13-4 (*Riemann*, 1857) *The derivatives and integrals of a normal integral of the first kind are expressible in terms of periods of normal integrals of the second and third kind as follows:*

$$\frac{2\pi i}{(n-1)!}\frac{d^nF^{(1)}}{d\zeta_P{}^n} = -\sum_{t=1}^{p}(\mathcal{Q}_t,dF^{(1)})(\mathcal{Q}'_t,dF_{P^n}{}^{(2)}) \tag{13-13a}$$

and keeping paths within a suitable canonical dissection of $\mathfrak{M}$,

$$-2\pi i\int_P^Q dF^{(1)} = -\sum_{t=1}^{p}(\mathcal{Q}_t,dF^{(1)})(\mathcal{Q}'_t,dF_{PQ}{}^{(3)}) \tag{13-13b}$$

The proofs in both cases follow from the Riemann period relations

$$\int_{\partial\mathfrak{M}^p} F\,dG = \sum_{t=1}^{P}\begin{vmatrix}(\mathcal{Q}_t,dF) & (\mathcal{Q}_t,dG)\\ (\mathcal{Q}'_t,dF) & (\mathcal{Q}'_t,dG)\end{vmatrix}$$

If we consider $\int F_{P^n}{}^{(2)}\,dF^{(1)}$, the right-hand side reduces to that of (13-13*a*) since $(\mathcal{Q}_t,dF_{P^n}{}^{(2)}) = 0$ by normalization. The left-hand side leads to (13-13*a*) if we use the Taylor series of $F^{(1)}$ and the singularity of $F_{P^n}{}^{(2)}$ shown in (13-11*a*).

If we consider $\int F^{(1)}\,dF_{PQ}{}^{(3)}$, on the other hand, the right-hand side of (13-13*b*) will check as before, but the left-hand side will check only after the following intermediate step which consists again of the use of the singularities shown in (13-11*b*):

$$\int_{\partial\mathfrak{M}^p} F^{(1)}\,dF_{PQ}{}^{(3)} = 2\pi i[F^{(1)}(P) - F^{(1)}(Q)] = -2\pi i\int_P^Q dF^{(1)}$$

EXERCISES

1 Take $\mathfrak{M}$ to be the period parallelogram in the z plane with side $2\omega_1$ (so that $\mathcal{Q}_1$ is the vector from 0 to $2\omega_1$) and side $2\omega_2$ (so that $\mathcal{Q}'_1$ is the vector from 0 to $2\omega_2$). Start with x and y as independent harmonic integrals of the first kind, and form ψ_1 and ψ_2, the normal (real) integrals of the first kind, by setting $\psi_1 = \alpha_1 x + \beta_1 y$, $\psi_2 = \alpha_2 x + \beta_2 y$. [It will be convenient to use the constants $2\omega_1 = 2\omega_{11} + i2\omega_{12}$, $2\omega_2 = 2\omega_{21} + i2\omega_{22}$ with ω_{ij} real and $\omega_{11}\omega_{22} - \omega_{12}\omega_{21} \neq 0$. Note $(\mathcal{K}_1,d\psi_1) = 2\alpha_1\omega_{11} + 2\beta_1\omega_{12} = 1$, etc.] Verify that the normal (complex) integral of the first kind is merely $2\pi iz/2\omega_1$.

2 Verify that the Riemann manifold for

$$w^2 = (z-a_1)\cdots(z-a_k) \qquad a_i \text{ distinct}$$

with $k = 2p+1$ or $k = 2p+2$, has as abelian integrals of the first kind the expressions $\int dz\cdot z^t/w$ where $t = 0, 1, 2, \ldots, p-1$.

3 Find those exponents t for which $\int dz\,(z-a_1)^t/w$ constitutes an abelian integral of the second kind.

4 Consider the Riemann manifold for

$$w^6 = z(z - \alpha)(z - \beta)^4 \qquad \alpha, \beta, 0 \text{ distinct}$$

(*a*) Find the local uniformizing parameters over $z = 0, \alpha, \beta, \infty$ (for example, $z = \zeta^6$ at $z = 0$). (*b*) How many points of $\mathfrak{M}$ lie above $z = 0, \alpha, \beta, \infty$ (for example, six points lie above $z = \infty$). (*c*) Find the total order of ramification b and the genus p by using Riemann's formula. (*d*) Show that the differential $(z - \beta)^s\, dz/w^t$ is of the first kind only when $t = 5, s = 3$ or $t = 4, s = 2$. (*e*) Verify that $(z - \beta)/w$ is an abelian integral of the second kind (in fact a meromorphic function) with two poles and two zeros, by using the local uniformizing parameters over $0, \alpha, \beta, \infty$.

5 Verify that the definition (13-12*b*) is independent of the chain of points (13-12*a*), by showing any two such definitions lead to integrals of the third kind which differ by an integral of the first kind in which $(\mathfrak{Q}_j, F^{(1)}) = 0$. Why is it necessary to speak of two chains in the same *dissected* manifold $\mathfrak{M}^p$?

13-2 Construction of the Function Field

We are considering a compact Riemann manifold $\mathfrak{M}$ over the z sphere conceived as an aggregate of triangles $\Delta_1, \ldots, \Delta_F$ in the usual fashion of Sec. 5-3. We wish to show there is a biunique (conformal) mapping of $\mathfrak{M}$ onto a Riemann manifold of some suitable algebraic curve, $\Phi(w_0, z_0) = 0$ over the z_0 sphere.

What we shall do is to define the meromorphic functions w_0, z_0 on $\mathfrak{M}$ in such a fashion that, say, the Riemann surface of w_0 over the z_0 sphere is equivalent to $\mathfrak{M}$.

We therefore consider general *meromorphic* functions F on $\mathfrak{M}$. A meromorphic function F is (trivially) a *normal* abelian integral of the second kind (indeed one whose periods are all zero).

We now consider the most general such F by prescribing what we shall call a *maximal* set of m singularities: Consider all functions F whose singularities are a subset (possibly none or all) of the following points of $\mathfrak{M}$:

$$P_1, \ldots, P_m \tag{13-14}$$

(We describe a multiple pole for F by listing the same singularity as many times as the multiplicity.) Thus a function F with all the poles in (13-14) will necessarily be of order m. We shall, however, ask only that the poles of F at any singularity be of *no higher order*† than the poles listed in (13-14) for that singularity.

For reasons which will become clear later on, the set of meromorphic functions with maximal set of singularities (13-14) will be called the *functional space with divisor*

$$D = P_1 \cdots P_m \tag{13-15}$$

(Thus we think of the divisor as the "maximum denominator.")

† Usually "degree" applies to a divisor, and "order" applies to a function, but the usage is not rigid.

This function space is a *linear* space over the complex numbers, since the set of functions with a *maximal* set of singularities is closed under addition and multiplication by complex constants. The linear space will be called $L(D)$. We are concerned ultimately with its complex *dimension* written $\dim L(D)$.

We are more immediately concerned, however, with proving the existence of nonconstant elements F in $L(D)$, since such an F is meromorphic on $\mathfrak{M}$. [We note that a zero or nonzero constant, say $F = 1$, is a member of $L(D)$, thus $\dim L(D) \geq 1$.] The *degree* of D is defined as m.

First of all let us note that every element of $L(D)$ is an abelian integral. Because of that it must have the form

(13-16) $$G = \sum_{t=1}^{m}{}^{*} \alpha_t F_{P_t}{}^{(2)}(\zeta) + \alpha_0 + \sum_{j=1}^{p} \beta_j F_j{}^{(1)}$$

Here the asterisk on the sum means that allowance is made for multiple poles [e.g., if $P_1 = P_2 = P_3 \neq P_4, \ldots$, then the first three terms are $\alpha_1 F_{P_1}{}^{(2)} + \alpha_2 F_{P_1{}^2}{}^{(2)} + \alpha_3 F_{P_1{}^3}{}^{(2)}$].

Clearly, the integrals G (of the second kind) which are really *functions* are those for which all $\beta_j = 0$ and

$$(\mathfrak{Q}_j', dG) = 0 \qquad \text{for } j = 1, 2, \ldots, p$$

[Note $(\mathfrak{Q}_j, dG) = 0$ automatically from $\beta_j = 0$ by use of normal integrals.] The conditions lead merely to

(13-17) $$\sum_{t=1}^{m}{}^{*} \alpha_t (\mathfrak{Q}_j', dF_{P_t}{}^{(2)}) = 0 \qquad j = 1, 2, \ldots, p$$

Here we note that if we have more unknowns in (13-17) than equations or if

(13-18) $$m \geq p + 1$$

then there will be a nonzero set of $\alpha_1, \ldots, \alpha_m$ for which (13-17) is solvable. Hence the set $L(D)$ will have a nonconstant function in it. In more elegant language, we have the following result:

THEOREM 13-1 (*Riemann*, 1857) *If the degree of the divisor D exceeds the genus of $\mathfrak{M}$, then the dimension of $L(D)$ exceeds* 1 (*or nonconstant meromorphic functions exist with D as the maximal set of poles*).

We recall that any F in $L(D)$, even if it is nonconstant, need not use all of the poles; i.e., it need not be of actual order m [since many of the α_t in (13-16) might vanish]. Let us consider such an F (and use m again to denote its order for convenience). Then $F(P)$ maps $\mathfrak{M}$ onto a Riemann manifold (indeed, a Riemann surface) over the F sphere which covers the F sphere m times (making due allowance for branch points).

Now we visualize $\mathfrak{M}$ as an m-sheeted Riemann surface over the F sphere. Thus every value of F (except for projections of branch points) corresponds to m points on $\mathfrak{M}$; for example, the points $F_1, \ldots, F_m$ on $\mathfrak{M}$ project into F.

LEMMA 13-5 *Let G be a meromorphic function on $\mathfrak{M}$ such that the m values $F_1, \ldots, F_m$ produce m distinct functions $G(F_1), \ldots, G(F_m)$ [that is, $G(F_i) \neq G(F_j)$ when $i \neq j$ except for a finite number of exceptions]. Then G is an algebraic function of F, and the Riemann surface of G over the F sphere is equivalent to $\mathfrak{M}$.*

First note that the precaution of using distinct functions $G(F_i)$ is necessary, since otherwise we could take $G = F$ as the second function only to find the Riemann surface of G over the F sphere is the F sphere itself (a trivial m-to-one projection of $\mathfrak{M}$).

We cannot employ the mechanism of Theorem 6-2 of Sec. 6-1 directly since $\mathfrak{M}$ is not yet known to be generated by an algebraic equation. Yet the method of forming m symmetric functions of $G(F_i)$ can be employed to form rational functions which are coefficients of the equation defining G (see Exercise 1). This equation is irreducible (Exercise 2), but this is not important. What is important is that m distinct values of G are definable for any F by analytic continuation (with a finite number of exceptional F), so that $G(F)$ defines the original Riemann manifold $\mathfrak{M}$.

LEMMA 13-6 *We can weaken the hypotheses of Lemma 13-5 to the local assumption that for some neighborhood $\mathfrak{N}$ on the F sphere, if $F_1, \ldots, F_m$ project into a common $F \in \mathfrak{N}$, then $G(F_i) \neq G(F_j)$ when $i \neq j$.*

PROOF For then if $F \in \mathfrak{N}$, we define n different values $G(F_i)$ if F_i projects into F. We triangulate $\mathfrak{M}$ so it consists of m sheets over the F sphere with branch points belonging to the vertices of the triangulation. Then $G(F)$ must be a different analytic function on each sheet, or else the values $G(F_i)$ will coincide by analytic continuation back to F_i lying over $\mathfrak{N}$. Q.E.D.

COMPLETION OF RIEMANN'S EXISTENCE THEOREM

We now set about using Lemma 13-6 by letting $G \in L(D)$ as described, with m ($\geq p + 1$) distinct poles $P_1, \ldots, P_m$, and $F \in L(D')$ where $D' = P_0{}^{p+1}$ (P_0 again distinct from the other poles). Then near $F = \infty$ (which occurs only at P_0) the values of $G_i(F)$ are the *distinct* values $P_1, \ldots, P_m$. Thus we invoke Lemmas 13-6 and 13-5 to complete Riemann's existence theorem†.

† By the remarks in Sec. 6-2, the function field of $\Phi(F,G) = 0$ *within automorphisms* lies in direct one-to-one correspondence with the manifold $\mathfrak{M}$ *within conformal equivalence*, but this is a matter of primary interest on the level of algebraic function theory.

GENUS ZERO AND UNITY

A rather satisfactory characterization can be made for compact Riemann manifolds of genus 0 and 1.

THEOREM 13-2 *A manifold* $\mathfrak{M}$ *is of genus* 0 *if and only if all* (*or any*) *of the following three properties hold:*

(*a*) $\mathfrak{M}$ *has a meromorphic function of order* 1.

(*b*) $\mathfrak{M}$ *is conformally equivalent to the sphere.*

(*c*) *There exists a meromorphic function* F *on* $\mathfrak{M}$ *with the property that every meromorphic function on* $\mathfrak{M}$ *is a rational function of* F.

PROOF We show that the following chain of implications hold:

$$\text{genus } 0 \to (a) \to (b) \to (c) \to (a) \to (b) \to \text{genus } 0$$

First of all, genus $0 \to (a)$ since the absence of a period structure enables us to obtain a rational function with any exact (preassigned) set of poles. In particular, by prescribing one pole we see a meromorphic function F exists of degree 1.

Next, $(a) \to (b)$, since $F(\mathfrak{M})$ maps $\mathfrak{M}$ onto the F sphere with every value taken once (according to Theorem 7-2, Corollary 1, in Sec. 7-1).

Also $(b) \to (c)$ because any function G meromorphic on $\mathfrak{M}$ is then meromorphic on $F(\mathfrak{M})$; thus G is defined single-valuedly, hence rationally in F.

To show $(c) \to (a)$, let us note that if F were not of order 1, then for two distinct points $P_1 \neq P_2$, $F(P_1) = F(P_2)$. We nevertheless can find a G whose poles are only at P_1 [for example, $G \in L(P_1{}^m)$ for m large enough]. Thus G is not a rational (single-valued) function of F [since $G(P_1) = \infty \neq G(P_2)$].

The final step, that $(a) \to (b) \to$ genus 0, is a trivial consequence of topological invariance of genus. (The sphere has genus 0.) Q.E.D.

THEOREM 13-3 *A manifold* $\mathfrak{M}$ *is of genus* 1 *if and only if both* (*or either*) *of the following properties are valid:*

(*a*) $\mathfrak{M}$ *is conformally equivalent to some period parallelogram.*

(*b*) $\mathfrak{M}$ *is conformally equivalent to some so-called "elliptic Riemann manifold,"*

$$w^2 = 4(z - e_1)(z - e_2)(z - e_3) \tag{13-19}$$

for suitably chosen distinct e_1, e_2, e_3.

PROOF It is clear that $(a) \to$ genus 1 and $(b) \to$ genus 1 by the topological definition of genus.

Next, let us show genus $1 \rightarrow (b)$. If $\mathfrak{M}$ is of genus 1, it possesses some meromorphic function $z(P)$ which is at most of order 2, by Theorem 13-1. Actually, $z(P)$ is really of order 2 [lest $z(\mathfrak{M})$ be the sphere, contrary to Theorem 13-2*b*]. There must also exist some other meromorphic function $w(P)$ which is *not* a rational function of z (see Theorem 13-2*c*). Since each z (with finite exceptions) corresponds to two points z_1, z_2 on $\mathfrak{M}$, the meromorphic functions $w_1 = w(z_1)$ and $w_2 = w(z_2)$ are root functions in the following sense: $w_1 + w_2 = R_1(z)$ and $w_1w_2 = R_2(z)$ are rational functions of z. Hence

$$w^2 - R_1(z)w + R_2(z) = 0 \tag{13-20}$$

If we solve (13-20), we see $w - R_1(z)/2$ is a radical in z, and by elementary transformations we can define a meromorphic function w on $\mathfrak{M}$ such that w^2 is a polynomial. Since the degree of this polynomial determines the genus [see (7-11)], we are led to (13-19) for suitable roots e_1, e_2, e_3.

The final step is to show $(b) \rightarrow (a)$, but this is a consequence of Hypotheses† 1 and 2 of Secs. 4-2 and 4-5. Q.E.D.

EXERCISES

1 Show the symmetric functions $g_1 = \Sigma G(F_i)$, $g_2 = \Sigma G(F_i)G(F_j)$, $(i \neq j)$, etc., are rational in F, and show G satisfies an algebraic equation, $G^m - g_1G^{m-1} + g_2G^{m-2} - \cdots = 0$.

2 Show the above equation is irreducible in G on the basis of the fact that the functions $G(F_i)$ are distinct (see Theorem 6-2 again).

3 Show that under the hypotheses of Lemma 13-6 the functions $G(F_i)$ and $G(F_j)$ can coincide only at a finite number of points.

† Note that for the last step we had to use the result on elliptic modular functions: Lacking a result of such depth, the best we could do is $(a) \rightarrow (b)$ which is essentially Lemma 4-3 of Sec. 4-5.

CHAPTER 14
ADVANCED RESULTS

In this section we consider two problems which arise directly from Riemann's existence theorem. We have shown all compact Riemann manifolds correspond exactly to the algebraic Riemann surfaces, but we still are led to ask how many different Riemann manifolds there are within conformal equivalence and how to describe with minimal effort the meromorphic functions on any manifold.

Neither of these questions can be answered with the same finality, as in the case of genus 0 or 1, when we study manifolds of genus $p \geq 2$. The important goal for us is not actually to answer these questions but to show how the best efforts at obtaining an answer have led to a new viewpoint embodying topology and algebra as a direct consequence.

14-1 Riemann-Roch Theorem

Given a divisor $D = P_1 \cdots P_m$ of degree m we consider the dimension of $L(D)$, the linear space of functions on $\mathfrak{M}$ (a compact manifold), whose poles are a subset of the points listed in the divisor (with appropriate multiplicity). We have previously considered the system of p equations in m unknowns

$$(14\text{-}1) \qquad \sum_{t=1}^{m}{}^{*} (\mathcal{Q}_j', dF_{P_t}{}^{(2)})\alpha_t = 0 \qquad j = 1, 2, \ldots, p$$

where the asterisk indicates a modification for multiple poles in D as previously described [see (13-16)].

We saw that if the complex vectors $(\alpha_1, \ldots, \alpha_p)$ satisfying (14-1) are a linear space of dimension q, then the dimension of $L(D)$ is $1 + q$ [1 takes into account *constant* functions in $L(D)$]. If we view (14-1) as p equations in m unknowns, then we can always write $q = m - p + i$ where $p - i$ $(\leq p)$ is the rank† of the $m \times p$ matrix of coefficients in (14-1).

† The *rank* is definable as the number of independent row (or column) vectors in the matrix.

Thus

(14-2) $$p - i = \text{rank}\,\|(\mathcal{Q}_j', dF_{P_t}{}^{(2)})\| \qquad 1 \le t \le m,\ 1 \le j \le p$$

Here i denotes the number of redundancies among the p restrictions on $(a_1, \ldots, a_m)$ contained in (14-1).

Thus we rewrite Theorem 13-1 of Sec. 13-2 as

(14-3) $$\dim L(D) = 1 + \deg D - p + i$$

Indeed, the theorem cited states only that $i \ge 0$ and nothing else!

The quantity $i(D)$ is called the "index of specialty" of D since it will be shown to vanish in the general case. Riemann's original investigations (1857) were carried to the extent of showing that $i = 0$ when $m \ge 2p - 1$ (see Lemma 14-2).

A vast improvement came in 1865 when Roch found an interpretation for $i(D)$ as the dimension of another linear space $L^*(D)$ (always over the complex constants), which we proceed to describe. From this, a Riemann-Roch theorem and, indeed, a Riemann-Roch "mystique" developed in which, by a process of reinterpretation, a principally analytic result (on functions with given poles) became highly algebraic.

First of all, we note that an abelian differential dF of first, second, or third kind on a compact manifold $\mathfrak{M}$ has a zero and pole structure (like a function) quite independent of uniformizing parameters. For instance, the *first* kind of differential, $dF^{(1)}$ (by definition) has no poles, but it can be said to have a *zero of order s at P*. If we follow the custom of using ζ_P as the local parameter so P corresponds to $\zeta_P = 0$, the zero of order s means

(14-4a) $$dF^{(1)} = \zeta_p{}^s\, d\zeta_p\, \phi(\zeta_p)$$

where $\phi(\zeta_p)$ is analytic and nonzero at $\zeta_p = 0$. Any other local parameter is given by $\zeta_p^* = c_1\zeta_p + c_2\zeta_p{}^2 + \cdots$ $(c_1 \ne 0)$ so that s is well defined by (14-4a). Alternately, we can define s by the conditions

(14-4b) $$\frac{d^{s+1}F^{(1)}}{d\zeta_p^{s+1}} \ne 0 \qquad \frac{d^sF^{(1)}}{d\zeta_p{}^s} = \cdots = \frac{dF^{(1)}}{d\zeta_p} = 0$$

(Note that here we regard $F^{(1)}$ as a local primitive function subjected to ordinary differentiation.)

Hence if we are given a divisor D, again, of degree m, we can consider *those differentials of first kind* which have *at least* as many zeros as indicated by D. Certainly, such differentials form a *linear space*. This means (for P^s a factor of D) that if the functions

$$\frac{dF_1{}^{(1)}}{\zeta_P{}^s\, d\zeta_P} \qquad \text{and} \qquad \frac{dF_2{}^{(1)}}{\zeta_P{}^s\, d\zeta_P}$$

are analytic at $\zeta_P = 0$ (but possibly vanishing), then the differential

$$dF_0^{(1)} = g_1 dF_1^{(1)} + g_2 dF_2^{(1)}$$

(for complex constants g_1,g_2) has the property that $dF_0^{(1)}/(\zeta_P^s\, d\zeta)$ is also analytic at $\zeta_P = 0$. Call this linear space $L^*(D)$.

THEOREM 14-1 (*Roch*) *The index of specialty* $i(D)$ *equals the (complex) dimension of* $L^*(D)$.

PROOF Let $D = P_1 \cdots P_m$ (where, say, multiple factors such as P_1^s are comprised through repetition of P_1). Then $L^*(D)$ is a subspace of the p-dimensional space of abelian differentials of the first kind

$$dF^{(1)} = \sum_{j=1}^{p} \beta_j\, dF_j^{(1)} \tag{14-5a}$$

as determined by m conditions (written for $D = P_1^s P_2 \cdots$); for example,

$$\frac{dF^{(1)}}{d\zeta_{P_1}} = \cdots = \frac{d^s F^{(1)}}{d\zeta_{P_1}{}^s} = \frac{dF^{(1)}}{d\zeta_{P_2}} = \cdots = 0 \tag{14-5b}$$

We now apply Lemma 13-4 [e.g., (13-13a)] ignoring constant factors. Then we see conditions (14-5b) amount to

$$\sum_{t=1}^{p} (\mathfrak{Q}_t, dF^{(1)})(\mathfrak{Q}'_t, dF^{(2)}_{P_1{}^n}) = 0 \qquad \text{for } n = 1, 2, \ldots, s, \text{ etc.} \tag{14-5c}$$

If we refer $dF^{(1)}$ to the normal integrals by (14-5a), then from (13-3) we see the summation in (14-5c) selects $t = j$ when $dF_j^{(1)}$ is used. We therefore get m conditions from (14-5c) of the type

$$\sum_{j=1}^{p} \beta_j(\mathfrak{Q}'_j, dF^{(2)}_{P_1{}^n}) = 0 \qquad \text{for } n = 1, 2, \ldots, s, \text{ etc.}$$

which we write in customary style [compare (14-1)], as

$$\sum_{j=1}^{p}{}^* \beta_j(\mathfrak{Q}'_j, dF_{P_t}{}^{(2)}) = 0 \qquad t = 1, 2, \ldots, m \tag{14-5d}$$

Here the asterisk denotes the usual modification for multiple points in D.

Certainly, the matrix of coefficients in (14-5d) is the transpose of the matrix in (14-1), and therefore it has the same rank $p - i$. Thus the abelian differentials of the first kind are reduced in dimension to $p - (p - i) = i$ under membership in $L^*(D)$. Q.E.D.

Now Riemann's theorem becomes the Riemann-Roch theorem

$$\dim L(D) = 1 + \deg D - p + \dim L^*(D) \tag{14-6}$$

DEGREE OF A DIFFERENTIAL

To make further use of Theorem 14-1, we need the following supplementary result:

LEMMA 14-1 *For any differential dF (of first, second, or third kind) on $\mathfrak{M}$, the degree of dF, or the number of zeros minus the number of poles, is $2p - 2$.*

PROOF First, we note that for any two differentials on $\mathfrak{M}$, namely dF and dG, the number of zeros minus the number of poles produces the same result. For $dF = dG \cdot (dF/dG)$ where $f = dF/dG$ is meromorphic on $\mathfrak{M}$ whence f produces as many zeros as poles.

Now consider just the case where the differential is simply dz for z a meromorphic function on $\mathfrak{M}$. Then $z(P)$ maps $\mathfrak{M}$ onto an n-sheeted Riemann surface over the z sphere. The zeros or poles which belong to dz can occur, of course, only for z a branch point or $z = \infty$ (since elsewhere the local uniformizing parameter ζ can be taken as z itself, and $dz/d\zeta = 1$ produces neither a zero nor a pole). For convenience, choose z so that at $z = \infty$ there is no branch point (see Exercise 1). Then $z = 1/\zeta$ expresses the parameter, and $dz = -d\zeta/\zeta^2$ introduces a pole of order 2 (for each of the n sheets). Likewise at a branch point over $z = \beta$, $z - \beta = \zeta^e$ and $dz = e\zeta^{e-1}\, d\zeta$, yielding a zero of order $e - 1$. Thus the total order of zeros of dz is $\Sigma(e - 1) = b$ (the total ramification) and zeros minus poles is $b - 2n = 2p - 2$ by Riemann's formula (see Theorem 7-4, Corollary 1, in Sec. 7-3). Q.E.D.

This lemma provides an alternate definition of genus independent of dimension of abelian integrals (or differentials) of the *first* kind, since only *one* abelian differential (of any kind) dF is required for the lemma to be applied!

LEMMA 14-2 (*Riemann*) *If* $\deg D \geq 2p - 1$, *then the index of specialty of D always vanishes; that is,* $i(D) = \dim L^*(D) = 0$.

PROOF From Lemma 14-1, any differential of the first kind ($\not\equiv 0$) (with no poles) has $2p - 2$ zeros and no more; hence, $L^*(D)$ consists only of the zero differential. Q.E.D.

There is another result that for the so-called *"general" divisors D of dimension* $\dim D \geq p$, we have $i(D) = 0$. A precise formulation requires that we think of $P_1, P_2, \ldots, P_m$ as m independent variables on $\mathfrak{M}$ so that D represents a point in m-dimensional complex space $\mathfrak{S}$. Then the configurations of $P_1, \ldots, P_m$ for which $i(D) = 0$ represent a lower dimensional manifold in $\mathfrak{S}$. The proof of such results is referred to the literature. We cannot do much better than general statements. For example, we shall see in Exercise 4 that for, say $p = 2$, it is possible to have $i(P_1P_2) > 0$ for infinitely many pairs P_1P_2.

WEIERSTRASS POINTS

An interesting consequence of Lemma 14-2 can be deduced from the following result:

LEMMA 14-3 *Consider the divisor $D = P^s$ corresponding to a multiple pole. Then*

$$\dim L(P^{s+1}) = \dim L(P^s) \quad or \quad \dim L(P^s) + 1 \tag{14-7}$$

PROOF First note $L(P^{s+1}) \supseteq L(P^s)$. Assume $L(P^{s+1}) \supset L(P^s)$. Then an element $f_0 \in L(P^{s+1})$ exists for which f has an actual pole or order $s + 1$ at P. If any other element $f \in L(P^{s+1})$, we can choose a constant β so as to subtract out the pole by using βf_0; that is, $f = \beta f_0 + g$ where $g \in L(P^s)$. Thus $\dim L(P^{s+1}) = 1 + \dim L(P^s)$. Q.E.D.

If $\dim L(P^{s+1}) = \dim L(P^s)$, we say that the power P^{s+1} or (the exponent $s + 1$) represents a *gap*. If $p = 0$, we have no gaps (rational functions). Let us review the gaps if $p > 0$:

$\dim L(P^0) = 1 \ldots$ (A constant *is* a function without poles)
$\dim L(P^1) = 1 \ldots$ (If $p > 0$, no function has order 1)
$\dim L(P^2) = 1$ or 2
. . . .

Now by Lemma 14-2 and (14-6),

$$\dim L(P^{2p-1}) = p$$
$$\dim L(P^{2p}) = p + 1$$
$$\cdots\cdots\cdots$$
$$\dim L(P^s) = s + 1 - p \qquad \text{if } s \geq 2p - 1 \tag{14-8}$$

Obviously if no gap existed at all [$\dim L(P^{s+1}) = \dim L(P^s) + 1$], then we should have $\dim L(P^s) = s + 1$. Thus from (14-8) there are p gaps somewhere. When $p = 1$, the one gap is obviously at P^1.

Let us then assume $p > 1$. The gaps must include P^e for $e = 1$ but they also contain other exponents e, and so a complete list would be

$$1 = e_1 < e_2 < \cdots < e_p \leq 2p - 1 \tag{14-9}$$

Now Weierstrass showed that *in general* (for all but a finite set of P), the gaps constitute the first p powers. Thus in general, if we want to construct a meromorphic function with a pole at P of order e, we cannot do it unless $e \geq p + 1$ (indeed, the index of specialty of P^p is accordingly zero for general values of P).

The exceptional values of P where $e_p > p$ are called "Weierstrass points." They are known to number between $2p + 2$ and $p^3 - p$ points, but little else is known in general. In particular, little is known about the nature of the gap structure generally (as in the case of the hyperelliptic manifold where the gaps are the p first *odd* powers $P, P^3, \ldots, P^{2p-1}$; see Exercise 2).

An important feature of the Weierstrass points is that they demonstrate the *heterogeneity* of a manifold of genus ≥ 2. Let us recall that on the sphere ($p = 0$) or parallelogram ($p = 1$) we can make any point transform into any other point by well-known transformations.† Here, however, as we shall soon see, we cannot transform the Weierstrass points into arbitrary points.

CONTINUOUS TRANSFORMATIONS

We are now prepared to review the problem of *self-transformations* of $\mathfrak{M}$ or conformal mappings of a compact Riemann manifold $\mathfrak{M}$ onto itself. We now know that $\mathfrak{M}$ can be taken equivalently to be a Riemann surface of, say, the irreducible algebraic function $\Phi(z,w) = 0$ over the z sphere.

Let $f(P)$ map $\mathfrak{M}$ onto itself. Then we can describe $f(P)$ as a rational function $Z = f(z,w)$ (see Sec. 6-1). Meanwhile, corresponding to Z there must be a unique root W of $\Phi(Z,W) = 0$, since $f(P)$ maps $\mathfrak{M}$ onto itself *biuniquely*. Thus the arbitrary one-to-one mapping of $\mathfrak{M}$ onto itself takes the form of polynomials in two variables (or algebraic functions)

$$\begin{aligned} Z &= f(z,w) \\ W &= g(z,w) \end{aligned} \tag{14-10}$$

subject to $\Phi(Z,W) = 0$ once more.

We say that a *continuous* self-transformation of $\mathfrak{M}$ exists when the algebraic functions on $\mathfrak{M}$, namely $Z = f(z,w,t)$ and $W = g(z,w,t)$, have a parameter t which takes arbitrarily small values ($\neq 0$). (Although f and g are analytic in z and w, we do not require that they be analytic in t, only continuous.)

LEMMA 14-4 *If* (14-10) *represents a continuous self-transformation of* $\mathfrak{M}$ *(with parameter t understood), then the individual normal differentials of the first kind are invariant; i.e., each $dF_j^{(1)}$ has the very same formula in z and w as it has in Z and W.*

PROOF The main idea is that all of the canonical cycles $\mathcal{Q}_1, \ldots, \mathcal{Q}_p, \mathcal{Q}_1', \ldots, \mathcal{Q}_p'$ are moved an arbitrarily small amount under (14-10) for t sufficiently small. Call $f(\mathcal{Q}_i)$ and $f(\mathcal{Q}_i')$ the images under f. Now each $\mathcal{Q}_i$ is then deformable into $f(\mathcal{Q}_i)$ [$\mathcal{Q}_i'$ into $f(\mathcal{Q}_i')$] by a topological transformation, hence the characterizing equations of $dF_j^{(1)}$, namely

$$(\mathcal{Q}_i, dF_j^{(1)}) = 2\pi i \delta_{ij} \tag{14-11}$$

† This is embodied in the "addition theorems." We elaborate this idea in (Schwarz') Theorem 14-2.

are preserved if Q_i is replaced by $f(Q_i)$, and $dF_j^{(1)}$ transforms under $f^{(P)}$ into a differential of the first kind *defined in the same manner as* $dF_j^{(1)}$.

Q.E.D.

Thus we see the only continuously parametrized mappings of a manifold of genus 1 onto itself are determined by the requirement that the abelian differentials of first kind (usually written du) be preserved. This implies $df(u) = du$ or $f(u) = u +$ constant $(= u + t)$, which leads us back to the addition theorem of elliptic functions (see Sec. 4-7).

THEOREM 14-2 (*Schwarz*, 1878) *A Riemann manifold of genus p has no continuous transformations onto itself if* $p \geq 2$. *If* $p = 0$ *or* 1, *the only continuous transformations are derived from the linear-fractional mappings of a sphere onto itself or from the addition theorem of elliptic functions. These transformations have three or one complex parameters, respectively.*

PROOF The only remaining case to be proved is the case where $p \geq 2$. Here we note that the ratio $dF_1^{(1)}/dF_2^{(1)} = h(z,w)$ is a meromorphic function invariant under (14-10) for sufficiently small t, by Lemma 14-4, or if (z,w) is transformed into (Z,W) by f, then

$$h(z,w) = h(Z,W) \tag{14-12}$$

Thus any point $P_0 = (z_0,w_0)$ can be transformed under $f(P)$ only to a root of $h(Z,W) = h(z_0,w_0)$, that is, only into one of a *finite* set of points as t varies. By continuity, the assumption of a continuous transformation implies that *any* P_0 must be a fixed point [that is, $f(P_0) = P_0$] for t small enough. This leads to a contradiction since the number of fixed points is finite when $f(P) \not\equiv P$. Indeed the fixed points are roots of a meromorphic function $f(P) - P$ $(= 0)$. Q.E.D.

Actually, when $p \geq 2$, the Weierstrass points provide a more concrete realization of Theorem 14-2 because of the following considerations: First of all, any transformation $f(P)$ of $\mathfrak{M}$ onto itself must map a Weierstrass point only onto itself or another Weierstrass point. To see this, let $h(P)$ denote the function with only e $(\leq p)$ poles all at P_0. Then $g(P) = h(f(P))$ produces a similar pole structure at $f(P_0)$; that is, $g(P)$ has the same pole structure at $P = f(P_0)$. Secondly, it can be shown that the group of self-transformations of $\mathfrak{M}$ is generated by the interchange $w \leftrightarrow -w$ for the hyperelliptic surface† (generated by $w^2 =$ polynomial in z) and by certain permutations of the Weierstrass points. The details unfortunately are too lengthy to give here, but it can be seen how to limit the number of self-transformations of $\mathfrak{M}$ to $2W!$, where W is the number of Weierstrass points.

† This transformation leaves unchanged each of the $2p + 2$ Weierstrass points which are described in Exercises 2 and 3.

MODULES

We now consider the problem of how many *complex parameters* (modules) are needed to specify a Riemann manifold $\mathfrak{M}$ once the genus p is specified. If $n \geq p + 1$, then there are generally† ∞^{n-p+1} functions f on $\mathfrak{M}$ which map $f(\mathfrak{M})$ onto an n-sheeted Riemann surface over the f sphere, so that n poles are assigned or the points $P_1, \ldots, P_n$ go into $f = \infty$. (This is true precisely because the index of specialty vanishes.) Since these points P_j are generally different, we have ∞^n ways to select the points and therefore ∞^k functions f mapping $\mathfrak{M}$ onto a fixed n-sheeted surface where we define $k = 2n - p + 1$.

We ask how many such $\mathfrak{M}$ are different? By Riemann's formula there must be $b = 2n + 2p - 2$ points of ramification, generally different. Now the structures (interlacings of sheets) at the various points represent a finite number of alternatives, therefore we can say that there are generally ∞^b different n-sheeted Riemann surfaces $\mathfrak{S}$ of genus p (when viewed as configurations over the f sphere).

We still must consider that geometrically different surfaces $\mathfrak{S}$ can still be conformally the same. To take into account the admissible continuous self-transformations for $p = 0$ or 1, let us define

$$q = \begin{cases} 3 & \text{if } p = 0 \\ 1 & \text{if } p = 1 \\ 0 & \text{if } p \geq 2 \end{cases} \tag{14-13}$$

Then there are generally‡ ∞^q mappings of $\mathfrak{S}$ onto itself. In other words, there are ∞^{b-q} different n-sheeted Riemann surfaces $\mathfrak{S}$ under conformal mapping, achieved by our ∞^k mappings of $\mathfrak{M}$.

Thus ∞^{b-q-k} points can generally be assigned arbitrarily when $\mathfrak{M}$ is mapped onto $\mathfrak{S}$. This leaves $b - q - k$ arbitrary (complex) parameters in the *equivalence* classes of $\mathfrak{S}$ under self-transformation. We evaluate $b - q - k = 3p - 3 + q$.

THEOREM 14-3 (*Riemann*) *A Riemann manifold of genus p has $3p - 3 + q$ continuous complex parameters under the definition* (14-13).

The proof sketched above is disconcerting in its abuse of the term "generally." A more acceptable proof would involve theorems of algebraic geometry on how a curve (corresponding to $\mathfrak{S}$) can be specified

† Here we can use Lemma 14-2 if $n = 2p - 1$ and eliminate one of the many uses of the word "generally"! We can take $n \geq 2p + 3$ to be safe on all the inequalities below.

‡ Even here we cannot escape from use of the word "generally," since a manifold of genus ≥ 2 (or even of genus 1) can have (discrete) self-transformations which cannot be imbedded in the parameters of the continuous self-transformations (see Exercise 8).

when certain points are fixed, but such matters must be referred as always to the literature.

Corollary 1 *An n-tuply-connected region is determined by* $3(n-2)+r$ *continuous real parameters for* $n \geq 1$, *where* $r = 3$ *for* $n = 1$, $r = 1$ *for* $n = 2$, *and* $r = 0$ *for* $n \geq 2$.

The proof involves information on canonical mappings in Sec. 9-3 combined with information on continuous mappings (see Exercise 9).

EXERCISES

1 Show that if Z is a meromorphic function on $\mathfrak{M}$, we can choose z as a linear function of Z so that $z = \infty$ is not a branch point.

2 Consider the hyperelliptic manifold for

$$w^2 = (z - e_1) \cdots (z - e_{2p+1})$$

of genus p (>1). Show that $z = e_1, \ldots, e_{2p+1}$ and ∞ are Weierstrass points by noting that, say, at $z = \infty$ the function z has a pole of order 2 ($<p+1$). What about $z = e_1, \ldots, e_{2p+1}$? Show the gaps are the odd powers $P, P^3, \ldots, P^{2p-1}$. (In fact, use $z = 1/\zeta^2$, $w = 1/\zeta^{2p+1} \cdots$ and note that the gaps lie between the powers $z, z^2, \ldots, z^p$.)

3 *A necessary and sufficient condition that a manifold be hyperelliptic is that it have a meromorphic function z of order* 2. The necessity is displayed by z itself (since z = constant corresponds to two points P_1,P_2 of $\mathfrak{M}$, one for each sheet). Prove the sufficiency by considering any other function w not rational in z and setting up $w(z(P_1))$ and $w(z(P_2))$ as in Theorem 13-3 of Sec. 13-2.

4 *Any manifold of genus* 2 *is hyperelliptic.* (*Hint:* Consider $z = dF_1^{(1)}/dF_2^{(1)}$ and use Lemma 14-1.) Show that (infinitely many) pairs of points can serve as poles of functions z, such as the one just described, and one of these may be chosen arbitrarily.

5 From the knowledge that Weierstrass points are finite in number, state a stronger form of Theorem 13-1.

6 Verify that for the hyperelliptic manifold in Exercise 2, the differentials of the first kind can be taken as $dz/w, z\,dz/w, \ldots, z^{p-1}\,dz/w$. Note that they have the property that at least one function (namely w) defined on $\mathfrak{M}$ is not rationally expressible in terms of the ratios of such differentials.

7 Show that the manifold $\mathfrak{M}$ for $w^4 + z^4 = 1$ is *not* hyperelliptic by noting that the ratios of differentials of the first kind determine both z and w (hence all functions on $\mathfrak{M}$). (*Hint:* Note that dz/w^3, $z\,dz/w^3$, dz/w^2 are suitable differentials of the first kind.) (In Sec. 7-3, Exercise 5, we saw $\mathfrak{M}$ is of genus 3.)

8 Find some self-transformations of the Riemann surface generated by $w^p + z^q = 1$, based on multiplying w or z by roots of unity. Show that for certain (p,q) there exist analytic transformations with a continuous parameter. Find all such pairs.

9 By using the Schottky double $\mathfrak{M}$ of an n-tuply-connected region $\mathfrak{R}$, show that a conformal mapping of $\mathfrak{R}$ onto itself constitutes a conformal mapping of $\mathfrak{M}$ onto itself by reflection, and from this (and other results) show that analytic mappings with continuous parameters of $\mathfrak{R}$ onto itself occur only for $n \leq 2$. Specify parameters.

10 Let $\mathfrak{F}(\mathfrak{M})$ be the field of functions meromorphic on $\mathfrak{M}$. Show that if z is any nonconstant element of $\mathfrak{F}(\mathfrak{M})$, another element w exists in $\mathfrak{F}(\mathfrak{M})$ such that $\mathfrak{F}(\mathfrak{M})$ consists of the rational functions of z and w. [*Hint:* Let w be a function with all poles at one single *nonramified* point of the Riemann surface of $z(\mathfrak{M})$.]

14-2 Abel's Theorem

Up until now we have used algebraic concepts no deeper than rank or dimension and topological concepts on the level of surface (or manifold) theory. In this concluding section we consider a classic theorem of Abel which has undergone an evolution in concept (not unrelated to the Riemann-Roch theorem). Abel's theorem deals with the arbitrary assignment of an exact number of poles and zeros of a meromorphic function on a compact manifold $\mathfrak{M}$ of genus p.

We recall that in the case of a period parallelogram in the u plane (or surface of genus 1), a necessary and sufficient condition was given in Theorem 4-4 of Sec. 4-8 for m points $P_1, \ldots, P_m$ to be zeros and for m points $Q_1, \ldots, Q_m$ to be poles of some elliptic function. The condition was

$$\sum_{t=1}^{m} u(Q_t) - \sum_{t=1}^{m} u(P_t) \equiv 0 \qquad (\text{mod } \boldsymbol{\Omega}) \tag{14-14}$$

where $\boldsymbol{\Omega}$ represents the module of periods. These periods $\boldsymbol{\Omega}$ were complex numbers of the type $n_1\Omega_1 + n_2\Omega_2$ where, say, $\Omega_1 = (\mathcal{Q}_1,du)$, $\Omega_2 = (\mathcal{Q}_1',du)$. (We previously wrote $2\omega_j$ for Ω_j.) More generally

$$\boldsymbol{\Omega} = (\mathcal{H},du) \tag{14-15}$$

where $\mathcal{H}$ is the general homology class $\mathcal{H} \sim n_1\mathcal{Q}_1 + n_1'\mathcal{Q}_1'$ (n_1,n_1' integral).

If we consider a Riemann manifold of genus p, then du has as generalization a linear space of dimension p (over the complex numbers) spanned by normal differentials of the first kind (defined in Sec. 13-1),

$$dF_1^{(1)}, \ldots, dF_p^{(1)}$$

A direct generalization of the module of complex periods $\boldsymbol{\Omega}$ would be the p-dimensional linear space (over integers) consisting of the vectors

$$\boldsymbol{\Omega}^{(p)} = [(\mathcal{H},dF_1^{(1)}), \ldots, (\mathcal{H},dF_p^{(1)})]$$

where the most general homology class is now the cycle

$$\mathcal{H} \sim n_1\mathcal{Q}_1 + n_2\mathcal{Q}_2 + \cdots + n_p\mathcal{Q}_p + n_1'\mathcal{Q}_1' + n_2'\mathcal{Q}_2' + \cdots + n_p'\mathcal{Q}_p' \tag{14-16a}$$

We can think of $\boldsymbol{\Omega}^{(p)}$ as a $2p$ periodic structure generated by integral combinations of $2p$ vectors† in p-dimensional complex space:

$$\left.\begin{array}{l} n_1[(\mathcal{Q}_1, dF_1^{(1)}), \ldots, (\mathcal{Q}_1, dF_p^{(1)})] \\ \cdots \qquad\qquad\qquad \cdots \\ + n_p'[(\mathcal{Q}_p', dF_1^{(1)}), \ldots, (\mathcal{Q}_p', dF_p^{(1)})] \end{array}\right\} (2p \text{ vectors}) \tag{14-16b}$$

Again, if we have two complex vectors $\boldsymbol{\xi} = [\xi_1, \ldots, \xi_p]$,

$$\boldsymbol{\eta} = [\eta_1, \ldots, \eta_p]$$

we say

$$\boldsymbol{\xi} \equiv \boldsymbol{\eta} \qquad (\text{mod } \boldsymbol{\Omega}^{(p)})$$

exactly when $\xi_j - \eta_j = (\mathcal{H}, dF_j^{(1)})$ for some cycle $\mathcal{H}$ ($j = 1, 2, \ldots, p$).

Now let us fix some base point P_0 on $\mathfrak{M}$ (its choice will be unimportant) and define the vector

$$\mathbf{h}_0(P) = \left(\int_{P_0}^{P} dF_1^{(1)}, \ldots, \int_{P_0}^{P} dF_p^{(1)}\right) \qquad (\text{mod } \boldsymbol{\Omega}^{(p)})$$

in p-dimensional complex space. [The paths of integration are taken the same for each component, but are not well defined so that $\mathbf{h}_0(P)$ is defined only modulo $\boldsymbol{\Omega}^{(p)}$.]

THEOREM 14-4 (*Abel*, 1829) *Let m points $P_1, \ldots, P_m$ be prescribed as zeros for a function on a manifold $\mathfrak{M}$ of genus p, and let m additional points $Q_1, \ldots, Q_m$ be prescribed as poles. [There may be repetitions constituting multiplicities in the usual fashion, but for now let $P_i \neq Q_j$ for any (i,j).] Then a necessary and sufficient condition that a meromorphic function exist on $\mathfrak{M}$ with the prescribed zeros and poles is that*

$$\sum_{t=1}^{m} \mathbf{h}_0(P_t) - \sum_{t=1}^{m} \mathbf{h}_0(Q_t) \equiv 0 \qquad (\text{mod } \boldsymbol{\Omega}^{(p)}) \tag{14-17a}$$

or equivalently (performing the subtraction in each of p components),

$$\sum_{t=1}^{m} \left(\int_{P_t}^{Q_t} dF_1^{(1)}, \ldots, \int_{P_t}^{Q_t} dF_p^{(1)}\right) \equiv 0 \qquad (\text{mod } \boldsymbol{\Omega}^{(p)}) \tag{14-17b}$$

for some path of integration $\mathcal{K}_t$ from P_t to Q_t.

PROOF Let f exist with the indicated description of zeros and poles. Then $d \log f - \sum_{t=1}^{m} dF_{P_tQ_t}^{(3)}$ must be an abelian differential of the first kind; i.e., for some (as yet unknown) $\alpha_1, \ldots, \alpha_p$

$$d \log f = \sum_{t=1}^{m} dF_{P_tQ_t}^{(3)} + \sum_{k=1}^{p} \alpha_k \, dF_k^{(1)} \tag{14-18}$$

† The set of vectors is generally called a *Riemann matrix*. It is an open question to decide when a given $2p \times p$ matrix is a Riemann matrix (i.e., when it arises from the periods of a manifold).

Of course, the definition of $F^{(3)}_{P_tQ_t}$ could depend on the path of continuation $\mathcal{K}_t$ from P_t to Q_t since P_t and Q_t are not close, but ambiguity is avoided by keeping $\mathcal{K}_t$ on a fixed canonical dissection $\mathfrak{M}^p$.

The condition that f is single-valued means $\log f$ has an ambiguity of $2\pi i$ times an integer or $(\mathcal{Q}_j, d \log f)/2\pi i$, and $(\mathcal{Q}_j, d \log f)/2\pi i$ are integers. Thus from (14-18), the first of the following two formulas is immediate:

$$\text{(14-19a)} \qquad 2\pi i n_j = (\mathcal{Q}_j, d \log f) = \sum 0 + \sum a_k 2\pi i \delta_{jk} = 2\pi i a_j$$

$$\text{(14-19b)} \qquad 2\pi i n_j' = (\mathcal{Q}_j', d \log f) = \sum_{t=1}^{m} (\mathcal{K}_t, dF_j{}^{(1)}) + \sum_{k=1}^{p} a_k(\mathcal{Q}_j', dF_k{}^{(1)})$$

In the latter formula we apply Lemma 13-4 with $dF^{(1)}$ taken as $dF_j{}^{(1)}$ in (13-13b) [so that the sum in (13-13b) singles out the term $t = j$].

Now, we prove in Exercise 2 that

$$\text{(14-20)} \qquad (\mathcal{Q}_j', dF_k{}^{(1)}) = (\mathcal{Q}_k', dF_j{}^{(1)})$$

We recall $2\pi i \delta_{kj} = (\mathcal{Q}_k, dF_j{}^{(1)})$. Therefore, by using (14-19a) in (14-19b) we find

$$\text{(14-19c)} \qquad \sum_{k=1}^{m} n_k'(\mathcal{Q}_k, dF_j{}^{(1)}) = \sum_{t=1}^{m} (\mathcal{K}_t, dF_j{}^{(1)}) + \sum_{k=1}^{p} n_k(\mathcal{Q}_k', dF_j{}^{(1)})$$

Hence if we set $\mathcal{K} \sim -\sum_{k=1}^{p} n_k\mathcal{Q}_k' + n_k'\mathcal{Q}_k$, we find (14-17b) holds in the form

$$\text{(14-21)} \qquad \sum_{t=1}^{m} (\mathcal{K}_t, dF_j{}^{(1)}) = (\mathcal{K}, dF_j{}^{(1)}) \qquad j = 1, 2, \ldots, p$$

The steps are reversible; i.e., if (14-21) holds, f is single-valued as defined in (14-18). Q.E.D.

GEOMETRIC APPLICATION

Abel's theorem is one of the most useful tools in the study of algebraic curves (with complex points). We can hardly do justice to it as we now cite the simplest example of its usage. Let $\mathfrak{M}$ be determined as the Riemann surface $\mathcal{S}$ of $\Phi(z,w) = 0$. Then let $\Psi(z,w)$ be any meromorphic function on $\mathcal{S}$ of order m. Consider the m (complex) roots of

$$\text{(14-22a)} \qquad \Psi(z,w) - k = 0 \qquad \Phi(z,w) = 0$$

namely $(z,w) = P_1, \ldots, P_m$.

Of course, the roots *individually* depend on k, but note that the poles $Q_1, \ldots, Q_m$ of $\Psi(z,w) - k$ are independent of k. Hence by (14-17b), regardless of k,

(14-22b) $$\sum_{t=1}^{m} F_j^{(1)}(P_t) = \text{const (say } \gamma_j) \qquad j = 1, \ldots, p$$

The result (14-22b) is quite powerful, but to see its application most simply, return to $p = 1$ (the cubic curve). We write

(14-23a) $$w^2 = 4z^3 - g_2 z - g_3 \qquad g_2{}^3 \neq 27 g_3{}^2$$

parametrized as usual by $z = \wp(u)$, $w = \wp'(u)$. Let us intersect this curve with a straight line, $w = Az + B$. The intersections are the roots of a meromorphic function Ψ of order 3, where

(14-23b) $$\Psi = -Az - B + w \qquad (\approx -2z^{3/2} \text{ at } z = \infty)$$

and A and B now both are variable.

Clearly the poles of Ψ are fixed (at ∞). Hence for the intersections, given as u_1, u_2, u_3 in the period parallelogram, we have the relation [like (14-22b)]

(14-24) $$u_1 + u_2 + u_3 = \text{const}$$

(Actually, the constant is 0 since the point at ∞ is parametrized by $u = 0$.) We see that Abel's theorem is equivalent to the addition theorem or that $[\wp(u_1),\wp'(u_1)]$, $[\wp(u_2),\wp'(u_2)]$, $[\wp(u_3),\wp'(u_3)]$ are on the *line* in the zw plane $\Psi = 0$. (Compare Exercise 1 of Sec. 4-8.)

Our next objective will be to build a generalized addition theorem out of Abel's theorem, but this requires a reexamination of the divisor concept.

DIVISOR CLASSES

We now modify the sense of Abel's theorem to permit the same point to serve as zero and pole with the meaning that if P is listed r times as a P_i (zero) and s times as a Q_j (pole), then P is a zero of order $r - s$ if $r > s$, P is a pole of order $s - r$ if $s > r$, and P is neither zero nor pole if $r = s$. Clearly the new interpretation follows immediately from the statement of the theorem.

We can now speak of the *group* of divisors. The previous expressions $D = P_1 \cdots P_m$, etc., are called "integral" divisors. By formal division we obtain the larger set

(14-25a) $$d = P_1{}^{e_1} \cdots P_k{}^{e_k} = \frac{A}{B}$$

where we use capital letters A, B, . . . for integral divisors and small letter d for the complete group so that e_i are positive, negative, or 0. (Formal cancellation in A and B will always be valid.) We use 1 to denote the unit divisor. We can then speak of a set of prescribed *zeros* P_i and *poles* Q_j as corresponding to the divisor (quotient of $A = P_1 \cdots$ and $B = Q_1 \cdots$)

$$d = \frac{A}{B} = \frac{P_1 \cdots P_m}{Q_1 \cdots Q_n} \tag{14-25b}$$

(m need not equal n).

We say that a nonzero function f (on $\mathfrak{M}$) is represented by a divisor d

$$(f) \doteq d \tag{14-26}$$

if the P_i are precisely the zeros and the Q_i are precisely the poles (as always we allow for multiple points). We shall use (f) to denote the divisor structure of f.

We call the *degree* of a divisor d the sum [in (14-25a)],

$$e_1 + \cdots + e_k = \deg d$$

Thus (14-26) is possible for given d only if $\deg d = 0$, or in (14-25b), if $n = m$. The converse is not true; it leads to Abel's theorem again, as we soon see.

We leave it to the reader to verify the fact that $\deg d_1 d_2 = \deg d_1 + \deg d_2$. (The only difficulty is to acknowledge the trivial effect of cancellations.)

We define $\mathbf{h}_0(d)$ for any divisor (14-25a) as

$$\mathbf{h}_0(d) = \sum_{j=1}^{k} \mathbf{h}_0(P_j)e_j \tag{14-27}$$

and as usual $\mathbf{h}_0(d)$ is determined modulo $\mathbf{\Omega}^{(p)}$, and

$$\mathbf{h}_0(d_1 d_2) = \mathbf{h}_0(d_1) + \mathbf{h}_0(d_2)$$

Abel's theorem now has a new form (considerably simpler!):

For a preassigned divisor d, we can find a meromorphic function f such that

$$(f) \doteq d \tag{14-28a}$$

if and only if $\deg d = 0$ *and*

$$\mathbf{h}_0(d) \equiv 0 \qquad (\text{mod } \mathbf{\Omega}^{(p)}) \tag{14-28b}$$

Now suppose d does *not* represent a meromorphic function f, or (14-28a) is invalid. Then we ask what *does* $\mathbf{h}_0(d)$ represent?

The answer is given by another definition: Let us say that two divisors d_1, d_2 belong to the *same class*, written

$$d_1 \sim d_2 \tag{14-29a}$$

when for some meromorphic function f (not identically zero),

$$\frac{d_1}{d_2} \doteq (f) \tag{14-29b}$$

We introduce the symbol δ_1 to denote the class of all divisors $d \sim d_1$ for a given d_1. All such d have the same degree since $d = (f)d_1$, with $\deg f = 0$. We call the common value $\deg \delta_1$.

The divisor class of any meromorphic function f, including a constant ($\neq 0$), is the same and written, say, $(f) \sim 1$. We generally have infinitely many divisor classes (see Exercises 5 to 7). We can form the *product of divisor classes*, for example, $\delta_1\delta_2$, by selecting $d_1 \in \delta_1$, $d_2 \in \delta_2$ and letting $d \sim d_1d_2$ determine δ. Then

$$\deg \delta_1\delta_2 = \deg \delta_1 + \deg \delta_2$$

Likewise, we form the quotient and

$$\deg \frac{\delta_1}{\delta_2} = \deg \delta_1 - \deg \delta_2$$

Then, more abstractly, *Abel's theorem states that two divisors d_1 and d_2 belong to the same class or $d_1 \sim d_2$* exactly when *they have the same degree and*

$$\mathbf{h}_0(d_1) \equiv \mathbf{h}_0(d_2) \qquad (\text{mod } \mathbf{\Omega}^{(p)})$$

RIEMANN-ROCH THEOREM

Now to achieve the next stage of abstraction we must return to the Riemann-Roch theorem. *Let us say that the linear space $L(d)$ represents all meromorphic functions f for which $(f)d$ is an integral divisor.* Thus, in keeping with our earlier terminology for $L(D)$ (in Sec. 14-1), $L(d)$ represents all f for which the poles are a subset of the points of the *numerator* and the zeros contain all the points of the *denominator*.† We again speak of $\dim L(d)$ but now we see $\dim L(d)$ depends only on the divisor class δ of d. Hence we write

$$\dim \delta = \dim L(d) \qquad d \in \delta$$

To see this, note that if $d_1/d_2 = (f_0)$, then $f \in L(d_1)$ exactly when $(f/f_0) \in L(d_2)$ so that the set $\{L(d_1)\} = f_0\{L(d_2)\}$.

LEMMA 14-5 *The value of* $\dim \delta > 0$ *if and only if the class contains an integral divisor, hence only if* $\deg \delta > 0$ *or* $\delta \sim 1$.

PROOF Consider $L(d)$ for $d \in \delta$. Then $\dim L(d) \neq 0$ if and only if $f \in L(d)$ (for $f \neq 0$ a meromorphic function). This is true if and only if $(f)d = A$ or $d \sim A$ for A *integral*. Clearly, $\deg d = \deg A \geq 0$ while if $\deg A = 0$, $A = 1$. Q.E.D.

† Some authors write $L(1/d)$ since it seems curious to obtain the poles from the numerator, etc.!

A further concept is the *canonical class* or the class of all abelian differentials denoted by $\mathcal{O}$. We saw that any two differentials (of any kind) on $\mathfrak{M}$, say dF, dG, are in the same class since they determine a meromorphic ratio. Hence consider the zero and pole structures (of $dF/d\zeta$) which determine (dF) as a divisor:

$$\frac{dF}{d\zeta} = \frac{P_1 \cdots P_s}{Q_1 \cdots Q_t} \qquad \text{written as } (dF)$$

(for ζ a locally uniformizing parameter at each P_i and Q_j). By Lemma 14-1, we saw

$$\deg \mathcal{O} = \deg (dF) = s - t = 2p - 2 \tag{14-30}$$

LEMMA 14-6 *If D is an integral divisor and dF is any fixed abelian differential,*

$$\dim L^*(D) = \dim L\left(\frac{dF}{D}\right) \tag{14-31}$$

PROOF We had defined $L^*(D)$ as the set of $dF^{(1)}$ (abelian integrals of *first* kind) for which the divisor $(dF^{(1)})$ has at least the points in D as zeros, or for which $(dF^{(1)})/D$ is integral. This set can be described as the set of differentials dG of *any* kind for which $(dG)/D$ is integral (since poles in dG are still excluded). The most general dG, however, is $h\,dF$ where dF is any fixed differential and h is the most general meromorphic function. Hence, $L^*(D)$ is the set of $h\,dF$ for which h varies so that $h(dF)/D$ is *integral*. Since now h varies and dF is fixed, the h in question are precisely the elements of $L(dF/D)$ under our general definition of $L(d)$ for d fractional. We now have shown the elements in $L^*(D)$ and $L(dF/D)$ have a one-to-one correspondence. Q.E.D.

We are prepared to reinterpret the theorems of Riemann and Roch [see (14-6)] in broader fashion. Using Lemma 14-6, we have

$$\dim L(D) = 1 + \deg D - p + \dim L\left(\frac{dF}{D}\right) \tag{14-32a}$$

If we reinterpret $2p - 2 = \deg dF$, we can write (14-32a) as

$$\dim L(D) - \tfrac{1}{2}\deg D = \dim L\left(\frac{dF}{D}\right) - \tfrac{1}{2}\deg\left(\frac{dF}{D}\right) \tag{14-32b}$$

The observation made by Dedekind (1880) was that (14-32b) is valid even when D is a *fractional* divisor. This gives us a more symmetric version of the Riemann-Roch theorem:

THEOREM 14-5 *If δ is any divisor class (fractional or integral) and $\mathcal{O}$ is the class of the differential, then*

$$\dim \delta - \tfrac{1}{2}\deg \delta = \dim \frac{\mathcal{O}}{\delta} - \tfrac{1}{2}\deg \frac{\mathcal{O}}{\delta} \tag{14-33}$$

PROOF We start with (14-32a) or (14-32b) when D is integral. Because of the symmetry with regard to δ and $\mathcal{O}/\delta$, the result can be considered proved, if either the class δ or $\mathcal{O}/\delta$ has an integral divisor or equivalently if either

$$\dim \delta > 0 \qquad \text{or} \qquad \dim \frac{\mathcal{O}}{\delta} > 0 \tag{14-34}$$

It remains to consider the case where both inequalities (14-34) are false or $\dim \delta = \dim \mathcal{O}/\delta = 0$. Then, substituting into (14-33) [or better still into (14-32a)], we have only to prove that $\deg \delta = p - 1$.

Thus we assume $\dim \delta = 0$, and we let δ contain $d = A/B$ for A, B integral divisors. We shall first obtain a contradiction from the assumption that $\deg d \geq p$. For then

$$\deg A \geq p + \deg B$$

Now $B \neq 1$. Otherwise d is integral and $\dim \delta > 0$ by Lemma 14-5. Thus by the original version of Riemann [see (14-3)],

$$\dim L(A) \geq \deg A - p + 1 \geq \deg B + 1$$

Therefore, the family of functions in the space $L(A)$ has enough complex linear parameters to satisfy the deg B different relations required in making a function f_1 ($\neq 0$) vanish at the points of B (in proper multiplicity). Hence if $d = A/B$ and $\deg d > p - 1$, we can find functions $f_1 \in L(d)$. Therefore, $\dim L(d) > 0$.

This is a contradiction to (14-34), therefore $\deg \delta \leq p - 1$; and likewise, $\deg \mathcal{O}/\delta \leq p - 1$ (symmetrically). This leads to $\deg \delta \geq p - 1$. Thus, $\deg \delta = p - 1$. Q.E.D.

We actually are most desirous of the following result:

COROLLARY 1 *If* $\deg A \geq p + \deg B$, *then* $\dim L(A/B) > 0$; *in other words, every divisor class* d $(=A/B)$ *of degree* $\geq p$ *has an integral representative.*

PROOF Apply (14-32a), substituting A/B for D, as is now legitimate, since D need not be integral. Then

$$\dim L\left(\frac{A}{B}\right) \geq 1 + \deg\left(\frac{A}{B}\right) - p \geq 1 \tag{14-35}$$

Q.E.D.

Note that if A/B is not integral, $L(A/B)$ must contain *nonconstant* functions (as zeros of B are mandatory); if A/B were integral, then (14-35) might only be evidenced by *constants* in $L(A/B)$ [for example, when $B = 1$ and $i(A) = 0$].

ADDITION THEOREM

Now we can represent any divisor class of degree 0 in reduced form

(14-36) $$d = \frac{A}{B} = \frac{P_1 \cdots P_m}{Q_1 \cdots Q_m}$$

as the quotient of integral divisors. Then from the corollary to (the Riemann-Roch) Theorem 14-5, if integral divisors A, B_1, B_2, satisfy

(14-37) $$\deg B_1 = \deg B_2 = \deg A = m \geq p$$

it follows that an integral divisor B_3 exists such that $\deg B_3 = m$ and

(14-38) $$\frac{B_1}{A} \cdot \frac{B_2}{A} \sim \frac{B_3}{A}$$

Take $d = B_1B_2/A$, and we find $L(d) > 0$ or an f exists such that fd shows no poles or $fd = B_3$. (Easily $\deg B_3 = \deg d = m$.) Now, by a seemingly algebraic argument we have achieved the classic† addition theorem of Abel from (14-28*a*) and (14-28*b*).

THEOREM 14-6 *Let any set of m ($\geq p$) points be fixed on a Riemann manifold forming a divisor $A = P_1^{(0)} \cdots P_m^{(0)}$. Then for any two sets of m points (written as divisors)*

$$B_1 = P_1^{(1)} \cdots P_m^{(1)} \qquad B_2 = P_1^{(2)} \cdots P_m^{(2)}$$

a sum set $B_3 = P_1^{(3)} \cdots P_m^{(3)}$ is determined (unique to within divisor class) such that

(14-39) $$\mathbf{h}_0\left(\frac{B_1}{A}\right) + \mathbf{h}_0\left(\frac{B_2}{A}\right) = \mathbf{h}_0\left(\frac{B_3}{A}\right)$$

Of course, (14-39) is a consequence of (14-38). We can write $\mathbf{h}(d)$ instead of $\mathbf{h}_0(d)$ since the value is independent of base point if $\deg d = 0$.

COROLLARY 1 *The set of vectors $\mathbf{h}_0(B/A)$ covers the whole p-dimensional complex space modulo $\mathbf{\Omega}$ as B varies, provided*

$$m = \deg A = \deg B = p$$

and the fixed divisor A is general, that is, $i(A) = 0$.

PROOF The choice of A is left as Exercises 9 and 10. Assume we have $i(A) = 0$. We let B be a variable divisor of degree p (on $\mathfrak{M}$). If B is sufficiently near A, we can define $\mathbf{h}_0(B/A)$ uniquely by taking short paths from B to A (lying in the neighborhood of A). First we notice

† The meaning of the addition theorem is quite deep, since it implies that if the functions f_1 and f_2 are represented by B_1/A and B_2/A, the function f_3 representing B_3/A is rationally related to f_1 and f_2 (because of the field). We must refer to the literature for more details.

that $\mathbf{h}_0(B/A)$ provides a mapping from B into some neighborhood of the origin of p-dimensional complex space. Actually, the mapping *covers* the origin. Consider $\mathbf{h}_0$ written out as components:

(14-40)

$$(h_1, \ldots, h_p) = \mathbf{h}_0\left(\frac{B}{A}\right) = \left(\sum_{t=1}^{p} \int_{P_t}^{Q_t} dF_1^{(1)}, \ldots, \sum_{t=1}^{m} \int_{P_t}^{Q_t} dF_p^{(1)}\right)$$

where $B = Q_1 \cdots Q_t$, $A = P_1 \cdots P_t$, and the paths of integration all lie in a small (parameter) neighborhood of each point $P_1, \ldots, P_t$. The Jacobian of $\mathbf{h}_0$ with respect to B has the components

(14-41)
$$a_{ij} = \frac{\partial h_i}{\partial \zeta_{Q_j}} = \frac{dF_i^{(1)}}{d\zeta_{Q_j}}$$

The determinant of $\|a_{ij}\|$ is nonvanishing for Q_j *at* P_j (hence for B close to A) by virtue of the choice of A [$i(A) = 0$].

Now by the addition theorem, if any neighborhood of the origin is covered in p-dimensional complex space by $\mathbf{h}_0(B/A)$, then for A suitably chosen, *twice* that neighborhood is covered (mod $\mathbf{\Omega}$). Thus $\mathbf{h}_0(B/A)$ *covers all of the space of* p complex variables (mod $\mathbf{\Omega}$) as B varies. Q.E.D.

THE JACOBI VARIETY

Let us now consider the manifold $\mathfrak{M}$ of genus p and form the (cartesian) product space $\mathfrak{M} \times \mathfrak{M} \times \cdots \times \mathfrak{M}$ (p times). Its general element is the ordered p-tuple $(P_1, \ldots, P_p)$ where $P_i \in \mathfrak{M}$. We define the *Jacobi variety* $\mathcal{J}$ as the manifold $\mathfrak{M} \times \mathfrak{M} \times \cdots \times \mathfrak{M}$ on which *points are identified if the corresponding products* $B = P_1P_2 \cdots P_p$ *are equivalent*† *divisors*. Thus the points of $\mathcal{J}$ are in one-to-one correspondence with divisor classes of degree p.

Among other things, this identification of points B permutes the ordering of the p-tuple $(P_1, \ldots, P_p)$. Actually, in general, $i(B) = 0$, so that only this trivial permutation causes identifications of integral divisors of degree p.

THEOREM 14-7 *The Jacobi variety $\mathcal{J}$ of a compact manifold $\mathfrak{M}$ of genus p is mapped by $\mathbf{h}_0(B)$ in one-to-one fashion onto the complex vectors in p-dimensional space modulo $\mathbf{\Omega}$ (the module of periods of abelian integrals of the first kind), if B represents an arbitrary divisor class of degree p on $\mathfrak{M}$.*

This result is seen to be a further restatement of Abel's theorem in view of the result that $\mathbf{h}_0(B/A)$ covers p-dimensional space (mod $\mathbf{\Omega}$) [while $\mathbf{h}_0(B)$ differs from $\mathbf{h}_0(B/A)$ by a constant vector].

† The set $\mathfrak{M} \times \mathfrak{M} \cdots \times \mathfrak{M}$ (p times) is a *manifold*, but it loses its general ($2p$) dimensionality when we identify points. It is therefore only a *variety* when $p > 1$.

This theorem is the strongest analog known for the one-to-one mapping of the compact manifold of genus 1 onto the period parallelogram. We are dealing with a "nonmanifold" of a rather esoteric type in the Jacobi variety. It is interesting that now we are no longer mapping an analytic (or geometric) construct such as a point. We are mapping a *divisor class*, an algebraic construct.

We can now look upon our geometric applications in a rather different light. We had just considered the set of points of intersection (14-22*a*) and (14-22*b*) of a fixed curve $\Phi(z,w) = 0$ (actually a Riemann surface $\mathfrak{M}$) with a variable one $\Psi(z,w) = k$, and we saw the intersection was essentially a divisor class on $\mathfrak{M}$ independent† of k. Now each divisor class represents a single point on the Jacobi variety, hence the Jacobi variety can be looked upon as a compendium of such intersection problems of the theory of algebraic curves.

EXERCISES

1 Show that the $2p$ vectors in (14-16*b*) have the form

$$n_1[2\pi i,0, \ldots 0] + n_2[0, 2\pi i, \ldots, 0] \cdots + n_p[0,0, \ldots, 2\pi i] + \sum_{j=1}^{p} n'_j[\tau_{j1},\tau_{j2}, \ldots, \tau_{jp}]$$

where $\tau_{jk} = (\mathcal{Q}'_j, dF_k^{(1)})$.

2 Show the matrix τ_{jk} is symmetric; that is, $\tau_{jk} = \tau_{kj}$, by applying Riemann's period integrals (Sec. 8-5) to $\int F_j^{(1)}\, dF_k^{(1)}$ over $\partial\mathfrak{M}^p$ (the canonically dissected manifold). *Hint:* The $\int = 0$ since $\partial\mathfrak{M}^p \sim 0$ and the determinants yield

$$0 = \sum_{t=1}^{m} \begin{vmatrix} (\mathcal{Q}_t, dF_j^{(1)}) & (\mathcal{Q}'_t, dF_j^{(1)}) \\ (\mathcal{Q}_t, dF_k^{(1)}) & (\mathcal{Q}'_t, dF_k^{(1)}) \end{vmatrix} = \sum_{t=1}^{m} 2\pi i[\delta_{tj}\tau_{tk} - \delta_{tk}\tau_{tj}]$$

3 Show the form $F(a_1, \ldots, a_p) = \Sigma a_t a_j \operatorname{Re} \tau_{tj}$ (in real a_j) is *negative definite;* that is, $F \leq 0$ and $F = 0$ only when each $a_t = 0$. [*Hint:* Show that if $F^{(1)} = \Sigma a_j F_j^{(1)}$, then $0 \leq \int (\operatorname{Re} F^{(1)})\, d(\operatorname{Im} F^{(1)}) = -2\pi F(a_1, \ldots, a_p)$, by Riemann's period integrals again.]

4 Show how the last result is related to the period parallelogram in the u plane: Here u is normal if $(\mathcal{Q}_1, du) = 2\pi i$, but previously (see Sec. 4-6) we required $\operatorname{Im} \Omega_2/\Omega_1 = \operatorname{Im} (\mathcal{Q}'_1, du)/(\mathcal{Q}_1, du)] > 0$.

5 Show that for $\mathfrak{M}$ of genus $p = 0$, all divisor classes are given by $\ldots, P^{-2}, P^{-1}, P, 1, P, P^2, \ldots$ where P is any point of $\mathfrak{M}$. (*Hint:* Zeros and poles are arbitrarily prescribable.)

† The same result would hold substantially if we intersected $\Phi = 0$ with a variable curve whose parameters are more arbitrarily disposed, say $\Psi(z,w;k_1, \ldots, k_t) = 0$. The set of intersections would also constitute a divisor class on $\mathfrak{M}$. Actually, the intersections are called a "linear series" and each linear series corresponds to a single divisor class (as the parameters vary).

6 Show that for $\mathfrak{M}$ of genus $p = 1$ there are infinitely many divisor classes of any given degree m but they are determined by one variable point (∞^1 classes), if $m \geq 1$. (Use Theorem 4-4 of Sec. 4-8.)

7 Show how to interpret (Jacobi's) Theorem 14-7 as determining the number of degrees of freedom in the divisor classes as ∞^p.

8 From Lemma 14-6, show $\{L^*(D)\} = dF\{L(dF/D)\}$ in the sense of a set-theoretic correspondence (for any dF).

9 Show that if $A = P_1 \cdots P_p$ for distinct points, then $i(A) = 0$ precisely when the determinant $\|a_{ij}\| \neq 0$ where

$$a_{ij} = \frac{dF_j^{(1)}}{d\zeta_{P_i}}$$

Hint: By Roch's theorem, $i(A) > 0$ precisely when some complex vector $(\beta_1, \ldots, \beta_p) \neq 0$ exists for which $dF^{(1)} = \Sigma\beta_j\, dF_j^{(1)}$ satisfies $dF^{(1)}/d\zeta = 0$ at all points of A. What modification is made if A has multiple points?

10 Show that any neighborhood of any set of points $P_1^0, \ldots, P_p^0$ on $\mathfrak{M}$ contains a point whose divisor A satisfies $i(A) = 0$. [Use the linear independence of $dF_j^{(1)}$ $(j = 1, \ldots, p)$.]

APPENDIX A
MINIMAL PRINCIPLES

We saw that there are two aspects to the existence theorems for the Dirichlet problem, the solution in a simply-connected region and the extension to Riemann manifolds. The first part of the problem is of more profound influence than previously indicated, particularly since it has served as a fountainhead for research in functional analysis.

Consider the ordinary Dirichlet problem for finding a (real) harmonic function $\psi(z)$, defined on a simply-connected finite region $\mathcal{R}$ with piecewise smooth boundary curves and smooth boundary values $g(Z)$ for Z on $\partial\mathcal{R}$. Historically, the existence of a solution was argued to be obvious from physical intuition (and the major part of the effort went into explicit solutions like Poisson's integral!). The reason given by Gauss, Kelvin, and others was that the harmonic function is produced by a static distribution of charges on $\partial\mathcal{R}$ in some manner (not specified) which will produce the smallest (potential) self-energy over $\mathcal{R}$. The energy is given (to within a constant factor) by

$$D_{\mathcal{R}}[u] = \iint_{\mathcal{R}} \left[\left(\frac{\partial u}{\partial x}\right)^2 + \left(\frac{\partial u}{\partial y}\right)^2\right] dx\, dy \tag{A-1}$$

In purely mathematical terms, what we wish to say is that the *known* solution $\psi(z)$ to the Dirichlet problem minimizes $D_{\mathcal{R}}[u]$, so that

$$D_{\mathcal{R}}[\psi] < D_{\mathcal{R}}[u] \tag{A-2}$$

if u is *any other* (twice) continuously differentiable function on $\mathcal{R}$ for which $u(Z) = g(Z)$, for Z on $\partial\mathcal{R}$.

The minimum property of ψ can be "supported" by the following manipulation [recall (8-40)]:

$$D_{\mathcal{R}}[u,v] = \iint_{\mathcal{R}} \left[\left(\frac{\partial u}{\partial x}\right)\left(\frac{\partial v}{\partial x}\right) + \left(\frac{\partial u}{\partial y}\right)\left(\frac{\partial v}{\partial y}\right)\right] dx\, dy \tag{A-3}$$

$$= -\iint_{\mathcal{R}} v\, \Delta u\, dx\, dy - \int_{\partial\mathcal{R}} v \frac{\partial u}{\partial n}\, ds \tag{A-4}$$

Then letting $u(Z) = g(Z)$ for Z on $\partial\mathcal{R}$, we can set $w = u - \psi$.

We expand

$$D_{\mathfrak{R}}[u] = D_{\mathfrak{R}}[\psi + w] = D_{\mathfrak{R}}[\psi] + 2D_{\mathfrak{R}}[\psi,w] + D_{\mathfrak{R}}[w] \tag{A-5a}$$

and we find $D_{\mathfrak{R}}[\psi,w] = 0$, since $\Delta\psi = 0$ in $\mathfrak{R}$ and $w = 0$ on $\partial\mathfrak{R}$. Thus

$$D_{\mathfrak{R}}[u] = D_{\mathfrak{R}}[\psi] + D_{\mathfrak{R}}[w] \tag{A-5b}$$

justifying (A-2), when $w \not\equiv 0$.

By our manipulation, however, we have *not* shown the solution ψ (to the Dirichlet problem) exists or even that a solution to the problem of minimizing $D_{\mathfrak{R}}[u]$ [$u = g(Z)$ on $\partial\mathfrak{R}$] exists. We have shown the following:

THEOREM A-1*a* *If a harmonic function ψ exists on $\overline{\mathfrak{R}}$ with given boundary values on $\partial\mathfrak{R}$, then $D_{\mathfrak{R}}[\psi]$ will be less than any other $D_{\mathfrak{R}}[u]$ where u is twice continuously differentiable on $\mathfrak{R}$ and u has the same given boundary values.*

Supposing our primary objective were to minimize $D_{\mathfrak{R}}[u]$ (rather than to find the harmonic function ψ), then we could rewrite the proof of Theorem A-1*a* slightly to obtain:

THEOREM A-1*b* *If $D_{\mathfrak{R}}[u]$ actually has a minimum over all functions u (with continuous second derivatives and smooth boundary values on $\partial\mathfrak{R}$), then the minimum will be achieved by a harmonic function ψ with given boundary data.*

The change in manipulations consists of letting u_0 be the minimizing function [for which $u_0(Z) = g(Z)$ on $\partial\mathfrak{R}$]. Then consider

$$f(\varepsilon) = D_{\mathfrak{R}}[u_0(z) + \varepsilon w(z)] = D_{\mathfrak{R}}[u_0] + 2\varepsilon D_{\mathfrak{R}}[u_0,w] + \varepsilon^2 D_{\mathfrak{R}}[w] \tag{A-6}$$

where $w(z)$ is a twice continuously differentiable function vanishing on $\partial\mathfrak{R}$ [so that $u_0(z) + \varepsilon w(z)$ has the boundary values $g(Z)$ on $\partial\mathfrak{R}$]. Then $f(\varepsilon)$ must be a minimum at $\varepsilon = 0$, and $f'(0) = 0$. Thus $0 = \frac{1}{2}f'(0) = D_{\mathfrak{R}}[u_0,w]$,

$$0 = -\iint_{\mathfrak{R}} w\,\Delta u_0\,dx\,dy - \int_{\partial\mathfrak{R}} 0 \cdot \frac{\partial u_0}{\partial n}\,ds \tag{A-7}$$

Actually, the condition $f'(0) = 0$ is only necessary for a minimum. (It would also prevail if we sought a maximum, but the facts would exclude the possibility.) The method here is called the method of the "first variation" [$f'(0)$] and the general subject is called the "calculus of variations."

It is now an elementary conclusion that if (A-7) holds for arbitrary w (vanishing on $\partial\mathfrak{R}$), then, necessarily, $\Delta u_0 = 0$. This proves Theorem A-1*b*.

Now the method of concluding on an a priori *basis* that $D_{\mathfrak{R}}[u]$ must have a minimum [subject to $u(Z) = g(Z)$ on $\partial\mathfrak{R}$, etc.] is called "Dirichlet's principle." It is *not* an a priori triviality, however. We could always say that $D_{\mathcal{R}}[u] \geq 0$ intrinsically, therefore $I = \inf D_{\mathcal{R}}[u]$ (≥ 0) exists, but we cannot even say that $I > 0$, much less can we say that a u_0 exists for which $D[u_0] = I$ [with $u_0(Z) = g(Z)$ on $\partial\mathcal{R}$]. Indeed, the best we can say is that a sequence $u_n(z)$ exists for which: $D_{\mathcal{R}}[u_1] \geq D_{\mathcal{R}}[u_2] \geq \cdots \geq I$ and indeed, that $D_{\mathcal{R}}[u_n] \to I$, but we cannot say *beforehand* that $u_n(z)$ must converge to a limit $u_0(z)$ uniformly (or any other way so as to produce a differentiable limit). This incorrect type of reasoning (assuming the limit beforehand) was a tempting mistake, made consistently by many including Riemann in 1851. It led Weierstrass and others to reexamine the foundations of analysis and even the foundations of irrational numbers.

In the meantime, pending a rigorous justification of the Dirichlet principle (as an a priori method), the existence theorems were established substantially by Schwarz and Neumann (ca. 1880) as done here. Ultimately, Hilbert, Weyl, Courant, and others (ca. 1908–1910) justified Riemann's "a priori reasoning" by a good deal of "a posteriori hindsight," opening new vistas in the theory of function spaces. We regretfully ignore their arguments which are too long to reproduce.

There are still three reasons† to be cognizant of the minimal methods:

(*a*) Intrinsic relevancy to abelian integrals

(*b*) Use of normal families of functions

(*c*) Other adaptations of extremal methods

INTRINSIC RELEVANCY TO ABELIAN INTEGRALS

We first note that according to Exercise 11, Sec. 8-3, $D_{\mathcal{R}}[u]$ is invariant under conformal mapping of $\mathcal{R}$. Specifically, if $u(z)$ is a given function (not necessarily harmonic) and if $z = z(\zeta)$ transforms $\mathcal{R}$ (in the z plane) into $\mathcal{R}'$ in the ζ plane in one-to-one fashion, then

(A-8) $$D_{\mathcal{R}}[u(z)] = D_{\mathcal{R}'}[u(z(\zeta))] = \iint_{\mathcal{R}'} \left[\left(\frac{\partial u}{\partial \xi}\right)^2 + \left(\frac{\partial u}{\partial \eta}\right)^2\right] d\xi\, d\eta$$

Likewise, $D_{\mathcal{R}}[u,v]$ is invariant (as it equals $\frac{1}{2}D_{\mathcal{R}}[u+v] - \frac{1}{2}D_{\mathcal{R}}[u] - \frac{1}{2}D_{\mathcal{R}}[v]$).

† We restrict ourselves to matters bearing on complex analysis. A more compelling *practical* reason is that the solutions to the Dirichlet problem can be approximated *numerically* by use of the Dirichlet principle. This is sometimes called the *Rayleigh-Ritz method;* it consists of parametrizing a family of functions with t parameters $\phi(x,y;a_1, \ldots, a_t)$ [analogously with the one-parameter family $u_0 + \varepsilon w$ in (A-6)] and forming:

$$f(a_1, \ldots, a_t) = D_{\mathcal{R}}[\phi(x,y;a_1, \ldots, a_t)]$$

[analogously with $f(\varepsilon)$ in (A-6)]. Then the "best" parameters $a_1, \ldots, a_t$ are determined by $\partial f/\partial a_1 = \cdots = \partial f/\partial a_t = 0$, and the suitability of ϕ as an approximation to the unknown harmonic function ψ is determined by how small $D[\phi]$ can be made. There is a vast literature on this method, but it bears little relevancy to our approach.

Thus $D_{\mathfrak{M}}[u]$ *can be defined invariantly for a Riemann manifold* $\mathfrak{M}$ (independently of parametrization) by summing $D_{\mathfrak{R}_n}[u]$ over all polygons $\mathfrak{R}_n$ of $\mathfrak{M}$. More important, if u is an integral with undetermined additive constant (when $\mathfrak{M}$ is not single-valued), $D_{\mathfrak{M}}[u]$ will be independent of such a constant; in other words, $D_{\mathfrak{M}}[u]$ will depend only on the differential $du = (\partial u/\partial x)\,dx + (\partial u/\partial y)\,dy$.

Finally, let us consider abelian integrals of the first kind when $\mathfrak{M}$ is of genus p with n (≥ 2) boundary curves. Then we can prescribe $\int_{Q_1} du = h_1, \ldots, \int_{Q_p'} du = h_p'$, for all $2p$ crosscuts and $u = k_t$ (constant) on all boundary curves $\mathfrak{C}_t$ ($1 \leq t \leq n$). The solution to the indicated problem can again be shown to minimize $D[u]$ over the indicated prescribed data.

The interesting point, however, is that we can also minimize $D_{\mathfrak{M}}[u]$ keeping $h_1, \ldots, h_p'$ (the $2p$ periods) fixed but *permitting* $k_1, \ldots, k_n$ *to vary.* Then the solution would turn out to be such as to minimize u exactly when u^* is *single-valued* around all $\mathfrak{C}_1, \ldots, \mathfrak{C}_n$; for example,

$$\int_{\mathfrak{C}_t} du^* = 0 \quad (t = 1, 2, \ldots, n)$$

Thus floating (or free) boundaries take care of themselves! In fact, if we permitted a boundary curve $\mathfrak{C}_t$ to be completely free (no prescription at all, not even u = undetermined constant), then we should automatically obtain ($\partial u/\partial n =$) $\partial u^*/\partial s = 0$ on $\mathfrak{C}_t$.

In effect, then, the Dirichlet problem simplifies the formulation of the natural boundary-value problems of hydrodynamics.

USE OF NORMAL FAMILIES OF FUNCTIONS

Now let us recall that the difficulties in the Dirichlet principle involve the fact that *real* functions u need not converge simply because $D[u]$ converges to, say, its minimum. In *complex* variable theory, however, we are dealing with functions which tend to converge more easily, as indicated by Harnack's theorems on harmonic functions (in Sec. 10-1). Hence let us consider the possibility of defining desired harmonic (or complex) functions by minimal procedures involving only *other* harmonic (or complex) functions. For convenience let us take *complex* functions as an example (rather than harmonic functions).

We define a *normal family* of functions $\mathfrak{F}$ (with Montel, 1927) as an aggregate of analytic functions $\{f(z)\}$ for z on, say, a finite region $\mathfrak{R}$, with the property that given any closed subdomain $\mathfrak{R}'$ and any infinite subsequence $\mathfrak{F}'$ of $\mathfrak{F}$, $\{f_n(z)\}$, then a subsequence of $\mathfrak{F}'$, $\{f_{n_t}(z)\}$ ($t = 1, 2, \ldots$), converges uniformly on $\mathfrak{R}'$.

THEOREM A-2 *An aggregate of (uniformly) bounded functions on a finite region* $\mathfrak{R}$ *is a normal family.*

PROOF Let us first consider the case where $\mathcal{R}$ is circular of radius R while $\mathcal{R}'$ is a closed concentric disk of radius R' ($<R$). Take the center to be the origin. We can then write for our sequence

$$f_n(z) = a_0^{(n)} + a_1^{(n)}z + \cdots + a_t^{(n)}z^t + \cdots \tag{A-9}$$

Since $|f_n(z)| <$ (say) M for $|z| \leq R$, then by the Cauchy estimate on derivatives,† $|a_t^{(n)}| \leq M/R^t$ and $f_n(z)$ is dominated term by term by the series formed from (A-9) by taking $|z| = R'$

$$\frac{M}{1 - (R'/R)} = M + M\frac{R'}{R} + M\left(\frac{R'}{R}\right)^2 + \cdots$$

Thus all series (A-9) [of the subsequence $f_n(z)$] converge uniformly. We therefore need only observe that the first coefficient $a_0^{(n)}$ (as n varies) is bounded and hence has a subsequence $n_1, n_2, \ldots$ for which $a_0^{(n_r)}$ converges to a limit point. Now from the subsequence $n_1, n_2, \ldots$, select a finer subsequence $n = m_1, m_2, \ldots$ for which $a_1^{(n)}$ (as well as $a_0^{(n)}$) converges to a limit point. This leads to a matrix of subsequences

$$\begin{array}{l} n_1, n_2, n_3, n_4, \ldots \\ m_1, m_2, m_3, m_4, \ldots \\ p_1, p_2, p_3, p_4, \ldots \\ q_1, q_2, q_3, q_4, \ldots \end{array}$$

each coming from the preceding. Form, next, the diagonal subsequence

$$n_1, m_2, p_3, q_4, \ldots$$

which we relabel $N_1, N_2, N_3, N_4, \ldots$. Then for any fixed t,

$$a_t^{(N_k)} \to a_t^{(0)} \qquad \text{as } k \to \infty \tag{A-10a}$$

for some limit $a_t^{(0)}$ (abs. $\leq M/R^t$). By uniform convergence of each of the functions $f_n(z)$ in $|z| \leq R'$, it follows that

$$f_{N_k}(z) \to f^{(0)}(z) \tag{A-10b}$$

where $f^{(0)}(z) = \sum_0^\infty a_t^{(0)}z^t$.

With the theorem established for circles, let us consider the more general region $\mathcal{R}$ and let $\mathcal{R}'$ be two overlapping disks interior to $\mathcal{R}$. For one disk, a subsequence can be chosen as indicated, and for the other disk a subsequence of the first subsequence can be chosen. Thus any finite union of disks is now a valid choice for $\mathcal{R}'$ ($\subset \mathcal{R}$). As seen earlier (e.g., Exercise 7 in Sec. 10-1), any $\mathcal{R}'$ which lies interior to $\mathcal{R}$ can be covered by such a finite set of circles, hence the theorem is proved. Q.E.D.

† Compare Exercise 5 in Sec. 1-7.

With this in mind we are prepared to prove anew the main conformal mapping theorem (of Sec. 10-2). Let $\mathfrak{R}$ be any simply-connected region of the z plane which we place interior to the unit disk $|z| < 1$ by the usual elementary transformations. Let $\mathfrak{F} = f(z)$ be the aggregate of functions such that $w = f(z)$ maps $\mathfrak{R}$ in univalent fashion on a region $\mathfrak{S}$ interior to the unit disk, keeping $f(0) = 0$. Clearly, $f(z)$ is a normal family, because $\mathfrak{F}$ is uniformly bounded by 1.

We next define (instead of $D[u]$) the functional

$$M_{\mathfrak{R}}[f] = \text{(minimum) distance from 0 to } \partial\mathfrak{S} \qquad f \in \mathfrak{F} \tag{A-11}$$

Clearly $M_{\mathfrak{R}}[f] \leq 1$. We *should like to show that for some* $f_0 \in \mathfrak{F}$ we have $M_{\mathfrak{R}}[f_0] = 1$. This would make f_0 a function mapping $\mathfrak{R}$ onto the unit disk.

THEOREM A-3 *Any simply-connected region whose boundary consists of more than one point can be mapped biuniquely onto the interior of a unit disk.*

PROOF The problem is to find a way of singling out $f_n(z)$, a subsequence of $\mathfrak{F}$ (as just defined), so that any convergent sequence within $f_n(z)$ must converge to the desired f_0. Take $f_n(z)$ an optimizing sequence for $M_{\mathfrak{R}}[f]$. We first must show $M_{\mathfrak{R}}[f_n] \to 1$. This, however, has already been accomplished by Caratheodory's construction (see Theorem 10-5). Then clearly, $f_n(z) \to f_0(z)$ for some $f_0(z)$ uniformly on any region interior to $\mathfrak{R}$. The function $f_0(z)$ must belong to $\mathfrak{F}$ (that is, it must be univalent) since it is the limit of a uniformly convergent sequence of such functions. (See Exercise 6, Sec. 1-8.) It then follows that $f_0(z)$ is the desired function.

Q.E.D.

Note that this proof, unlike the earlier one, involving Green's function does not require that the boundary of $\mathfrak{R}$ be piecewise smooth, even as a matter of convenience.

OTHER ADAPTATIONS OF EXTREMAL METHODS

At the other extreme we can use unsmooth functions advantageously in minimal problems to bring out properties of analytic functions. This is a more modern and more powerful development than those just described, but we shall present only one illustration of outstanding elegance, originated by Grötzsch (1928).

In considering the module problem, we ask whether or not two regions (of, say, the plane) with like connectivity are conformally equivalent.

Let $\mathfrak{R}$ be a region of connectivity n (with boundary curves

$$\partial\mathfrak{R} = \mathfrak{C}_1 + \mathfrak{C}_2 + \cdots + \mathfrak{C}_n)$$

Let us define an *extremal metric* for each homology class $\mathfrak{H}$ as follows: We consider the aggregate of all *continuous* nonnegative functions $\rho(x,y)$

defined in $\mathcal{R}$, subject to normalization† by

(A-12*a*) $$\iint_{\mathcal{R}} \rho^2\, dx\, dy = 1$$

For each ρ (and each homology class $\mathcal{H}$) we define

(A-12*b*) $$m_{\mathcal{R}}[\rho,\mathcal{H}] = \inf_{\mathcal{C}\sim\mathcal{H}} \int_{\mathcal{C}} \rho|dz|$$

where the infimum is taken over all $\mathcal{C}$ homologous with $\mathcal{H}$ (in the closure $\overline{\mathcal{R}}$).

We then define the *extremal length* of the curves homologous with $\mathcal{H}$ as:

(A-13) $$M_{\mathcal{R}}\{\mathcal{H}\} = \sup_{\rho} m[\rho,\mathcal{H}]$$

where now the supremum is taken over all ρ. If an optimal ρ exists, it is called the *extremal weight function.*

Let us now observe that the values $M\{\mathcal{H}\}$ are *conformally invariant,* for if $\mathcal{R}$ is mapped onto $\mathcal{R}'$, then $dx\, dy$ equals $|dz/dz'|^2\, dx'\, dy'$ and $|dz|$ equals $|dz/dz'|\,|dz'|$; hence each ρ is merely changed to $\rho/|dz'/dz|$.

Of course, it is not evident that an optimizing curve $\mathcal{C}$ exists for (A-12*b*) or that an optimizing ρ exists for (A-13).

Nevertheless, we could see that $M_{\mathcal{R}}\{\mathcal{H}\} > 0$, for if we take

$$\rho_0 = \text{const} = 1/\sqrt{A}$$

for A the area of $\mathcal{R}$, then $m_{\mathcal{R}}[\rho_0,\mathcal{H}] = l\{\mathcal{H}\}/\sqrt{A}$ where $l\{\mathcal{H}\}$ is the minimum euclidean length of chains homologous with $\mathcal{H}$ [for example, if $\mathcal{H} \sim \mathcal{C}_1 + \mathcal{C}_2$, then if $\mathcal{C}_1$ and $\mathcal{C}_2$ are convex, $l\{\mathcal{H}\} = \text{length}\,\mathcal{C}_1 + \text{length}\,\mathcal{C}_2$]. Clearly then, $M_{\mathcal{R}}\{\mathcal{H}\} \geq m_{\mathcal{R}}[\rho_0,\mathcal{H}] > 0$.

The strength of this method lies in our ability to do elementary things with elegance. For instance, we can generalize on the result that two annuli are inequivalent if one lies inside the other, without directly using harmonic functions! (In fact, we do not need to know the major existence theorems.)

THEOREM A-4 *Let $\mathcal{R}$ and $\mathcal{R}'$ be two regions of the same connectivity n (>0) and bounded by circles $\mathcal{C}_n$ and $\mathcal{C}'_n$ on the outside. Let $\mathcal{R} \supset \mathcal{R}'$ in such a fashion that $\mathcal{C}'_n$ lies wholly interior to $\mathcal{R}$. Then $\mathcal{R}$ and $\mathcal{R}'$ cannot be conformally equivalent in such a fashion as to identify $\mathcal{C}_n$ and $\mathcal{C}'_n$.*

PROOF Let us take any continuous nonnegative ρ in $\mathcal{R}$, the larger region. Then this defines a function ρ' (assumed not identically 0) in $\mathcal{R}'$, the smaller region, as $\rho/\sqrt{\lambda[\rho]}$ where $\lambda[\rho]$ is defined by

$$\lambda[\rho] = \iint_{\mathcal{R}'} \rho^2\, dx\, dy \leq \iint_{\mathcal{R}} \rho^2\, dx\, dy = 1$$

† We do not require that ρ be bounded as $z \to \partial\mathcal{R}$. (In advanced treatments ρ need not even be continuous.)

Using $\rho' = \rho/\sqrt{\lambda[\rho]}$ in $\mathcal{R}'$, we find

$$m_{\mathcal{R}'}[\rho',\mathcal{C}_n] = \inf_{\mathcal{C}\sim\mathcal{C}_n} \int_{\mathcal{C}} \rho'|dz| \qquad \text{for } \mathcal{C} \subset \mathcal{R}'$$

Now for each $\mathcal{C}(\sim\mathcal{C}_n) \subset \mathcal{R}' \subset \mathcal{R}$

$$\int_{\mathcal{C}} \rho'|dz| = \frac{1}{\sqrt{\lambda[\rho]}} \int_{\mathcal{C}} \rho|dz|$$

with *more* curves $\mathcal{C}$ $(\sim\mathcal{C}_n)$ lying in $\mathcal{R}$ than $\mathcal{R}'$. Now the minimum is greater if taken over fewer items, so

$$\inf_{\mathcal{C}\subset\mathcal{R}'} \int_{\mathcal{C}} \rho'|dz| = \frac{1}{\sqrt{\lambda[\rho]}} \inf_{\mathcal{C}\subset\mathcal{R}'} \int_{\mathcal{C}} \rho|dz| \geq \frac{1}{\sqrt{\lambda[\rho]}} \inf_{\mathcal{C}\subset\mathcal{R}} \int_{\mathcal{C}} \rho|dz|$$

(A-14)
$$m_{\mathcal{R}'}[\rho',\mathcal{C}_n'] \geq \frac{m_{\mathcal{R}}[\rho,\mathcal{C}_n]}{\sqrt{\lambda[\rho]}}$$

Now if ρ were actually the (optimum) extremal weight function for $\mathcal{R}$, we *could* show without trouble that $0 < \lambda[\rho] < 1$. Then from (A-14),

(A-15*a*)
$$m_{\mathcal{R}'}[\rho',\mathcal{C}_n'] \geq \frac{M_{\mathcal{R}}\{\mathcal{C}_n\}}{\sqrt{\lambda[\rho]}}$$

and therefore the supremum on the left would yield

(A-15*b*)
$$M_{\mathcal{R}'}\{\mathcal{C}_n'\} > M_{\mathcal{R}}\{\mathcal{C}_n\}$$

Now (A-15*b*) would contradict conformal equivalence, proving Theorem A-4.

Indeed, although it can be shown that the extremal values are achieved for both (A-12*b*) and (A-13) by more advanced treatments of the theory, we should like to obtain (A-15*b*) without such deep results. We wish to exclude from consideration the possibility that $\lambda[\rho]$ is near 1 or 0.

To exclude $\lambda[\rho]$ near 1, we refer to Exercise 7, which shows that if $\rho_t(x,y)$ is a sequence of functions for which the integral over $\mathcal{R}'' = \mathcal{R} - \overline{\mathcal{R}'}$ (the difference region) satisfies

(A-16)
$$\iint_{\mathcal{R}''} \rho_t^2\, dx\, dy \to 0 \qquad \text{as } t \to \infty$$

then for some (circular) $\mathcal{C} \sim \mathcal{C}_n$ in $\mathcal{R}$, $\int_{\mathcal{C}} \rho_t|dz| \to 0$. (Here $\mathcal{C}$ depends on t.) Hence, we can exclude the possibility that $\lambda[\rho_t] \to 1$ for any maximizing sequence $\rho_t(x,y)$ for which $m_{\mathcal{R}}[\rho_t,\mathcal{C}_n] \to M_{\mathcal{R}}\{\mathcal{C}_n\}$. In a similar way we show (A-16) cannot hold for integrals over $\mathcal{R}'$, thus $\lambda[\rho]$ is not near 0 either.

Q.E.D.

Thus we have proved Theorem A-4 without explicitly using any existence theorems for harmonic functions or extremal metrics! We have hereby learned about modules of conformal equivalence by elementary means. It is no wonder that extremal methods have been applied so successfully toward this end (as evidenced in the literature).

EXERCISES

1 Let $\mathfrak{R}_1$ be the unit disk $|z| < 1$, and let the boundary data be

$$g(Z) = \frac{a_0}{2} + \sum_1^\infty (a_n \cos n\theta + b_n \sin n\theta) = f(\theta)$$

for $Z = e^{i\theta}$, and let $\Sigma |a_n| + |b_n| < \infty$ so that $f(\theta)$ is continuous by uniform convergence. Then the Poisson solution exists and equals

$$\psi(z) = \frac{a_0}{2} + \sum_1^\infty r^n(a_n \cos n\theta + b_n \sin n\theta)$$

Verify that $D_{\mathfrak{R}_\rho}[\psi]$, as defined by $\iint\limits_{\mathfrak{R}_\rho} \left[\left(\frac{\partial \psi}{\partial r}\right)^2 + \frac{1}{r^2}\left(\frac{\partial \psi}{\partial \theta}\right)^2\right] r\, dr\, d\theta$ over $\mathfrak{R}_\rho$ of radius ρ (<1), equals

$$\sum_1^\infty (a_n^2 + b_n^2) n\pi\rho^{2n}$$

Show that $D_{\mathfrak{R}_\rho}[\psi]$ can approach ∞ as $\rho \to 1$, even under the assertion of uniform convergence of the series for $f(\theta)$. Show that if $f(\theta)$ has continuous second derivatives, then a_n and b_n are $< k/n^2$ and $D_{\mathfrak{R}_1}[\psi]$ exists. (*Hint:* Hadamard's famous counterexample is $a_n = 1/m^2$ for $n = m!$ whereas the other a_n and b_n vanish.)

2 (*Weierstrass*) Let $u(x)$ be a continuous and differentiable function in $0 \le x \le 1$ with boundary conditions $u(0) = 0$, $u(1) = 1$. Define $D[u] = \int_0^1 \sqrt{|u'(x)|}\, dx$. Show $\inf D[u] = 0$ as u varies over all such functions while $D[u] = 0$ would not lead to any smooth function (where du/dx is continuous). (*Hint:* Try $u = x^n$.)

3 In the preceding exercise, set $D[u] = \int_0^1 [u'(x)]^2\, dx$ subject to $u(0) = 0$, $u(1) = 1$. Show that the minimum is achieved for $u_0 = x$ by showing that analogously with (A-5*b*), if $w(x) = 0$ at $x = 0$ and 1, then $D[u_0 + w] = D[u_0] + D[w]$.

4 Likewise consider $D[u] = \int_0^1 |u'(x)|\, dx$ subject to $u(0) = 0$, $u(1) = 1$, and show $\min D[u] = 1$, but the u_0 for which $D[u_0] = 1$ is not unique. What about $\int_0^1 \sqrt{1 + [u'(x)]^2}\, dx$?

5 Verify that for a Riemann manifold $\mathfrak{M}$ with boundary curves $\mathcal{C}_1, \ldots, \mathcal{C}_n$ and crosscuts $\mathcal{Q}_1, \mathcal{Q}_1', \ldots, \mathcal{Q}_p, \mathcal{Q}_p'$, the analogs of Theorems A-1*a* and A-1*b* involve minimizing $D[u]$ with $u(Z)$ prescribed on $\partial\mathfrak{M}$ with periods $u\{\mathcal{Q}_t\} = (\mathcal{Q}_t, du) = h_t$, etc. Show (A-4) is replaced by

$$D_{\mathfrak{M}}[u,v] = -\iint\limits_{\mathfrak{M}^*} v\, \Delta u\, dx\, dy - \sum_{t=1}^n \int_{\mathcal{C}_t} v \frac{\partial u}{\partial n}\, ds + \sum_{t=1}^p \left\{\int_{Q_t} v\{\mathcal{Q}'\} \frac{\partial u}{\partial n}\, ds - \int_{Q_t'} v\{\mathcal{Q}_t\} \frac{\partial u}{\partial n}\, ds\right\}$$

From this, give the analogs of Theorems A-1a and A-1b for a Riemann manifold of p handles and n boundaries.

6 Show that for the Riemann manifold $\mathfrak{M}$ *if a harmonic function* ψ *exists* on $\mathfrak{M}$ with prescribed periods and boundary data on some segments $\mathcal{Z}_1$ of $\partial\mathfrak{M} = \mathcal{C}_1 + \cdots + \mathcal{C}_n$, but with $\partial\psi/\partial n = 0$ on other (free) segments $\mathcal{Z}_2$ of $\partial\mathfrak{M} = \mathcal{Z}_1 + \mathcal{Z}_2$, then ψ will minimize $D_{\mathfrak{M}}[u]$ over all smooth u with given periods and boundary data on $\mathcal{Z}_1$ (whereas u is unrestricted on $\mathcal{Z}_2$). (The existence of such a ψ is referred to the literature.)

7 In Theorem A-4, assume $\mathcal{C}_n$ is the circle $|z| = 1$ and ($\mathfrak{R}'$ and) $\mathcal{C}_n'$ lies inside the circle $|z| = r_0$ (<1). Verify that by the Schwarz' inequality,

$$\int_{r_0}^{1} dr \left(\int_0^{2\pi} 1 \cdot \rho r \, d\theta\right)^2 \leq 2\pi \int_0^{2\pi} \int_{r_0}^{1} \rho^2 r^2 \, dr \, d\theta \leq 2\pi \int_0^{2\pi} \int_{r_0}^{1} \rho^2 r \, dr \, d\theta$$

hence show that if $\int_0^{2\pi} \int_{r_0}^{1} \rho_t^2 r \, dr \, d\theta \to 0$ for a sequence of functions ρ_t, then for a circular $\mathcal{C}(\sim\mathcal{C}_n)$, depending on t, $\int_{\mathcal{C}} \rho_t \, |dz| \to 0$.

8 Consider the Riemann manifold consisting of $0 \leq x \leq a$, $0 \leq y \leq b$ with identification of sides by $z + a \to z$. Show the extremal weight function is $\rho_0 = 1/\sqrt{ab}$ for curves homologous with the (closed) curve $\mathcal{C}_1$: $0 \leq x \leq a$ (y = constant). [*Hint:* From ρ_0 we verify $m[\rho_0,\mathcal{C}_1] = a\rho_0 = \sqrt{a/b}$.] Hence, starting with any other ρ satisfying (A-12a), note

$$\int_0^b \int_0^a (\rho - \rho_0)^2 \, dx \, dy > 0$$

hence, show

$$\frac{1}{\rho_0} \geq \int_0^b dy \int_0^a \rho \, dx \geq bm[\rho,\mathcal{C}_1]$$

and thus

$$m[\rho_0,\mathcal{C}_1] \geq m[\rho,\mathcal{C}_1] \qquad \text{with equality only when } \rho = \rho_0$$

9 Consider the extremal weight function for the ring $r_1 < |z| < r_2$ and the class of circles centered at the origin, by using the above result.

10 Consider the Riemann manifold consisting of the (torus) parallelogram structure $z \to z + \omega_1$, $z \to z + \omega_2$ (where ω_2/ω_1 is not real). Let $\mathcal{C}_1$ represent the curves homologous to ω_1 and $\mathcal{C}_2$ represent the curves homologous to ω_2. Show that for either $\mathcal{C}_1$ or $\mathcal{C}_2$ the extremal weight ρ_0 is the constant $|\text{Im } \omega_1\bar{\omega}_2|^{-1/2}$, while

$$m\{\mathcal{C}_1\} = \frac{|\omega_1|}{|\text{Im } \omega_1\bar{\omega}_2|^{1/2}} \qquad m\{\mathcal{C}_2\} = \frac{|\omega_2|}{|\text{Im } \omega_1\bar{\omega}_2|^{1/2}}$$

APPENDIX B
INFINITE MANIFOLDS

Finite manifolds, taken by themselves, constitute a neat subject with existence theorems which hardly can admit improvement (except for modular problems which are almost unapproachable!). The physically important problems seem to all correspond to finitely triangulated manifolds. Why then should we want to consider infinite manifolds?

The answer goes back to the reason we studied differentials in preference to integrals (e.g., in Sec. 5-2). The reason was that the differentials on, say, a finite manifold $\mathfrak{M}$ are *actually defined on* $\mathfrak{M}$, whereas the integrals of these differentials were defined only on the canonical dissection $\mathfrak{M}^p$. But integrals are really defined on a (generally infinite) manifold $\mathfrak{M}^c$ with no artificial boundary cuts, the so-called "(universal) covering surface" of $\mathfrak{M}$.

ANNULUS

Let us illustrate the idea of a covering surface by taking for $\mathfrak{M}$ the annulus

$$\mathfrak{M}: \quad r_1 < |z| < r_2 \tag{B-1}$$

The expression $g(z)\,dz$ is a (meromorphic) differential on $\mathfrak{M}$ (as we recall) when $g(z)$ is meromorphic on $\mathfrak{M}$; indeed, if $g(z) = dz/2\pi iz$, then $g(z)\,dz$ was seen to be an *abelian* differential.

Suppose we next turn our attention to integrals; for example, $G(z) = \int g(z)\,dz$ or more specially,

$$G(z) = 1/(2\pi i) \log z + \text{const}$$

Clearly $G(z)$ is not defined on $\mathfrak{M}$ (inasmuch as it is multivalued), but $G(z)$ is defined on $\mathfrak{M}^c$, the Riemann surface for $\log z$ or the spiral region covering $\mathfrak{M}$. This region $\mathfrak{M}^c$ is shown in Fig. B-1. It has no necessary connection with $\log z$ since it is indeed a geometrical entity which could have been constructed even if we were ignorant of any abelian differentials on $\mathfrak{M}$.

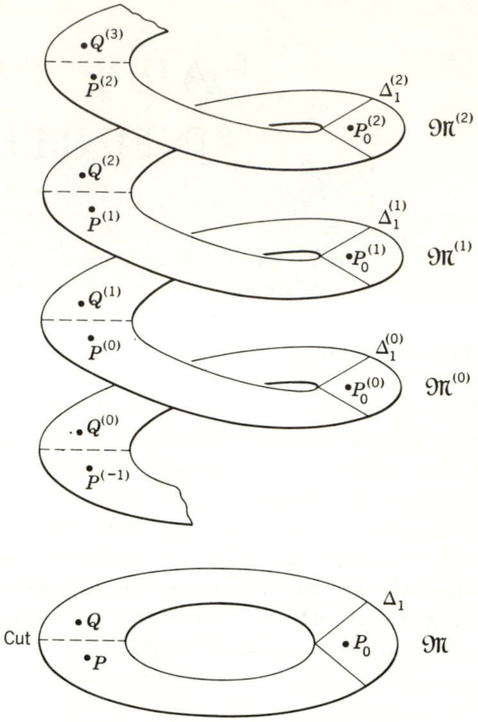

Fig. B-1 Riemann surface of log z. (Only one triangle Δ_1 is shown in the annulus.)

For example, if $\mathfrak{M}$ is triangulated as $\Delta_1, \Delta_2, \ldots, \Delta_F$, then $\mathfrak{M}_c$ consists of infinitely many replicas of each Δ_i, namely

(B-2) $$\mathfrak{M}^{(k)}: \quad \Delta_1{}^{(k)}, \Delta_2{}^{(k)}, \ldots, \Delta_F{}^{(k)}$$

each corresponding to a value of k in

(B-3) $$G(z) = \frac{1}{2\pi i} \log z + k$$

where, say, log z denotes the principal value. We might say, in effect, that there are infinitely many replicas of $\mathfrak{M}$, namely $\mathfrak{M}^{(k)}$; but that these are cut and pasted in the fashion of Fig. B-1 to form $\mathfrak{M}^c$, *so that any integral* $G(z) = \int g(z)\,dz$ [not just $G(z)$ in (B-3)] *is uniquely defined* by analytic continuation on $\mathfrak{M}^c$.

Referring to Sec. 3-5, we might say that the function $w = \log z$ maps the covering surface of the annulus $\mathfrak{M}^c$ onto an infinite strip ($\log r_1 < \operatorname{Re} w < \log r_2$) in biunique fashion.

GENERAL DEFINITION

Let us render the above situation more general. To begin with, the primary definition of Riemann manifold $\mathfrak{M}$ (finite or infinite) is as a collection of objects $[P,\mathcal{Z}_P,\Phi_P,\mathfrak{R}_P]$ (points P with cyclic neighborhoods $\mathcal{Z}_P$, and parametrizations Φ_P with plane neighborhoods $\mathfrak{R}_P$). This collection must be accompanied by mappings of intersections $\mathcal{Z}_P \cap \mathcal{Z}_Q$ and $\mathfrak{R}_P \cap \mathfrak{R}_Q$ which assure connectivity between any two points. (See Sec. 5-1.)

Now start with any *base point* P_0 [essentially the lower limit of $G(z) = \int_{P_0}^{P} g(z)\,dz$] and consider all paths $\mathcal{C}$ lying in $\mathfrak{M}$ and connecting P_0 with an arbitrary P. Thus $G(z)$ is dependent on $\mathcal{C}$; for example, $G(z) = (\mathcal{C},g\,dz)$.

We next divide the paths $\mathcal{C}$ from P_0 to P into the *homotopy* equivalence classes such that two paths $\mathcal{C}_0$, $\mathcal{C}_1$ are *equivalent* (or homotopic) exactly when $\mathcal{C}_0$ can be continuously deformed into $\mathcal{C}_1$ always within $\mathfrak{M}$. Thus the closed path $\mathcal{C}_0 - \mathcal{C}_1$ is *zero-homotopic;* i.e., it can be continuously deformed to some point of $\mathfrak{M}$, taken as P_0 for convenience.

Here we need the result that *if* $\mathcal{C}_0$ *and* $\mathcal{C}_1$ *are homotopic, they are homologous.* We must refer the reader to standard texts in topology on this point, but the intuitive reasoning is that some region $\mathfrak{R}$ is swept out in the deformation of $\mathcal{C}_0$ into $\mathcal{C}_1$, and $\mathcal{C}_0 - \mathcal{C}_1$ is the boundary of such a region $\mathfrak{R}$. (In effect, if $\mathcal{C}_0$ crosses $\mathcal{C}_1$, the "region" $\mathfrak{R}$ will become a combination of regions, e.g., a 2-chain, and the proof will become more elaborate.) Conversely, however, homologous curves are not homotopic necessarily, as cited in Exercise 1 of Sec. 1-6.

For example, if $\mathfrak{M}$ is simply-connected, any closed curve is deformable to a point hence there is only *one* equivalence class of paths joining two specified points P_0, P. Thus $\mathfrak{M} \equiv \mathfrak{M}^c$.

Let us denote each equivalence class of paths from P_0 to P by $\widehat{P_0P}^{(k)}$ where the index k enumerates such paths for each terminal P. For each $\widehat{P_0P}^{(k)}$ we associate a $P^{(k)}$ and we construct the replicas $[P^{(k)},\mathcal{Z}_P{}^{(k)}]$ and the parametrizations $\Phi_{P^{(k)}}(\zeta_P{}^{(k)})$ on $\mathfrak{R}_P{}^{(k)}$. We have only to set up the matching of intersections $\mathcal{Z}_P \cap \mathcal{Z}_Q$ and the Riemann manifold $\mathfrak{M}^c$ is completely described (without use of abelian integrals).

Clearly if $\mathcal{Z}_P$ intersects $\mathcal{Z}_Q$, we desire that some $\mathcal{Z}_P{}^{(k)}$ intersect some $\mathcal{Z}_Q{}^{(k')}$ but this is not wholly trivial since k and k' might be different! For instance, in Fig. B-1, the replica $\mathfrak{M}^{(0)}$ meets $\mathfrak{M}^{(1)}$ at the "cuts" ($\arg z = \pi$ and 3π), thus some $\mathcal{Z}_P{}^{(0)}$ will meet some $\mathcal{Z}_Q{}^{(1)}$. Hence if $\mathcal{Z}_P$ intersects $\mathcal{Z}_Q$ on $\mathfrak{M}$, we let $\mathcal{C}$ be a path in the class $\widehat{P_0P}^{(k)}$ and we let $\mathcal{C}_*$ be a path from P to Q through $\mathcal{Z}_P \cap \mathcal{Z}_Q$. Now $\mathcal{C} + \mathcal{C}_*$ must be a path from P_0 to Q in some new equivalence class $\widehat{P_0Q}^{(k')}$. (Although we hope generally to have $k = k'$, we would have a triviality, $\mathfrak{M}^c \equiv \mathfrak{M}$, unless $k \neq k'$ somewhere, or two replicas meet somewhere!) Hence $Z_P^{(k)} \cap Z_Q^{(k')}$ is parametrized exactly by the same relations as those which parametrize $Z_P \cap Z_Q$. The definition of $\mathfrak{M}^c$ is now complete (if we accept the loose description of the joining of $Z_P^{(k)}$ and $Z_Q^{(k')}$ on the cuts).

It is useful to refer to $\mathfrak{M}$ as the *projection* of $\mathfrak{M}^c$ and P as the projection of $P^{(k)}$, Z_P as the projection of $Z_P^{(k)}$, etc.

We can think of the definition in terms of triangulation; that is,

$$\mathfrak{M} = \Delta_1 + \Delta_2 + \cdots + \Delta_F \qquad \text{(if } \mathfrak{M} \text{ is finite)} \tag{B-4}$$

Then if $P_0 \in \Delta_1$, say, and $P \in \Delta_2$, say, the number of classes $\widehat{P_0P}$ does not exceed the number of finite connected sequences of adjacent triangles commencing with Δ_1 and terminating with Δ_2 (actually enumerable even if $\mathfrak{M}$ is enumerally infinite). This idea permits us to think of pasting sheets together, where the sheets are replicas of $\Delta_t^{(k)}$ (shown in Fig. B-2) or even more elaborate polygons composed of the Δ_t.

TRIPLY-CONNECTED REGION

For example, consider the triply-connected region $\mathfrak{M}$ in Fig. B-2 formed by cutting disks around $z = 0$, $z = 1$, and $z = \infty$ out of the z sphere. If we cut out $\mathcal{C}_{01}$ and $\mathcal{C}_{\infty 0}$ (the slits shown), we have a simply-connected region $\mathfrak{M}'$. We then construct replicas of $\mathfrak{M}'$ labeled in such a fashion as to indicate the analytic continuation from P_0 across $\mathcal{C}_{01}$ and $\mathcal{C}_{\infty 0}$. For example, $\mathfrak{M}'$ is pasted to fit with four replicas, say, $\mathfrak{M}'^{(1)}$ and $\mathfrak{M}'^{(2)}$ across $\mathcal{C}_{01}$ (above and below) and $\mathfrak{M}'^{(3)}$ and $\mathfrak{M}'^{(4)}$ across $\mathcal{C}_{\infty 0}$ (above and below). (See Fig. B-2.) Thus any finite sequence of $\mathfrak{M}'^{(1)}$, $\mathfrak{M}'^{(2)}$, $\mathfrak{M}'^{(3)}$, $\mathfrak{M}'^{(4)}$ (abbreviated as digits 1, 2, 3, 4) will determine a unique replica of $\mathfrak{M}'$, provided the sequence does not have self-cancelling successors such as 12, 21, 34, 43, e.g., $k = 142311$. The aggregate of such $\mathfrak{M}'^{(k)}$ constitutes $\mathfrak{M}^c$. (The pasting process accounts for the pattern of intersections $Z_P \cap Z_Q$).

TORUS

An even more elegant example of a model produced by pasting sheets is provided by the torus. If we consider the torus to be the period

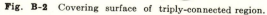

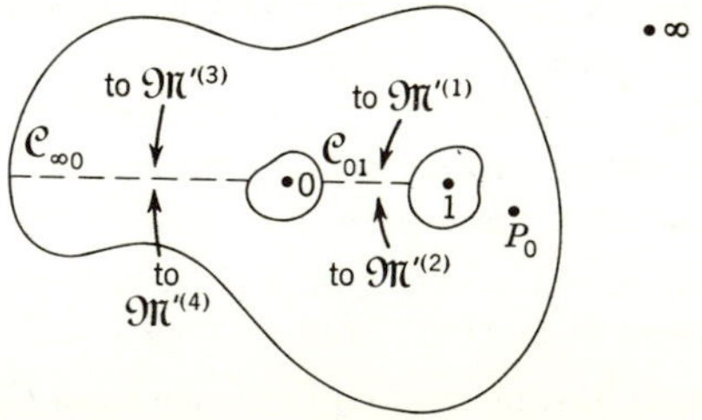

Fig. B-2 Covering surface of triply-connected region.

parallelogram of the u plane formed by identifying opposite sides, then the covering surface is the finite u plane. Thus if u_P is the coordinate of P, then the replicas of P can be parametrized by $P_{(m_1,m_2)}$, whose value is $u = u_P + 2\omega_1 m_1 + 2\omega_2 m_2$. The intersections $\mathcal{Z}_P \cap \mathcal{Z}_Q$ are the natural ones in the u plane. (See Fig. 4-3*a*.)

THEOREM B-1 *The (universal) covering surface is simply-connected.*

For a quick sketch of a proof, let us see how we would deform to P_0 an arbitrary closed curve $\mathcal{K}^c$ on $\mathfrak{M}^c$. Let us deform $\mathcal{K}^c$ (by retraction to P_0) so we can think of it as joining the base point P_0 with itself. It is clear that the curve $\mathcal{K}^c$ must be zero-homotopic or else it would join P_0 to one of its replicas on $\mathfrak{M}^c$ (but not to P_0). The process of shrinking $\mathcal{K}^c$ to a point (say P_0) causes it to sweep out an area which it bounds. Thus $\mathcal{K}^c \sim 0$.

COROLLARY 1 *The integrals on $\mathfrak{M}$ can be defined uniquely on $\mathfrak{M}^c$ by analytic continuation in a natural way. The values of integrals on different replicas $\Delta_i^{(k)}$ of a triangle Δ_i differ by constants.*

Here we see the role of the *universal* covering surface (based on *homotopy equivalence*), namely that the Weierstrass monodromy theorem (Sec. 3-1) provides for analytic continuation. If we were interested only in *abelian integrals* we could have used a *special* covering surface (based on *homology equivalence*).

OTHER INFINITE MANIFOLDS

Now we can turn our attention to other infinite manifolds (than $\mathfrak{M}^c$) which rise in problems incidental to finite manifolds. The punctured disk, for example,

$$\mathfrak{M}: \quad 0 < |z| < 1 \tag{B-5}$$

has been conspicuously ignored because of the "simplicial complex" concept of the finite Riemann manifold. (We wanted a one-dimensional boundary.) Actually, we can treat (B-5) as an infinite manifold (see Exercise 6, Sec. 5-3).

A more significant problem is the simply-connected triangulated manifold. If it is compact (hence finite), it is conformally equivalent to a sphere

$$\mathfrak{M}_{(0)}: \quad z \text{ sphere} \qquad (\text{elliptic}\dagger) \tag{B-6a}$$

If the manifold is noncompact and finite, it is conformally equivalent to

† The designations "elliptic," "parabolic," and "hyperbolic" refer to classical forms of noneuclidean geometry on each manifold (see Bibliography). They are related to the designations of the linear transformations in Sec. 2-7.

a disk

(B-6b) $$\mathfrak{M}_{(1)}: \quad |z| < 1 \qquad \text{(hyperbolic)}$$

An infinite manifold might be conformally equivalent to $\mathfrak{M}_{(1)}$, but it could also be equivalent to the finite z plane

(B-6c) $$\mathfrak{M}_{(2)}: \quad |z| < \infty \qquad \text{(parabolic)}$$

Both $\mathfrak{M}_{(1)}$ and $\mathfrak{M}_{(2)}$ might be presented to us as an infinite aggregate of triangles, however. So if we look at the *triangulations*, we could not tell immediately that $\mathfrak{M}_{(1)}$ and $\mathfrak{M}_{(2)}$ were conformally inequivalent (e.g., they are topologically equivalent by Exercise 1 of Sec. 2-1). We could tell that $\mathfrak{M}_{(1)}$ and $\mathfrak{M}_{(2)}$ are conformally inequivalent, however, by Liouville's theorem [$f(\mathfrak{M}_{(2)}) = \mathfrak{M}_1$ is impossible]. Therefore, triangulations of $\mathfrak{M}_{(1)}$ and $\mathfrak{M}_{(2)}$ can be put together to correspond topologically but not conformally.

Let us define a simply-connected infinite Riemann manifold as *parabolic* if it is equivalent to $\mathfrak{M}_1$ and *hyperbolic* if it is equivalent to $\mathfrak{M}_2$. It is a deep result that these possibilities are exhaustive (compare Exercise 2).

THEOREM B-2 (*a*) *The covering surface of an annulus is hyperbolic.*

(*b*) *The covering surface of the doubly-punctured sphere is parabolic.*

(*c*) *The covering surface of a triply-punctured sphere is hyperbolic.*

(*d*) *The same is true for the covering surface of a general triply-connected region.*

A typical method of proof is to explicitly display the mapping function. For instance, for (*b*) the function $w = \log z$ maps the covering surface of the doubly-punctured z sphere $0 < |z| < \infty$ onto the (finite) w plane ($|w| < \infty$).

The most difficult part of the theorem is (*c*) (the rest is left as exercises below). We refer to $\lambda(\tau)$ defined in Sec. 6-3. The exact value of $\lambda(\tau)$ is not important (indeed, we can define it anew). What is important is the property: *The function* $\lambda = \lambda(\tau)$ *maps the upper half* λ *plane onto the region*

(B-7) $$-1 < \operatorname{Re} \tau < 0 \qquad \tfrac{1}{2} < |\tau - \tfrac{1}{2}|$$

(*the left-hand shaded part of* $\mathfrak{M}'$ *in Fig. B-3*), *so that* $\tau = 0$ *becomes* $\lambda = 1$, $\tau = -1$ *becomes* $\lambda = 0$, *and* $\tau = i\infty$ *becomes* $\lambda = \infty$.

With $\lambda(\tau)$ satisfying the above property, then by reflection, the triply-punctured plane leads to a covering surface which fills the upper half plane.

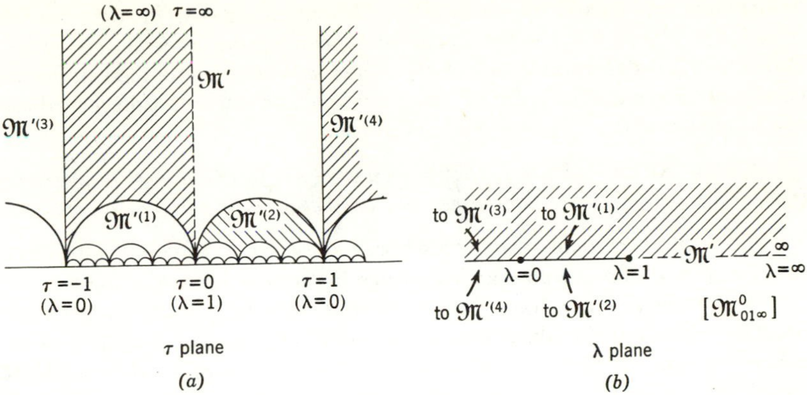

Fig. B-3 Mapping of the covering surface of the triply-punctured plane onto the upper half plane. Note that $\mathfrak{M}'$ corresponds to the slit plane (not to a half plane). (Compare Fig. 6-1.) The labeling corresponds with Fig. B-2.

Thus a biunique function $\lambda = \lambda(\tau)$ is obtained which maps the upper half τ plane biuniquely onto $\mathfrak{M}^c_{01\infty}$, the covering surface for the triply-punctured λ plane $(\lambda \neq 0,1,\infty)$.

Actually, we are more interested, however, in the inverse of $\lambda(\tau)$

(B-8) $$\tau = \tau(\lambda)$$

which maps $\mathfrak{M}^c_{01\infty}$ onto upper half τ plane biuniquely.

PICARD'S THEOREM

THEOREM B-3 *Let $\mathcal{R}$ be a simply-connected region on the z sphere, and let $w = \Phi(z)$ be a function on $\mathcal{R}$ with values lying in $\mathfrak{M}$ a subregion of the w sphere of arbitrary connectivity. Let $\mathfrak{M}^c$ be the covering surface of $\mathfrak{M}$ (in the w sphere), and let $\gamma(w)$ be a (single-valued) meromorphic function defined on $\mathfrak{M}^c$. Then (by analytic continuation) we can define $\gamma(\Phi(z)) = \Gamma(z)$, a (single-valued) meromorphic function on $\mathcal{R}$.*

The simplest (and most frequent) application is that if $\Phi(z) \neq 0$ in $\mathcal{R}$, a *simply-connected* region, we have a unique function $\log \Phi(z) = \Gamma(z)$ defined in $\mathcal{R}$. (Here $\mathfrak{M}$ is the region $0 < |w| < \infty$.) If $\mathcal{R}$ were not simply-connected, the conclusion could fail [e.g., if $\Phi(z) = z$ in $\mathcal{R}$ the punctured plane $0 < |z| < \infty$, then $\log \Phi(z)$ is not uniquely defined in $\mathcal{R}$].

For proof, let us consider the function element $w = \Phi(z)$ mapping z_0 into w_0. Then take w_0 as the base point of $\mathfrak{M}^c$. We wish to show that for any $z \in \mathcal{R}$ we obtain a value of $\gamma(\Phi(z))$ by analytic continuation which is unique regardless of the path connecting z with z_0. Let $\mathcal{C}_1$ and $\mathcal{C}_2$ be two paths connecting z to z_0. Since $\mathcal{R}$ is simply-connected, $\mathcal{C}_1$ is deformable into $\mathcal{C}_2$ and by continuity of the mapping $\Phi(\mathcal{C}_1)$ is deformable into $\Phi(\mathcal{C}_2)$, where $\Phi(\mathcal{C}_i)$ is the image of $\mathcal{C}_i$, connecting w_0 with $w = \Phi(z)$. But if $\Phi(\mathcal{C}_1)$ is deformable into $\Phi(\mathcal{C}_2)$, then $w = \Phi(z)$ as defined under $\mathcal{C}_i$ corresponds to a unique point of $\mathfrak{M}^c$ under analytic continuation regardless of the choice of $\mathcal{C}_i$. Thus $\gamma(\Phi(z))$ is unique.

COROLLARY 1 (*Picard*, 1879) *An analytic function $w = \Phi(z)$ in the (finite) z plane can omit to assume only one value unless $\Phi(z)$ is a constant.*

Thus $\exp z = w$ illustrates this result since $\exp z$ is never 0 but $\exp z$ takes any other finite value at least once for some finite z ($= \log w$).

To prove this corollary, let us assume that $\Phi(z)$ is a nonconstant function which fails to take the values α and β. {Assume $\alpha = 0$ and $\beta = 1$, otherwise we use $[\Phi(z) - \alpha]/[\beta - \alpha]$.} Then, by Theorem B-3, we construct $\mathfrak{M}^c_{01\infty}$, the covering surface of the triply-punctured w sphere ($w \neq 0$, 1, ∞), and set $\Gamma(z) = \tau(\Phi(z))$ where $\tau(\lambda)$ is defined in (B-8) for λ on $\mathfrak{M}^c_{01\infty}$. Now for all z, $\Gamma(z)$ is bounded away from, say, $-i$ (recall $\operatorname{Im} \tau > 0$). Thus, contrary to Liouville's theorem, $(\Gamma(z) + i)^{-1}$ would be bounded for all z unless $\Gamma(z)$ [and therefore $\Phi(z)$] were a constant.

EXERCISES

1 Construct a covering surface for the punctured torus, e.g., the period parallelogram with point $u = 0$ deleted.

2 We can define Green's function with pole P_0 on an infinite simply-connected manifold $\mathfrak{M}$ by considering all finite submanifolds $\mathfrak{N} \subset \mathfrak{M}$ which contain P_0 and calling $g(\mathfrak{N},P_0)$ the Green's function.

$$g(\mathfrak{N},P_0) = -\log |u - u_{P_0}| + \text{reg func}$$
$$g(\mathfrak{N},P_0) = 0 \qquad \text{on } \partial\mathfrak{N}$$

Then let $\mathfrak{N}$ increase in any monotonic sequence $\mathfrak{N}_1 \subseteq \mathfrak{N}_2 \subseteq \mathfrak{N}_3 \subseteq \cdots$ such that each triangle Δ_i of $\mathfrak{M}$ belongs to $\mathfrak{N}_t$ for some $t(i)$. Then either

$$g(\mathfrak{N}_t,P) \to \infty \qquad \text{as } t \to \infty$$

or else

$$g(\mathfrak{N}_t,P) \to g(\mathfrak{M},P) \qquad \text{as } t \to \infty$$

(for all $P \neq P_0$)

Show this alternative is independent of the choice of the sequence $\mathfrak{N}_t$.

3 Show that the existence of the Green's function leads to the hyperbolic case and the nonexistence of the Green's function leads to the parabolic case.

4 Show that in accordance with Exercise 3, $\mathfrak{M}$ can be mapped onto either $\mathfrak{M}_{(1)}$ or $\mathfrak{M}_{(2)}$ in (B-6*b*) and (B-6*c*), hence the result is independent of choice P_0.

5 Show that a submanifold of a hyperbolic manifold is hyperbolic.

Table 1 SUMMARY OF EXISTENCE AND UNIQUENESS PROOFS

<table>
<tr><th></th><th>First Dirichlet Problem</th><th>Second Dirichlet Problem</th></tr>
<tr><td>Physical justification (Sec. 9-1)</td><td>Hydrodynamics</td><td>Electrostatics</td></tr>
<tr><td rowspan="4">Prescribed data for unknown $d\psi$ (Sec. 8-4)</td><td>(8-44a) $d\psi^* = 0$ on $\partial\mathfrak{M}$</td><td>(8-46e) ψ given on $\partial\mathfrak{M}$ after canonical dissection $\mathfrak{M}_p$ of $\mathfrak{M}$</td></tr>
<tr><td colspan="2">$(\mathcal{Q}_i,d\psi)$, $(\mathcal{Q}'_i,d\psi)$ given ($2p$ values), but</td></tr>
<tr><td>$(\mathcal{C}_j,d\psi)$ arbitrary (8-44b)
for $1 \leq j \leq n - 1$
($j = n$ follows Lemma 8-4)</td><td>$(\mathcal{C}_j,d\psi)$ vanishes (8-46a)
for $1 \leq j \leq n$</td></tr>
<tr><td colspan="2">(8-44c) (8-44d) Singularities prescribed with consistency conditions based on conservation laws (8-46c) (8-46d)</td></tr>
<tr><td>Uniqueness proofs (Sec. 8-5)</td><td>Assume $\partial\psi/\partial x$ and $\partial\psi/\partial y$ are well behaved on $\partial\mathfrak{M}$. Use Riemann's period relation and Dirichlet integral.</td><td>Assume only that $\int d\psi$ locally has limiting value on $\partial\mathfrak{M}$ (and is bounded near discontinuity, Sec. 11-1). Use maximum-minimum principle.</td></tr>
<tr><td>Logical sequence</td><td>Reduced by Schottky double (Sec. 11-3) to compact case ($n = 0$) and then to second problem. First Dirichlet problem implies the Riemann existence theorem for $\mathfrak{M}$ compact (Chap. 13).</td><td>Proved by alternating procedures (Chap. 12), first for noncompact manifold. This implies proof for compact manifold where two types of Dirichlet problems are the same.</td></tr>
</table>

BIBLIOGRAPHY

AHLFORS, L., "Complex Analysis," McGraw-Hill, New York, 1953.

——— and L. SARIO, "Riemann Surfaces," Princeton, Princeton, N.J., 1960.

APPELL, P., and E. GOURSAT, "Theorie des fonctions algebriques," Gauthier-Villars, Paris, 1929.

ARTIN, E., Algebraic numbers and algebraic functions (mimeographed notes), New York University, New York, and Princeton, Princeton, N.J., 1951.

BEHNKE, H., and F. SOMMER, "Theorie der analytischen Funktionen einer komplexen Veränderlichen," Springer, Berlin, 1962.

BERGMAN, S., "The Kernel Function," American Mathematics Society, New York, 1950.

BIEBERBACH, L., "Einführung in die Konforme Abbilfung," Göschen, Berlin, 1915.

CAIRNS, S. S., "Introductory Topology," Ronald, New York, 1961.

CARATHEODORY, C., "Conformal Representation," Cambridge, London, 1932.

CARTAN, H., Algebraic topology (mimeographed notes), Harvard, Cambridge, Mass., 1949.

——— "Theorie des fonctions analytiques," Herman, Paris, 1961.

CONFORTO, F., "Abelschen Funktionen und algebraische Geometrie," Springer, Berlin, 1956.

COPSON, E. T., "Introduction to the Theory of Functions of a Complex Variable," Oxford, London, 1935.

COURANT, R., "Dirichlet's Principle, Conformal Mapping and Minimal Surfaces," Interscience, New York, 1950 (with appendix by M. Schiffer, *q.v*).

——— and A. HURWITZ, "Vorlesungen uber allgemeine Funktionentheorie und elliptische Funktionen," Springer, Berlin, 1929.

EILENBERG, S., and N. STEENROD, "Foundations of Algebraic Topology," Princeton, Princeton, N.J., 1952.

FLANDERS, H., "Differential Forms," Academic, New York, 1963.

FRICKE, R., and F. KLEIN, "Vorlesungen über die Theorie der elliptischen Modulfunktionen," Leipzig, 1890–2.

GOLDBERG, S. I., "Curvature and Homology," Academic, New York, 1962.

HENSEL, K., and G. LANDSBERG, "Theorie der algebraischen Funktionen einer Veränderlichen," Teubner, Leipzig, 1902.

HILLE, E., "Analytic Function Theory," vol. I, Ginn, Boston, 1959.

HILTON, P. J., and S. WYLIE, "Homology Theory," New York, 1962.

HODGE, W. V. D., "Theory and Application of Harmonic Integrals," Cambridge, New York, 1959.

KNOPP, K., "Theory of Functions," vols. I and II, De Gruyter, Berlin, 1937.

KOBER, H., "Dictionary of Conformal Representations," Dover, New York, 1952.

LEHNER, J., "Discontinuous Groups and Automorphic Functions," American Mathematical Society, Providence, R. I., 1964.

MACLANE, S., "Homology," Springer, Berlin, 1963.

MILNE-THOMSON, L. M., "Theoretical Hydrodynamics," Macmillan, New York, 1955.

MONTEL, P., "Leçons sur les familles normales," Gauthier-Villars, Paris, 1927.

NEHARI, Z., "Conformal Mapping," McGraw-Hill, New York, 1952.

NEVANLINNA, R., "Uniformisierung," Springer, Berlin, 1953.

OSGOOD, W. F., "Lehrbuch der Funktionentheorie," vols. I and II (2), Teubner. Leipzig, 1928, 1932.

PFLUGER, A., "Theorie der Riemannschen Flächen," Springer, Berlin, 1957.

PICARD, E., "Traité d'Analyse," II, Gauthier-Villars, Paris, 1893.

de Rham, G., "Varietés differentiables (Actualités scientifiques)," Hermann, Paris, 1955.
Riemann, B., "Collected Works," Dover, New York, 1953.
Schiffer, M., "Some Recent Developments in the Theory of Conformal Mapping" (appendix to Courant, R., "Dirichlet's Principle," *q.v.*), 1950.
——— and D. C. Spencer, "Functionals of Finite Riemann Surfaces," Princeton, Princeton, N.J., 1954.
Seifert, H., and W. Threlfall, "Lehrbuch der Topologie," Teubner, Leipzig, 1934.
Severei, F., "Lezioni die Geometria Algebrica," Draghi, Padova, 1908.
Springer, G., "Introduction to Riemann Surfaces," Addison-Wesley, Reading, Mass., 1957.
Stoilow, S., "Leçons sur les principes topologiques de la theorie des fonctions analytiques," Gauthier-Villars, Paris, 1938.
Weyl, H., "Die Idee der Riemannschen Fläche," Teubner, Leipzig, 1913.
Whittaker, E. T., and G. N. Watson, "Modern Analysis," Cambridge, London, 1940.
Whyburn, G. T., "Topological Analysis," Princeton, Princeton, N.J., 1964.

SPECIAL SOURCE MATERIAL

References to books are given by authors (cited in the Bibliography). The references are minimal in each case.

PART I. Ahlfors, Hille, Knopp
Chapter 1: Sec. 1-4 *Jordan Curve Theorem*, Hille; Sec. 1-5 *Topological Remarks*, Seifert and Threlfall.
Chapter 2: Sec. 2-6 *Möbius Geometry*, Caratheodory.
Chapter 3: Sec. 3-5 *Special Mappings*, Kober.

PART II. Courant and Hurwitz, Weyl, Fricke and Klein
Chapter 4: Sec. 4-3 *Circular Pendulum*, P. Appell (Comptes Rendus, vol. 87, 1878); Secs. 4-4 to 4-6 *Special Functions*, Copson; Sec. 4-7 *Addition Theorems*, Whittaker and Watson.
Chapter 5: Sec. 5-3 *Rado's Theorem*, T. Rado (Acta Szeged, vol. 2, 1925), Nevanlinna, Springer.
Chapter 6: Sec. 6-3 *Modular Functions*, Fricke and Klein; Lehner.

PART III. Ahlfors and Sario, Behnke and Sommer, Courant
Chapter 7: *Topological Models*, Cairns.
Chapter 8: *Differentials*, Hodge, de Rham, Goldberg; *Harmonic Functions*, Picard, Ahlfors and Sario.
Chapter 9: *Canonical Mappings*, Courant, Schiffer, Nehari, Ahlfors, Bergman.

PART IV. Picard, Behnke and Sommer, Nevanlinna
Chapter 10: *Potential Functions*, Osgood; Sec. 10-2 *Topological Arguments*, Ahlfors.
Chapter 11: *Boundary Behavior*, Courant; C. Caratheodory (Math. Ann., vol. 73, 1913); *Schottky Double*, F. Schottky (Crelle's Journal, vol. 83, 1877), Nevanlinna.
Chapter 12: *Alternating Principle*, Picard.

PART V. Behnke and Sommer, Springer, Appell and Goursat
Chapter 14: *Divisors*, R. Dedekind, and H. Weber (Crelle's Journal, vol. 82, 1882), Hensel and Landsberg, Artin.

Appendix A: Courant, Schiffer, Ahlfors and Sario
Appendix B: *Covering Surfaces*, Ahlfors and Sario; *Picard's Theorem*, Copson; *Non-euclidean Geometry*, Nevanlinna.

INDEX

INDEX

Abel, N. H., 71
  addition theorem of, 292
  theorem of, 284
Abelian differential, 157, 177, 188
Abelian integral, 9, 71, 157, 299, 307
  [*See also* Normal (abelian) differential; Integral]
Addition theorem of Abel, 292
Addition theorem of Euler, 90
Ahlfors, *viii*, 12
Algebraic function, 108, 272
Algebraic manifold, 113, 121
Alternating procedures, 242–262
Analytic boundary, 228
Analytic continuation, 47
Analytic function, 4
Analytic manifold, 96
Annulus, 11
  covering surface of, 307
  mapping theorem for, 196
Argument, 9
  principle of, 23, 199, 211

Basis of differentials and integrals, 192, 266
Basis of homology classes, 154*n*., 238, 265
Basis of period module, 78, 122
Bergman, S., *viii*
Betti numbers, 155
Bicontinuous function, 15
Bicontinuous transformation, 15
Bilinear operation, 6, 170
Birational equivalence, 120
Boundary of region, 7, 11
  (*See also* Floating boundary; Free boundary)
Boundary operator, 11, 17, 152
Branch point, 54, 99, 111

Canonical (differential) class, 290
Canonical dissection, 136–141
Canonical domain, 196–199
Canonical mapping, 196–199
Canonical model, 137, 139
Caratheodory, C., 212, 220
Cauchy, A. L., 7, 19, 171, 184
  integral theorem of, 7, 17
  kernel of, 19
  representation theorem of, 19
  residue theorem of, 23
Cauchy-Riemann equations, 4*n*., 171
Chain, 6, 152
Christoffel, E. B., 232
Circular pendulum, 74
Circulation, 187
Coboundary, 161
Cocycle, 161
Coherent neighborhood (structure), 98

Cohomology, 160
Compact manifold, 106, 120, 260
Compactification, 25
(*See also* Schottky double)
Complex multiplication, 127
Conformal mapping, 57, 172, 214
(*See also* elementary mappings)
Conjugate crosscut, 137
Conjugate form, 171, 173, 176
(*See also* Schottky double)
Courant, R., *viii*, 201, 299
Covering surface, 307, 311
Crosscut, 137
Cross ratio, 36, 74, 83
Curl, 165
(*See also* circulation)
Curve, closed and simple, 7
(*See also* Jordan theorem)
Cycle, 112, 153
Cyclic neighborhood, 54, 61, 69, 97

Degree (or order), of divisor, 271
of divisor class, 288
of equation, function, 28, 108, 116, 135, 272
Differential (form), 6, 33, 157, 160, 276
of Abelian type, 157, 177, 191
class of, 290
of first, second, and third kind, 157, 191
on manifold, 101
Differential form, 161
(*See also* Normal differential; Meromorphic differential)
Dipole, 42, 191, 197
Dirichlet, P. G. L., 177, 201, 299
ordinary problem of, 203, 242
problem of, 177
Dirichlet principle, 297
Disk, 11
mapping theorem for, 214
Divergence, 165
Divisor, 31, 119, 270, 287
degree of, 271, 288
dimension of, 271, 276
Divisor classes, 287
representation of function by, 288
(*See also* Gaps)
Domain, 4*n*.
(*See also* Fundamental domain)

Doubly periodic function, 73
Doubly periodic structure, 73, 122
(*See also* Variety, of Jacobi)

Eisenstein series, 81, 84
Electrostatic analogy, 185, 315
Elementary mappings, 60
Elliptic functions, 67, 77, 80, 90, 92, 102
(*See also* Modular function)
Elliptic integral, 71, 74
Elliptic region (or manifold), 311
Elliptic transformation, 40
Equivalence of fields and manifolds, 119
Euler, L., addition theorem of, 90
Euler-Poincaré theorem, 142
Exterior differential calculus, 162
Extremal length, 303
Extremal metric, 303

Field (algebraic), 77, 116, 270
First existence theorem, 177, 315
Fixed point, 39
Floating boundary, 185, 300
Flux, 185
Folium, 114, 144
Form [*see* Conjugate form; Differential (form)]
Free boundary, 185, 300
Function element, 47
Fundamental domain, 85, 102, 125–128

Gaps (in divisors), 279
Gauss, C. F., 57, 166, 297
mean-value theorem of, 204
Genus, 137
Riemann's formula for, 143
Genus 0, 273
Genus 1, 273
Genus 2, 283
Germ, 47*n*.
Global uniformizing parameter, 97*n*.
Goursat, E., 7
Gradient, 165
Graph, 142
Grassman algebra, 162
Green's function, 188, 211, 216, 314
of complex (hydrodynamic) type, 189
Green's theorem, 166*n*.
Grötzsch, H., 302

Hadamard, J., 305
Harmonic differential, 172
Harmonic function, 172
Harmonic integrals, 172
[*See also* Normal (abelian) differential; Integral; *and* Basis of differentials and integrals]
Harnack's theorem, 208, 247
Hilbert, D., 201, 299
(*See also* Kernel)
Holomorphic function, 4
Homeomorphism, 12*n*., 15
Homology, 17, 153
Homology basis, 154*n*., 238, 265
Homothetic transformation, 40
Homotopy, 18*n*., 309
Hydrodynamic analogy, 186, 315
Hyperbolic region (or manifold), 312
Hyperbolic transformation, 40
Hyperelliptic surfaces, 143, 283
Hypothesis 1, 73
Hypothesis 2, 81

Ideal boundary (point), 26, 106
Image, 37, 41, 63, 193, 230
Implicit functions, 51
Index of specialty, 276
Infinity, 25
(*See also* Ideal boundary)
Inner normal, 10
Integral, 5
of first, second, and third type, 71, 157, 265–269
(*See also* Abelian integral; Abelian differential)
Interior mapping, 52
Invariants, of elliptic function fields, 122
under linear mapping, 34
(*See also* Module of double period)
Isogonal mapping, 57

Jacobi's theorem on double periods, 79
(*See also* Variety, of Jacobi)
Jordan, C., theorem of, 11
Jordan manifold, 136

Kernel, of Cauchy, 19
of Hilbert, 209
of Poisson, 205
Klein, F., 86, 127

Lagrange, J. L., 20, 129
Legendre, A. M., 71, 95
Linear (fractional) transformations, 30, 34, 39
Linear mapping, 61
Linear series, 294*n*.
Liouville's theorem, 22, 79, 119, 136, 312, 314
Local uniformizing parameter, 54, 86*n*., 97
illustrations of, 102*n*., 112

Manifold, 12, 65, 96
equivalence (class) of, 101, 117
finite triangulation of, 103, 107, 120
infinite triangulation of, 106, 107, 307
(*See also* Algebraic manifold; Analytic manifold; Compact manifold)
Mapping, 46, 57, 101, 196, 210
(*See also* Homeomorphism; Canonical mapping; Conformal mapping; Interior mapping; Isogonal mapping)
Matrix (*see* Modular matrix; Riemann matrix)
Maximum-minimum principle, 56, 182
Meromorphic differential, 101, 177
Meromorphic function, 28, 29, 101, 117
Möbius circles, 36
Möbius geometry, 36–38
Möbius strip, 13, 136
Model (*see* Canonical model; Polygonal model; Simplicial model; Rectangular model)
Modular function, 84, 102, 122, 129, 312
Modular matrix, 192
Module of conformal equivalence, 123, 192, 196, 282, 302
Module of double period, 78, 122, 284
Monodromic function, 48
Monodromy substitution, 111
Montel, P., 300
Morera's theorem, 8
Multiform function, 49
Multivalued function, 49

Neighborhood, on left side, 10
structure of, 97
(*See also* Coherent neighborhood; Cyclic neighborhood)

Neumann, C., 201, 299
Normal (abelian) differential, 191, 197, 265
of first, second, and third types, 265–269
Normal family, 300
Notations (*recurring frequently*),
$\mathcal{C}_i$, 10
$J(\tau)$, 84
$\mathcal{K}_j$, 138
$\lambda(\tau)$, 129
$l(\mathcal{C})$, 7
$L(D)$, 271
$L^*(D)$, 277
$\mathfrak{M}^p$, $\mathfrak{M}^*$, 136–138
$\mathfrak{M}^0$, $\hat{\mathfrak{M}}$, 235–237
$N(\mathcal{C}; a)$, 9
$\mathcal{Q}_i$, $\mathcal{Q}'_i$, 137
res $f(a)\ dz$, 23, 135
$\wp(u)$, 80
$\Omega_{m,n}$, $\mathbf{\Omega}$, 78
$(\mathcal{C}, f(z)\ dz)$, 6
(σ, ω), 167

Order, of branch point and ramification, 54, 143
Order (*see* Degree)
Orientability of manifold, 13, 105, 134
Orientation, of curve, 5
of point, 17, 152
of region, 17

Pairing (operation), 170
Parabolic region (or manifold), 312
Parabolic transformation, 42
Period of abelian integral, 71
Periods, parallelogram of, 76, 102, 158, 177, 178
(*See also* Variety, of Jacobi)
Peripheral sequence, 146
[*See also* Simple boundary (property)]
Picard, E., *viii*, 201
Poincaré, H., 125
Poincaré-de Rham theorem, 169*n.*
(*See also* Euler-Poincaré theorem)
Poisson's integral, 205, 224
Poles, 22, 101
prescription of, 31, 92, 119, 271
Polydromic function, 49
Polygonal model, 138, 249
Positive orientation of curve, 11
Projection of manifold, 96

Rado's theorem, 106
Rank of matrix, 275*n.*
Rational (field of) functions, 28, 67, 102, 116, 270
Rayleigh-Ritz method, 299*n.*
Rectangular model, 136, 250
Reflection principle, 63, 229, 237
Region, 4
Regular function, 4
Residue, 23
at infinity, 32
on manifold, 135
de Rham G., (*see* Poincaré-de Rham theorem)
Riemann, B., 4*n.*, 22, 26, 46, 58, 63, 116, 121, 201, 271, 278, 299
fundamental theorem of, 121, 271
genus formula of, 143, 278
manifold of, 96
mapping theorem of, 63, 214
matrix of, 285*n.*
period relation of, 180, 269
sphere of, 26, 102
Riemann-Roch theorem, 277, 290
(*See also* Cauchy-Riemann equations)
Roch theorem, 277

Schiffer, M., *viii*
Schlicht mapping, 60
Schottky conjugate, 235, 315
Schottky double, 235, 315
Schwarz, H. A., 63, 92, 201, 208, 281, 299
inequality of, 220
Schwarz-Christoffel mapping, 232
Second Existence theorem, 178, 315
Simple boundary (property), 146
Simple connectivity, 11, 167, 311
Simple curve, 7
Simplicial structure, 145
Singularity, of essential or removable type, 22
of harmonic functions, 222, 225
planting of, 256
Specialty, 276
Stereographic projection, 26
Stokes' theorem, 18, 166
Strong reflection principle, 229
Symmetric functions, 113, 118

Topological mapping, 15
Torus, 14, 76
  conformal invariants of, 123
Triangulation, 103
  (*See also* Manifold)

Unicursal curve, 94
Uniform function, 49
Uniformization, 68, 82
  (*See also* Global uniformizing parameter; Local uniformizing parameter)

Variety, 13
  of Jacobi, 293
Vector analysis, 164

Weierstrass, K., 22, 46, 80, 89, 279, 299
  elliptic functions of, 80, 93, 94
  gap theorem of, 279
  monodromy theorem of, 49, 311
  normal form of cubic, 82
Weierstrass points (of manifold), 279
Weyl, H., 65, 103, 201, 299
Whyburn, G. T., 4*n*.
Winding number, 9

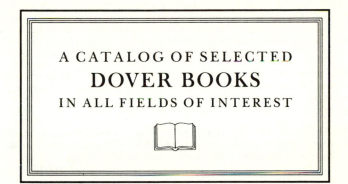

THE BLUE FAIRY BOOK, Andrew Lang. The first, most famous collection, with many familiar tales: Little Red Riding Hood, Aladdin and the Wonderful Lamp, Puss in Boots, Sleeping Beauty, Hansel and Gretel, Rumpelstiltskin; 37 in all. 138 illustrations. 390pp. 5⅜ × 8½. 21437-0 Pa. $5.95

THE STORY OF THE CHAMPIONS OF THE ROUND TABLE, Howard Pyle. Sir Launcelot, Sir Tristram and Sir Percival in spirited adventures of love and triumph retold in Pyle's inimitable style. 50 drawings, 31 full-page. xviii + 329pp. 6½ × 9¼. 21883-X Pa. $6.95

AUDUBON AND HIS JOURNALS, Maria Audubon. Unmatched two-volume portrait of the great artist, naturalist and author contains his journals, an excellent biography by his granddaughter, expert annotations by the noted ornithologist, Dr. Elliott Coues, and 37 superb illustrations. Total of 1,200pp. 5⅜ × 8.
Vol. I 25143-8 Pa. $8.95
Vol. II 25144-6 Pa. $8.95

GREAT DINOSAUR HUNTERS AND THEIR DISCOVERIES, Edwin H. Colbert. Fascinating, lavishly illustrated chronicle of dinosaur research, 1820's to 1960. Achievements of Cope, Marsh, Brown, Buckland, Mantell, Huxley, many others. 384pp. 5¼ × 8¼. 24701-5 Pa. $6.95

THE TASTEMAKERS, Russell Lynes. Informal, illustrated social history of American taste 1850's–1950's. First popularized categories Highbrow, Lowbrow, Middlebrow. 129 illustrations. New (1979) afterword. 384pp. 6 × 9. 23993-4 Pa. $6.95

DOUBLE CROSS PURPOSES, Ronald A. Knox. A treasure hunt in the Scottish Highlands, an old map, unidentified corpse, surprise discoveries keep reader guessing in this cleverly intricate tale of financial skullduggery. 2 black-and-white maps. 320pp. 5⅜ × 8½. (Available in U.S. only) 25032-6 Pa. $5.95

AUTHENTIC VICTORIAN DECORATION AND ORNAMENTATION IN FULL COLOR: 46 Plates from "Studies in Design," Christopher Dresser. Superb full-color lithographs reproduced from rare original portfolio of a major Victorian designer. 48pp. 9¼ × 12¼. 25083-0 Pa. $7.95

PRIMITIVE ART, Franz Boas. Remains the best text ever prepared on subject, thoroughly discussing Indian, African, Asian, Australian, and, especially, Northern American primitive art. Over 950 illustrations show ceramics, masks, totem poles, weapons, textiles, paintings, much more. 376pp. 5⅜ × 8. 20025-6 Pa. $6.95

SIDELIGHTS ON RELATIVITY, Albert Einstein. Unabridged republication of two lectures delivered by the great physicist in 1920–21. *Ether and Relativity* and *Geometry and Experience*. Elegant ideas in non-mathematical form, accessible to intelligent layman. vi + 56pp. 5⅜ × 8½. 24511-X Pa. $2.95

THE WIT AND HUMOR OF OSCAR WILDE, edited by Alvin Redman. More than 1,000 ripostes, paradoxes, wisecracks: Work is the curse of the drinking classes, I can resist everything except temptation, etc. 258pp. 5⅜ × 8½. 20602-5 Pa. $3.95

ADVENTURES WITH A MICROSCOPE, Richard Headstrom. 59 adventures with clothing fibers, protozoa, ferns and lichens, roots and leaves, much more. 142 illustrations. 232pp. 5⅜ × 8½. 23471-1 Pa. $3.95